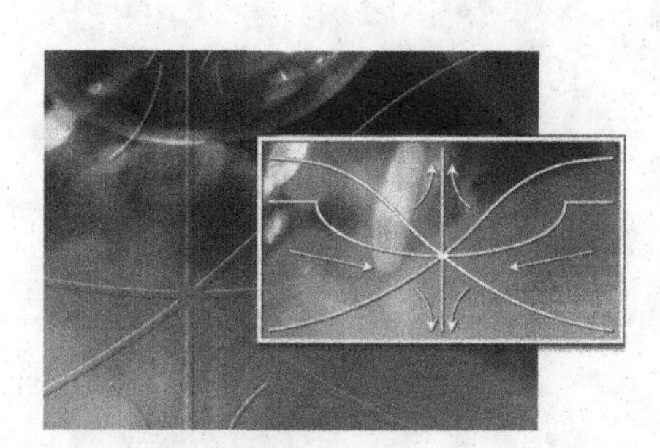

METHODS OF WAVE THEORY IN DISPERSIVE MEDIA

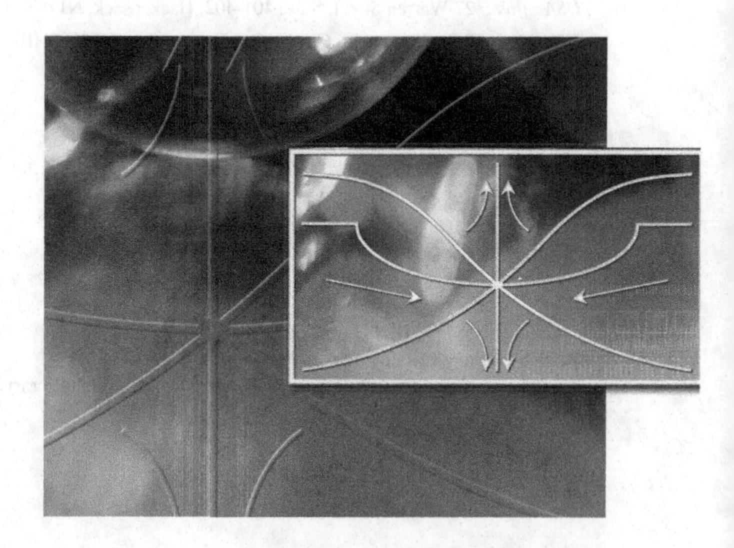

METHODS OF WAVE THEORY IN DISPERSIVE MEDIA

M V Kuzelev

Moscow State University, Russia

A A Rukhadze

Russian Academy of Sciences, Russia

World Scientific

NEW JERSEY • LONDON • SINGAPORE • BEIJING • SHANGHAI • HONG KONG • TAIPEI • CHENNAI

Published by

World Scientific Publishing Co. Pte. Ltd.
5 Toh Tuck Link, Singapore 596224
USA office: 27 Warren Street, Suite 401-402, Hackensack, NJ 07601
UK office: 57 Shelton Street, Covent Garden, London WC2H 9HE

British Library Cataloguing-in-Publication Data
A catalogue record for this book is available from the British Library.

METHODS OF WAVE THEORY IN DISPERSIVE MEDIA

ISBN-13 978-981-4261-69-2
ISBN-10 981-4261-69-6

Printed in Singapore.

Synopsis

The monograph presents main analytic mathematical methods and general problems in the theory of linear waves in dispersive media and systems, including nonequilibrium ones. To show how the general theory can be applied in practice, a unified description is given of important physical systems that are traditionally studied in the mechanics of continuous media, electrodynamics, plasma physics, electronics, and physical kinetics. An analysis is made of the interaction of waves in coupled systems, the propagation and evolution of localized wave perturbations, and the emission of waves in dispersive media under the action of external sources moving in a prescribed manner. A general theory of instabilities of linear systems is presented in which the criteria for absolute and convective instabilities are formulated and compared, and Green's functions for some nonequilibrium media are calculated. Special attention is paid to problems in the theory of linear electromagnetic waves in plasmas and plasmalike media. The monograph also contains a number of original results of the present-day wave theory that have been published by now in scientific journals only.

The book is aimed at researchers and experts, as well as students and postgraduates, who specialize in such fields as the electrodynamics and mechanics of continuous media, physical electronics, and radiophysics.

Synopsis

Introduction

The monograph presents mathematical methods of description and general physical results in the theory of linear waves in dispersive media and systems, including nonequilibrium ones. In essence, it gives formulations and solutions of problems for nth order linear partial differential equations and also physical interpretations of the solutions and their practical applications. Since the literature (manuals, monographs, reviews, etc.) on the theory of linear waves is now so extensive that it seems to be exhaustive, the question naturally arises of whether it is expedient to publish books like the one you are reading. We think, however, that books of this kind are still needed. The main reasons are twofold. First, wave theory is traditionally presented in the context of particular physical objects, such as optical waves, radiowaves, plasma waves, waves in fluids, and acoustic waves. But mathematically, wave theory can be constructed and presented irrespective of the physical nature of the wave process. And second, the very important subjects of modern natural sciences are nonequilibrium physical systems, for which wave theory plays a secondary role and is merely part of such original branches of physics as physical kinetics, plasma physics, microwave electronics, to name but a few. Yet, there is clearly a need for an original theory of waves in nonequilibrium media. In our monograph, the general theory of linear waves is presented just as a branch of mathematical physics that describes the dynamics of linear waves in equilibrium and nonequilibrium dispersive media and systems, irrespective of their physical nature. The practical application of the general theory is illustrated by considering fairly simple but important physical systems that are traditionally studied in the mechanics of continuous media, electrodynamics, plasma physics, and electronics. The practical problems are solved by a unified approach presented in the mathematical part of the wave theory. Along with traditional information, the monograph contains a number of new original results that we have obtained in studying nonequilibrium and resonant phenomena in plasmalike media and that have found practical applications in electronics and radiophysics. In studying linear waves in dispersive media and systems, we proceed from the general to the special, and we hope that our theoretical study will be of interest to both beginners (students and postgraduates) and experts in the physics of wave processes. Note finally that the monograph is based on the course of lectures given by the authors to senior students at Moscow State University (Faculty of Physics, Division of Physical Electronics).

Contents

Chapter 1

Linear Harmonic Waves in Dispersive Systems. Initial-Value Problem and Problem with an External Source

1. Harmonic Waves in Dispersive Systems

Assume that small perturbations of the equilibrium state of a one-dimensional physical system satisfy the following set of first-order linear homogeneous partial differential equations:

$$\frac{\partial \psi_s}{\partial t} + \sum_{j=1}^{n} \left(A_{sj} \frac{\partial \psi_j}{\partial z} + B_{sj} \psi_j \right) = 0, \quad s = 1, 2, \ldots, n. \tag{1.1}$$

Here, $\{\psi_1(t,z), \psi_2(t,z), \ldots, \psi_n(t,z)\} \equiv \boldsymbol{\Psi}(t,z)$ is the vector of small perturbations of the equilibrium, called the state vector of the system. The number of components $\psi_s(t,z)$ of the state vector is equal to the number n of equations in set (1.1), and A_{sj} and B_{sj} are square $n \times n$ matrices with constant elements. Among the equations that are reduced to equations of the form (1.1), we can mention acoustic equations, hydrodynamic equations, equations for the electromagnetic field in various material media, linear plasma electrodynamic equations, linearized equations of theoretical microwave electronics, and many others.

We seek a solution to Eqs. (1.1) in the form

$$\boldsymbol{\Psi}(t,z) = \boldsymbol{\Phi}(\omega,k) \exp(-i\omega t + ikz), \tag{1.2a}$$

$$\boldsymbol{\Phi}(\omega,k) = \{\phi_1(\omega,k), \phi_2(\omega,k), \ldots, \phi_n(\omega,k)\}. \tag{1.2b}$$

Here, $\boldsymbol{\Phi}(\omega,k)$ is the complex state vector; $\phi_s(\omega,k)$, with $s = 1, 2, \ldots, n$, are the state vector components; ω is the frequency; and k is the wavenumber. The state vector $\boldsymbol{\Psi}(t,z)$ is a physical quantity and as such is real. Consequently, only the real part of complex function (1.2a) has a physical meaning. It is convenient, however, to perform linear operations on the state vector in its complex form without restriction and to switch to its real part only in the final result.

Each component of the state vector (1.2a) is a plane harmonic wave, which is characterized by the time period

$$T = \frac{2\pi}{\omega} \tag{1.3}$$

and the spatial period, or wavelength,

$$\lambda = \frac{2\pi}{k} . \tag{1.4}$$

An important parameter of a plane harmonic wave is its phase velocity, i.e., the propagation velocity of the constant-phase points (planes) in space. This is the velocity at which an "observer" should move along the z axis in order for the state vector (1.2) to be constant. The phase velocity is obviously determined from the relationship

$$\omega t - kz = \text{const} , \tag{1.5}$$

which indicates that the phase of a plane wave is constant. Differentiating relationship (1.5) with respect to time and taking into account the fact that the observer's speed is dz/dt yields the definition of the phase velocity:

$$V_{ph} = \frac{\omega}{k} . \tag{1.6}$$

We substitute solution (1.2) into homogeneous equations (1.1) to arrive at the following set of linear homogeneous algebraic equations for the components $\phi_s(\omega, k)$ of the complex state vector:

$$-i\omega\phi_s(\omega, k) + \sum_{j=1}^{n}(ikA_{sj} + B_{sj})\phi_j(\omega, k) = 0, \quad s = 1, 2, \ldots, n . \tag{1.7}$$

The number of equations in set (1.7) and the number of unknowns $\phi_s(\omega, k)$ are both equal to n. Of course, we are interested only in nontrivial solutions to Eqs. (1.7), i.e., in such sets of state vector components $\phi_1(\omega, k)$, $\phi_2(\omega, k), \ldots, \phi_n(\omega, k)$ in which at least one is nonzero. Otherwise, the state vector (1.2) would be identically zero, a physically uninteresting case. From linear algebra, it is known that a set of linear homogeneous algebraic equations has a nontrivial solution if and only if its determinant is zero. For the set of Eqs. (1.7), this condition can be written as

$$D(\omega, k) \equiv \det(-i\omega\delta_{sj} + ikA_{sj} + B_{sj}) = 0, \quad s, j = 1, 2, \ldots, n , \tag{1.8}$$

where δ_{sj} is the Kronecker symbol. Relationship (1.8) is called the dispersion (characteristic) relation for determining the spectra of eigenmodes. The function of two variables $D(\omega, k)$ is called the dispersion function.

Dispersion relation (1.8) is a relationship between two independent quantities — frequency ω and wavenumber k. Consequently, this dispersion relation can be solved either with respect to frequency (in order to determine the dependence $\omega = \omega(k)$) or with respect to wavenumber (in order to find the function $k = k(\omega)$). The first approach yields a solution to the so-called initial-value problem. The second approach is used to solve the boundary-value problem. In the present monograph, we will only consider initial-value problems in which dispersion relation (1.8) is solved with respect to frequency and the frequency spectra of the eigenmodes, $\omega = \omega(k)$, are determined.

Dispersion relation (1.8) usually has more than one solution, i.e., $\omega = \omega_m(k)$, with $m = 1, 2, \ldots$. In this case, a physical system is said to have several branches of eigenmodes with eigenfrequencies $\omega_m(k)$. From Eqs. (1.1) and (1.7) we can see that dispersion relation (1.8) is an nth order algebraic equation for the frequency ω. In algebra courses, it is proved that such an equation has n roots, each corresponding to its own branch of eigenmodes. Hence, the number of different solutions to dispersion relation (1.8), or equivalently the number of different branches of eigenmodes, is equal to n. But it should be noted that some of the solutions to dispersion relation (1.8) can be trivial ($\omega_m = 0$) and therefore should be excluded from consideration,[1] in which case the number of eigenmode branches is in fact less than the number of equations in set (1.1). In addition, the dispersion relation can have coincident (multiple) roots — a so-called degenerate case that requires a separate analysis.

Corresponding to each eigenfrequency $\omega_m(k)$ there is a state eigenvector $\mathbf{\Psi}_m(t, z)$. The complex state eigenvector $\mathbf{\Phi}_m(\omega, k)$ that corresponds to the vector $\mathbf{\Psi}_m(t, z)$ is found from the set of algebraic equations (1.7) but with the eigenfrequency in place of an arbitrary frequency ω. In this case, the frequency ω is no longer an independent variable, so, in expression (1.2b), we can introduce the notation

$$\begin{aligned}
&\mathbf{\Phi}_m(\omega, k) = \mathbf{\Phi}(\omega_m(k), k) \equiv \mathbf{\Phi}_m(k)\,,\\
&\quad\phi_s(\omega, k) = \phi_s(\omega_m(k), k) \equiv \phi_s^{(m)}(k)\,,\\
&\quad\mathbf{\Phi}_m(k) = \{\phi_1^{(m)}(k), \phi_2^{(m)}(k), \ldots, \phi_n^{(m)}(k)\} \equiv \{\phi_1(k), \phi_2(k), \ldots, \phi_n(k)\})(m)\,.
\end{aligned} \tag{1.9}$$

Since the solution to a set of homogeneous algebraic equations is defined to within a constant factor, the vector $A_m\mathbf{\Phi}_m(k)$, with A_m being an arbitrary constant, also satisfies the set of Eqs. (1.7). Hence, in solving the initial-value problem, the state vector of a physical system that is described by linear differential equations (1.1) has the form

$$\mathbf{\Psi}_m(t, z) = A_m\mathbf{\Phi}_m(k)\exp[-i\omega_m(k)t + ikz]\,. \tag{1.10}$$

Moreover, there are as many such vectors as there are eigenmode branches, i.e., $m = 1, 2, \ldots, n$. And finally, keeping in mind the superposition principle, which implies in particular that a sum of solutions to a linear equation is also its solution, we write the solution to the initial-value problem for a set of linear homogeneous differential equations (1.1) as the sum over all eigenmode branches:

$$\mathbf{\Psi}(t, z) = \sum_{m=1}^{n} \mathbf{\Psi}_m(t, z) = \sum_{m=1}^{n} A_m\mathbf{\Phi}_m(k)\exp[-i\omega_m(k)t + ikz]\,. \tag{1.11}$$

Solution (1.11) contains the wavenumber k and constant factors A_m, which are called complex amplitudes. In order to determine the wavenumber and amplitudes,

[1] This concerns only harmonic waves; on the other hand, such solutions correspond to constant, but spatially nonuniform, fields.

additional conditions are required. How to formulate these additional conditions and how to use them will be described in Sec. 2.

Let us consider the phase velocity (1.6) for a particular harmonic eigenmode (1.10) of a certain physical system:

$$V_{ph}^{(m)} = \frac{\omega_m(k)}{k}. \tag{1.12}$$

If the phase velocity (1.12) is independent of the wavenumber k, then, according to the terminology adopted in the theory of linear waves, the eigenmode is said to have no dispersion. If the phase velocity $V_{ph}^{(m)}$ is a function of the wavenumber k, the eigenmode is called dispersive. Systems (media) in which there are dispersive eigenmodes are referred to as systems with dispersion or dispersive systems (media). For purely harmonic waves, the notion of dispersion is meaningless. But for more complicated, nonharmonic wave formations, the notion of wave dispersion plays an important role.

In accordance with what was said above, the eigenmode is nondispersive if its eigenfrequency is given by the formula

$$\omega_m(k) = \alpha k, \quad \alpha = \text{const} \tag{1.13}$$

Indeed, the phase velocity (1.12) in this case is independent of the wavenumber k. In wave theory, frequency spectra of the form (1.13) are called acoustic-like spectra.

Historically, the notion of dispersion has come to wave theory from optics. Since we are dealing with waves of quite a general nature, we extend the notion of dispersion as follows. A wave is considered to be nondispersive if its eigenfrequency has the form

$$\omega_m(k) = \alpha k + \beta, \quad \alpha = \text{const}, \quad \beta = \text{const} \tag{1.14}$$

For $\alpha = 0$, spectrum (1.14) is called optical. For $\beta \neq 0$, the phase velocity of a wave with the frequency (1.14) depends on the wavenumber k. But from the standpoint of the dynamics of nonharmonic wave formations, the frequency spectra (1.13) and (1.14) are equivalent, as will be shown later. We stress that, in spectra (1.13) and (1.14), the symbol "const" implies independence on the wavenumber k. For

$$\frac{d^2\omega_m(k)}{dk^2} \neq 0, \tag{1.15}$$

the eigenfrequency cannot be represented in the form (1.14) and the wave is dispersive. Inequality (1.15) is a mathematical criterion of whether the wave is dispersive or not.

Spatially harmonic solution (1.11) to the initial-value problem contains important information about the state of a physical system (medium). The solutions to dispersion relation (1.8) are generally complex,

$$\omega_m(k) = \omega_m'(k) + i\omega_m''(k), \tag{1.16}$$

so it is convenient to rewrite solution (1.11) as

$$\Psi(t,z) = \sum_{m=1}^{n} \Psi_m(t,z) = \sum_{m=1}^{n} A_m \Phi_m(k) \exp[\omega_m''(k)t] \exp[-i\omega_m'(k)t + ikz] \,. \quad (1.17)$$

If, for all m (i.e., for all the branches of eigenmodes), the imaginary parts are negative, $\omega_m''(k) < 0$, then all the terms in solution (1.17) decrease exponentially with time t. In this case, the negative imaginary part of the frequency is called the damping rate of the wave. On sufficiently long time scales, only the term with the minimum absolute value of the damping rate is important in solution (1.17). If one of the roots of the dispersion relation has a zero imaginary part, $\omega_m''(k) = 0$, then the corresponding term of the sum in solution (1.17) is not damped with time and describes an undamped eigenmode. And finally, if at least one of the roots has a positive imaginary part, $\omega_m''(k) > 0$, then the corresponding eigenmode grows with time. This is the case only when the system (medium) is in an unstable nonequilibrium state. The positive imaginary part of the frequency is called the growth rate of the wave or the instability growth rate.

In what follows, we will primarily focus on systems for which dispersion relations (1.8) are algebraic equations with real coefficients (except in Secs. 6–8, 10, 12, 17, 25). It is known that, if a certain complex number $\omega' + i\omega''$ is a root of such an equation, then its complex conjugate, $\omega' - i\omega''$, is a root too. It is also known that an algebraic equation with real coefficients can have no roots at all. That is, either we have $\omega_m''(k) = 0$ for all m, in which case the system is in a stable state, or there is an eigenmode branch such that $\omega_m''(k) > 0$, in which case the system is unstable. But it is somewhat incorrect to speak of wave damping in systems described by dispersion relations with real coefficients. Indeed, for any eigenmode branch with $\omega_{m_1}'' < 0$, there is a complex-conjugate branch with $\omega_{m_2}'' = -\omega_{m_1}'' > 0$ — a circumstance implying that the system is unstable. In actuality, wave damping always results from the dissipation of perturbation energy. A dispersion relation with real coefficients describes a nondissipative system.

Conceiving the wave phase velocity as the propagation velocity of constant-phase (but not constant-amplitude) points is also meaningful for complex frequencies. It is only necessary to rewrite formula (1.12) as

$$V_{ph}^{(m)} = \frac{\mathrm{Re}\,\omega_m(k)}{k} = \frac{\omega_m'(k)}{k} \,. \quad (1.18)$$

It is also obvious that introducing the notion of the wave period (1.3) is meaningful only when the imaginary part of the frequency is much less than its real part.

2. Initial-Value Problem. Eigenmode Method

In order to complete an investigation of the problem of excitation of harmonic eigenmodes in a system described by differential equations (1.1), it is necessary to find the wavenumber k and constant complex amplitudes A_m in the general solution

(1.11). To do this in the most illustrative way, it is convenient to change the notation system, i.e., to pass over from row vectors (1.2) to column vectors. Thus, we write the harmonic solution (1.11) to Eqs. (1.1) as (see also (1.9))

$$\mathbf{\Psi}(t,z) = \sum_{m=1}^{n} \mathbf{\Psi}_m(t,z) = \sum_{m=1}^{n} A_m \begin{pmatrix} \phi_1^{(m)}(k) \\ \phi_2^{(m)}(k) \\ \vdots \\ \phi_n^{(m)}(k) \end{pmatrix} \exp[-i\omega_m(k)t + ikz]\,. \tag{2.1}$$

Let us consider the structure of the column vector in solution (2.1) in more detail. The components $\phi_s^{(m)}(k)$ of the complex state vector satisfy the set of algebraic equations (1.7) with $\omega = \omega_m(k)$. Since Eqs. (1.7) are homogeneous, the components $\phi_s^{(m)}(k)$ can be found in the following way. The terms containing one of the components, say $\phi_1^{(m)}$ for definiteness, are moved to the right-hand side of Eqs. (1.7) and are treated as being known. The set of Eqs. (1.7) is then solved in a conventional manner (by linear algebra methods) for the remaining components $\phi_2^{(m)}, \phi_3^{(m)}, \ldots, \phi_n^{(m)}$. The resulting solutions are linear in $\phi_1^{(m)}$:

$$\phi_s^{(m)}(k) = L_s(\omega_m(k), k) \cdot \phi_1^{(m)}(k) \equiv L_{sm}(k)\phi_1^{(m)}(k), \quad s = 2, 3, \ldots, n\,, \tag{2.2}$$

where L_s are functions of the coefficients of Eqs. (1.7). As for the components $\phi_1^{(m)}$, they are arbitrary and can be chosen to be, e.g., unity. In so doing, the dimension should be accounted for as follows. When the complex component $\phi_1^{(m)}$ of the state vector is dimensional, it is convenient to assign its dimension to the complex amplitudes A_m, i.e., in effect, to make the redefinition $A_m\phi_1^{(m)}(k) \equiv A_m(k)$.

The last point deserves some clarification. After the terms with $\phi_1^{(m)}$ have been moved to the right-hand side of Eqs. (1.7), the number of unknowns becomes $n-1$, while the number of equations remains equal to n — a situation that poses no mathematical difficulty, however. Since ω_m is a root of dispersion relation (1.8), the determinant of the set of Eqs. (1.7) is zero. Consequently, one (any one) of the equations is a consequence of the remaining equations and thus can be dropped from the set. Hence, the number of unknowns and the number of equations are in fact the same. An approach for finding the complex amplitudes $A_m(k) = A_m\phi_1^{(m)}(k)$ and the functions $\phi_s^{(m)}(k)$ (2.2) that is presented below is called the eigenmode method.

Assume that, at the initial time $t = 0$, the spatially harmonic state vector of the system is given by the formula

$$\mathbf{\Psi}(0, Z) = \begin{pmatrix} b_1(\chi) \\ b_2(\chi) \\ \vdots \\ b_n(\chi) \end{pmatrix} \exp(i\chi z)\,, \tag{2.3}$$

where χ and $b_s(\chi)(s = 1, 2, \ldots, n)$ are known (prescribed) constant quantities. Vector relationship (2.3) is an initial condition for differential equations (1.1). Specifically, Eqs. (1.1) supplemented with relationships (2.3) constitute a so-called initial-value problem or a problem with initial conditions. The problem at hand is an initial-value problem with harmonic initial conditions. Let us consider the main steps in finding its solution.

At subsequent times ($t > 0$), the state vector satisfies Eqs. (1.1) and is therefore described by formula (2.1) (or (1.11)). Substituting $t = 0$ into formula (2.1) and equating the result to the initial state vector (2.3) yields the relationship

$$\sum_{m=1}^{n} A_m \begin{pmatrix} \phi_1^{(m)}(k) \\ \phi_2^{(m)}(k) \\ \vdots \\ \phi_n^{(m)}(k) \end{pmatrix} \exp(ikz) = \begin{pmatrix} b_1(\chi) \\ b_2(\chi) \\ \vdots \\ b_n(\chi) \end{pmatrix} \exp(i\chi z), \tag{2.4}$$

which should be satisfied identically for any $z \in (-\infty, +\infty)$. This is clearly the case only when $k = \chi$. Hence, the wavenumber k in solution (2.1) (and in (1.11)) is determined by the structure of the initial perturbation of the state vector that is harmonic in the spatial variable z. The case of a nonharmonic perturbation will be considered below.

Taking into account the equality $k = \chi$ and cancelling the common exponential factor in relationship (2.4), we obtain the set of linear algebraic equations

$$\sum_{m=1}^{n} A_m \begin{pmatrix} \phi_1^{(m)}(k) \\ \phi_2^{(m)}(k) \\ \vdots \\ \phi_n^{(m)}(k) \end{pmatrix} = \begin{pmatrix} b_1(k) \\ b_2(k) \\ \vdots \\ b_n(k) \end{pmatrix}, \tag{2.5a}$$

in which, by virtue of relationships (2.2), the complex state vector components $\phi_s^{(m)}$ are knowns. From the set of Eqs. (2.5a) we can determine the unknown complex amplitudes $A_m = A_m(k)$.

With relationships (2.2), we introduce the new notation $A_m\phi_1^{(m)}(k) \equiv \mathcal{A}_m(k)$ to rewrite Eqs. (2.5a) as

$$\sum_{m=1}^{n} \mathcal{A}_m(k) \begin{pmatrix} 1 \\ L_2(\omega_m(k), k) \\ \vdots \\ L_n(\omega_m(k), k) \end{pmatrix} \equiv \sum_{m=1}^{n} \mathcal{A}_m(k) \begin{pmatrix} L_{1m} \\ L_{2m} \\ \vdots \\ L_{nm} \end{pmatrix} = \begin{pmatrix} b_1(k) \\ b_2(k) \\ \vdots \\ b_n(k) \end{pmatrix}, \quad L_{1m} \equiv 1. \tag{2.5b}$$

It is in this form that the equations are usually used to solve particular initial-value problems.

Concerning the set of Eqs. (2.5), some points need to be clarified. If the number of equations in set (2.5) is equal to the number of unknowns, then the equations can be solved unambiguously by linear algebra methods. The number of equations is equal to the number n of state vector components, and the number of unknowns is equal to the number of wave branches, i.e., to the number of solutions to dispersion relation (1.8). Generally, the number of solutions to dispersion relation (1.8) is also equal to n. But the dispersion relation can have trivial roots, which are to be discarded. It might seem that the number of equations in set (2.5) is greater than the number of unknowns, but this is not so. In all such cases, the "redundant" components of the state vector are linear combinations of its remaining components, provided that the problem is well-posed. That is why the number of unknowns and the number of equations in set (2.5) are in fact always the same (for details on this issue, see Chapter 2, Secs. 7, 9, 10).

3. Characteristic Function of the State Vector. Dispersion Operator

Mathematically, harmonic solution (1.2) to the set of linear homogeneous differential equations (1.1) is the simplest possible solution. More complicated linear equations, such as homogeneous pseudodifferential and integrodifferential ones, also have harmonic solutions. Acting by differential and integral operators upon complex functions (1.2) reduces to their multiplication by constants. In complex (exponential) form, the rules for acting on trigonometric functions are written as

$$\begin{gathered}\frac{\partial}{\partial t} \to -i\omega, \qquad \frac{\partial}{\partial z} \to ik, \\ \int (\ldots)dt \to i\omega^{-1}, \quad \int (\ldots)dz \to -ik^{-1},\end{gathered} \tag{3.1}$$

as may be verified by directly inserting solution (1.2). Hence, in the class of solutions of the form (1.2), the differential and integrodifferential problems for the state vector $\mathbf{\Psi}(t,z)$ are reduced to an algebraic problem of determining the frequency $\omega(k)$ and complex state vector $\mathbf{\Phi}(\omega, k)$.

But harmonic waves (1.2) are not the only most general solutions to Eqs. (1.1) and similar equations. Commonly, the solution is a superposition of harmonic waves. In addition, perturbations of the state vector can be produced not only by the initial deviation of the system from an equilibrium state but also by external sources (forces), which are described by the nonzero right-hand sides of Eqs. (1.1). That is, in the most general case, it is necessary to begin with the following basic set of linear inhomogeneous partial differential equations:

$$\frac{\partial \psi_s}{\partial t} + \sum_{j=1}^{n}\left(A_{sj}\frac{\partial \psi_j}{\partial z} + B_{sj}\psi_j\right) = f_s(t,z), \quad s = 1, 2, \ldots, n\,, \tag{3.2}$$

which should be supplemented with certain additional (e.g., initial) conditions. Functions (1.2) are not the only solution to Eqs. (3.2).

The sets of Eqs. (1.1) or (3.2) can be solved for the state vector $\mathbf{\Psi}(t,z) = \{\psi_1, \psi_2, \ldots, \psi_n\}$. But it is more convenient to apply another approach — that based on generalizing the notions of dispersion relation and state vector. The approach described below is analogous to the "rules" that are used in quantum mechanics to pass over from the de Broglie wave function for the free electron to the Schrödinger equation for the wave function. We begin with formula (2.1) for the state vector,

$$\mathbf{\Psi}(t,z) = A \begin{pmatrix} \phi_1(\omega,k) \\ \phi_2(\omega,k) \\ \vdots \\ \phi_n(\omega,k) \end{pmatrix} \exp(-i\omega t + ikz)\,, \tag{3.3}$$

in which we take into account the contribution of only one branch of the eigenmodes (i.e., in writing formula (3.3), we have omitted the number m of the eigenmode branch). Using formula (2.2) for the components $\phi_s(\omega,k)$ of the complex state vector and setting $\phi_1 = 1$, we rewrite formula (3.3) as

$$\mathbf{\Psi}(t,z) = \begin{pmatrix} 1 \\ L_2(\omega,k) \\ \vdots \\ L_n(\omega,k) \end{pmatrix} A(\omega,k)\exp(-i\omega t + ikz)\,. \tag{3.4}$$

Here, $L_s(\omega,k)$ are known functions of the frequency and wavenumber, ω and k, and also of the coefficients of Eqs. (1.1). We also use the following obvious relationship:

$$D(\omega,k)A(\omega,k)\exp(-i\omega t + ikz) = 0\,, \tag{3.5}$$

where $D(\omega,k)$ is the left-hand side of dispersion relation (1.8). Relationship (3.5) reflects an important, although quite obvious, fact: in the class of harmonic waves (1.2), specifically, $\mathbf{\Psi} \sim A(\omega,k)\exp(-i\omega t+ikz)$, Eqs. (1.1) have nontrivial solutions ($A \neq 0$) only when the frequency ω and wavenumber k are related by the dispersion relation $D(\omega,k) = 0$.

Let us now generalize relationships (3.4) and (3.5) to arbitrary perturbations. To do this, we make the replacement

$$A(\omega,k)\exp(-i\omega t + ikz) \to A(t,z)\,, \tag{3.6}$$

where $A(t,z)$ is an arbitrary function of time and coordinates, called the characteristic function of the state vector. Replacement (3.6) implies that, in relationships (3.4) and (3.5), it is necessary to switch from the frequency ω and wavenumber k to the corresponding operators. This is done by generalizing the first two of formulas (3.1), namely, by introducing the frequency and wavenumber operators, $\hat{\omega}$ and $\hat{k}$:

$$\hat{\omega} = i\frac{\partial}{\partial t}\,, \quad \hat{k} = -i\frac{\partial}{\partial z}\,. \tag{3.7}$$

Note that a harmonic wave $\varphi(t,z) = \mathrm{Const}\cdot\exp(-i\omega t + ikz)$ is an eigenfunction of operators (3.7) by virtue of the relationships

$$\hat{\omega}\varphi = \omega\varphi, \quad \hat{k}\varphi = k\varphi\,. \tag{3.8}$$

In what follows, we will also use operators inverse to (3.7). These are obtained by inverting the last two of formulas (3.1):

$$\hat{\omega}^{-1} = -i \int (\ldots) dt, \quad \hat{k}^{-1} = i \int (\ldots) dz\,. \tag{3.9}$$

Making replacement (3.6) in relationships (3.4) and (3.5), i.e., switching from the frequency ω and wavenumber k to the corresponding operators, we obtain the differential relationships

$$\boldsymbol{\Psi}(t, z) = \begin{pmatrix} 1 \\ L_2(\hat{\omega}, \hat{k}) \\ \vdots \\ L_n(\hat{\omega}, \hat{k}) \end{pmatrix} A(t, z)\,, \tag{3.10}$$

$$D(\hat{\omega}, \hat{k}) A(t, z) = 0\,. \tag{3.11}$$

Vector formula (3.10) allows the state vector to be calculated from its characteristic function. As for relationship (3.11), it is the basic equation in the general theory of linear waves in dispersive systems (media). The operator function (the function of operators) $D(\hat{\omega}, \hat{k})$ is called the dispersion operator. Note that, since the operators $D(\hat{\omega}, \hat{k})$ and $L_s(\hat{\omega}, \hat{k})$ are commutative, Eq. (3.1) is satisfied by any component of the state vector (3.10) and also by an arbitrary linear combination of the components.

We assume that the dispersion function $D(\omega, k)$ is a polynomial of finite degree in its arguments. Consequently, the dispersion operator $D(\hat{\omega}, \hat{k})$ corresponding to this function is a differential operator. It is known, however, that, for certain media and systems (such as a "kinetic" hot plasma, systems that are inhomogeneous in a direction transverse to the Z axis, and so on), the dispersion function $D(\omega, k)$ is a transcendental function. Hence, in the most general case, the function $D(\hat{\omega}, \hat{k})$ is a pseudodifferential operator.

If there are external perturbing sources and the basic equations are inhomogeneous equations (3.2), then Eq. (3.11) is also inhomogeneous,

$$D(\hat{\omega}, \hat{k}) A(t, z) = F(t, z)\,. \tag{3.12}$$

Here, the external force $F(t, z)$ is expressed in terms of the functions $f_s(t, z)$ (see (3.2)) and their derivatives.

In writing formulas (3.3) and (3.4), we have omitted the number of the branch of the eigenmodes of the system. The reason is that the formulation of the problem in the language of differential equations (3.11) and (3.12) does not at all imply introducing such notions as waves, frequencies, and wavenumbers. Equation (3.11) and relationship (3.10) are applicable to describing arbitrary perturbations of a physical system, in which case, instead of different eigenmode branches, one must speak of linearly independent solutions to differential equation (3.11).

4. Laplace Transform Method

An efficient method of solving problems in mathematical physics and linear wave theory is that based on integral Laplace transforms. Let us outline the required information on this transform method and on its main properties. Consider a piece-wise smooth function $\varphi(t)$ of a real variable t that satisfies the following conditions:

(1) $\varphi(t) \equiv 0$ for $t < 0$; and
(2) for $t \to \infty$ the function $\varphi(t)$ has a finite order of growth, i.e., $|\varphi(t)| \leq C\exp(\alpha t)$, with C and α being constants. The constant α is called the growth order exponent of the function $\varphi(t)$.

The Laplace transform of a function $\varphi(t)$ of a real variable t is a transform that relates the function $\varphi(t)$ to a function $\varphi(\omega)$ of a complex variable ω, defined by the integral

$$\varphi(\omega) = \int_0^\infty \varphi(t)\exp(i\omega t)dt\,. \tag{4.1}$$

Since the function $\varphi(t)$ has a finite order of growth, integral (4.1) converges in the region $\operatorname{Im}\omega = \omega'' > \alpha$ and the function $\varphi\omega$, called the Laplace transformed function $\varphi(t)$, is an analytic function of the complex variable ω in this region. Note that we are using the same notation for the original function and for its Laplace transform. But no confusion will result, because we will always indicate the corresponding argument: for instance, in the initial-value problem, the original function is written with the argument t and the transformed function, with the argument ω.

Let us list the properties of the transformed function that are important for further analysis.

1. If $\varphi(\omega)$ is the transformed function $\varphi(t)$, then the transformed derivative $\varphi'(t)$ is defined by the formula

$$\varphi'(\omega) = -i\omega\varphi(\omega) - \varphi(t=0)\,. \tag{4.2}$$

2. If $\varphi(\omega)$ is the transformed function $\varphi(t)$, then the transformed nth-order derivative $\varphi^{(n)}(t)$ is defined by the following formula (which is a generalization of the previous property):

$$\varphi^{(n)}(\omega) = (-i\omega)^n\left[\varphi(\omega) - \sum_{q=1}^{n}\frac{\varphi^{(q-1)}(t=0)}{(-i\omega)^q}\right]. \tag{4.3}$$

3. If $\varphi(\omega)$ is the transformed function $\varphi(t)$, then the transformed integral

$$\phi(t) = \int_0^t \varphi(\tau)d\tau \tag{4.4}$$

is defined by the formula

$$\phi(\omega) = \frac{i}{\omega}\varphi(\omega)\,. \tag{4.5}$$

4. If $\varphi_1(\omega)$ and $\varphi_2(\omega)$ are the transformed functions $\varphi_1(t)$ and $\varphi_2(t)$, respectively, then the transformed integral

$$S(t) = \int_0^t \varphi_1(\tau)\varphi_2(t-\tau)d\tau = \int_0^t \varphi_1(t-\tau)\varphi_2(\tau)d\tau\,, \tag{4.6}$$

called the convolution, is defined by the formula

$$S(\omega) = \varphi_1(\omega)\varphi_2(\omega)\,. \tag{4.7}$$

In Laplace transform theory, the main formula is the Mellin formula, with which to determine the original function $\varphi(t)$ from its Laplace transform $\varphi(\omega)$:

$$\varphi(t) = \frac{1}{2\pi}\int_{C(\omega)} \varphi(\omega)\exp(-i\omega t)d\omega\,. \tag{4.8}$$

Here, $C(\omega)$ is a contour of integration that lies in the upper half of the complex frequency plane ω and passes above all the singularities in the integrand. Since the integrand is an analytic function in the region $\mathrm{Im}\,\omega = \omega'' > \alpha$, the integration contour $C(\omega)$ may be any straight line $\mathrm{Im}\,\omega = \omega'' = \sigma > \alpha$ that is parallel to the real axis $\mathrm{Re}\,\omega$ in the complex frequency plane ω (Fig. 1). This straight line lies above all the singularities in the Laplace transform $\varphi(\omega)$.

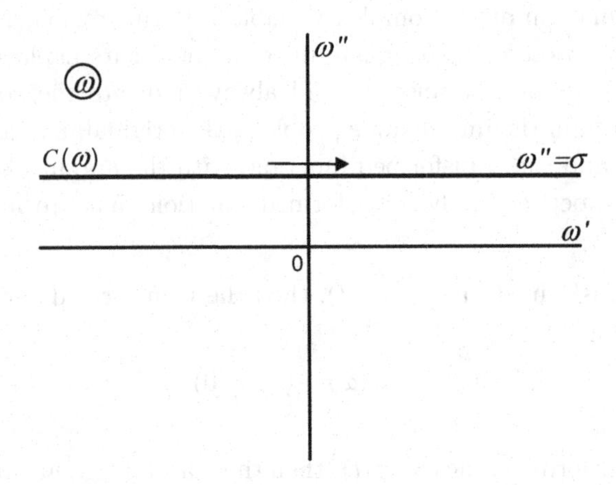

Fig. 1. Contour of integration $C(\omega)$ for calculating the Mellin integral. The arrow shows the direction of integration along the contour.

Integral (4.8) is calculated using Jordan's lemma, which states that, if a function $\varphi(\xi)$ of a complex variable ξ is analytic in the upper half-plane $\mathrm{Im}\,\xi > 0$, (the lower half-plane $\mathrm{Im}\,\xi < 0$) everywhere except at a finite number of isolated singular points and if this function in the upper (lower) half-plane approaches zero uniformly with respect to $\arg\xi$ as $|\xi| \to \infty$, then, for $\beta > 0$ ($\beta < 0$), the following relationship holds:

$$\lim_{R\to\infty}\int_{C(R)} \varphi(\xi)\exp(i\beta\xi)d\xi = 0\,, \tag{4.9}$$

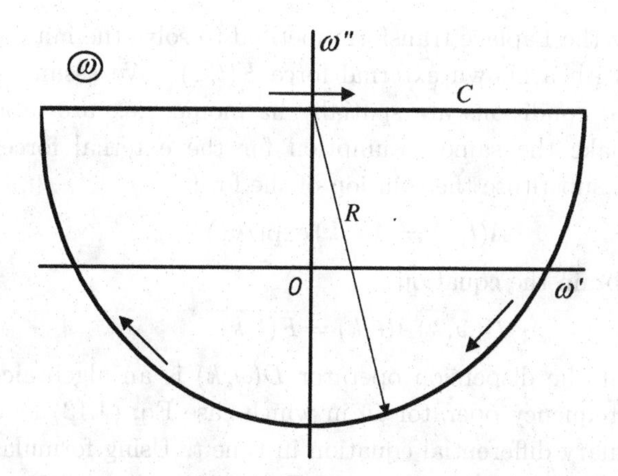

Fig. 2. Contour of integration C $(R \to \infty)$ in formula (4.11).

where $C(R)$ is the arc of the semicircle $|\xi| = R$ in the upper (lower) half-plane of the complex variable ξ.

Let us apply Jordan's lemma to calculate integral (4.8). Setting $\beta = -t < 0$ and $\xi = \omega$, we analytically continue the function $\varphi(\omega)$, which is initially defined in the region $\operatorname{Im}\omega = \omega'' > \alpha$, into the entire complex plane ω and assume that, in the region $\operatorname{Im}\omega = \omega'' < \alpha$, the analytic continuation of the function $\varphi(\omega)$ satisfies the Jordan's lemma conditions. Since $\beta = -t < 0$, relationship (4.9) can be reduced to the form

$$\lim_{R\to\infty} \int_{C(R)} \varphi(\omega)\exp(-i\omega t)d\omega = 0\,, \tag{4.10}$$

where $\tilde{C}(R)$ is the arc of the semicircle $|\omega - i\sigma| = R$ in the lower half-plane of the complex variable ω. With relationship (4.10), formula (4.8) can be represented as the following loop integral:

$$\varphi(t) = \frac{1}{2\pi}\oint_C \varphi(\omega)\exp(-i\omega t)d\omega = -i\sum_{m=1}^{n} \operatorname{Res}[\varphi(\omega)\exp(-i\omega t), \omega_m]\,. \tag{4.11}$$

Here, C is a closed contour composed of the contour $C(\omega)$ (see (4.8)) and a closing semicircle of infinite radius lying in the lower half-plane of the ω complex plane (Fig. 2). In formula (4.11), the integral over the closed contour has been calculated by Cauchy's theorem. The rest of the notation is as follows: ω_m (with $m = 1, 2, \ldots, n$) are the singular points (poles) of the Laplace transform $\varphi(\omega)$. The minus sign arises from the negative (clockwise) direction of integration along the contour C (see Figs. 1, 2) — a direction that is traditionally (almost always) chosen in wave theory (in the theory of functions of a complex variable, the Mellin integral is written in another form, namely, that with the opposite (counterclockwise) direction of integration along the contour and with the replacement $i\omega \to -p$.

Let us now apply the Laplace transform method to solve the initial-value problem for Eq. (3.12) with a known external force $F(t,z)$. We assume for the moment that the initial conditions are spatially harmonic, i.e., are proportional to $\sim \exp(ikz)$, and make the same assumption for the external force, $F(t,z) = F(t,k)\exp(ikz)$. We substitute the solution of the form

$$A(t,z) = A(t,k)\exp(ikz) \tag{4.12}$$

into Eq. (3.12) to obtain the equation

$$D(\hat{\omega},k)A(t,k) = F(t,k)\,. \tag{4.13}$$

We also assume that the dispersion operator $D(\hat{\omega},k)$ is an algebraic polynomial of degree n in the frequency operator $\hat{\omega}$, in which case Eq. (4.13) is an nth-order inhomogeneous ordinary differential equation in time t. Using formulas (3.10) and (4.12), we can write the initial conditions for Eq. (4.13) as (see (2.5))

$$\begin{pmatrix} 1 \\ L_2(\hat{\omega},k) \\ \vdots \\ L_n(\hat{\omega},k) \end{pmatrix} A(t,k)|_{t=0} = \begin{pmatrix} b_1(k) \\ b_2(k) \\ \vdots \\ b_n(k) \end{pmatrix}. \tag{4.14}$$

Since the problem given by Eq. (4.13) and initial conditions (4.14) is linear, homogeneous equation (4.13) ($F(t,k) \equiv 0$) with nonzero initial conditions (4.14) and inhomogeneous equation (4.13) with zero initial conditions can be solved separately. Let us first solve the homogeneous equation. We multiply the equation $D(\hat{\omega},k)A(t,k) = 0$ by $\exp(i\omega t)$ and integrate over t from zero to infinity. Taking into account property 2 of the Laplace transform (see (4.3)) and the definition of the frequency operator in (3.7), we reduce the homogeneous differential equation to the algebraic relationship

$$D(\omega,k)A(\omega,k) = P_{n-1}(\omega,k)\,. \tag{4.15}$$

Here, $D(\omega,k)$ is the dispersion function, $A(\omega,k)$ is the Laplace transformed characteristic function $A(t,k)$, and $P_{n-1}(\omega,k)$ is a polynomial of degree at most $n-1$ in ω (see formula (4.3), whose right-hand side contains derivatives of order at most $n-1$). The coefficients of the polynomial $P_{n-1}(\omega,k)$ depend on the coefficients of the basic equation (4.13), as well as on the function $A(t=0,k) \equiv A_0^{(0)}(k)$ and its derivatives $A^{(q)}(t=0,k) \equiv A_0^{(q)}(k)$ (with $q = 1,\ldots,n-1$)), at $t = 0$.

From relationship (4.15) we obtain the following expression for the Laplace transformed characteristic function:

$$A(\omega,k) = \frac{P_{n-1}(\omega,k)}{D(\omega,k)}\,. \tag{4.16}$$

The original function, or the inverse Laplace transform of function (4.16), can be calculated from the Mellin formula (4.8):

$$A(t,k) = \frac{1}{2\pi}\int_{C(\omega} \frac{P_{n-1}(\omega,k)}{D(\omega,k)} \exp(-i\omega t)d\omega\,, \tag{4.17}$$

where $C(\omega)$ is a straight line that passes above all the singularities in the integrand (Fig. 1). Since function (4.16) is the ratio of two polynomials, the singular points of the integrand in expression (4.17) can only be the roots of the dispersion relation $D(\omega,k) = 0$. In formula (4.16), the degree of the numerator as a function of frequency ω is less, by at least one, than that of the denominator. Consequently, for $t > 0$, function (4.16) satisfies the Jordan's lemma conditions in the lower half-plane $\operatorname{Im}\omega = \omega'' < \alpha$. In this case, expression (4.17) reduces to (see (4.11))

$$A(t,k) = -i\sum_{m=1}^{n} \operatorname{Res}\left[\frac{P_{n-1}(\omega,k)}{D(\omega,k)}\exp(-i\omega t), \omega_m\right], \tag{4.18}$$

where $\omega_m = \omega_m(k)$ are the roots of the dispersion relation $D(\omega,k) = 0$. If all the roots are simple (recall that the case of multiple roots will be considered separately), then, representing the dispersion function as

$$D(\omega,k) = \prod_{m=1}^{N}(\omega - \omega_m(k)) \tag{4.19}$$

and substituting representation (4.19) into expression (4.18), we arrive at the following final solution to the homogeneous equation $D(\hat{\omega},k)A(t,k) = 0$:

$$A(t,k) = \sum_{m=1}^{n}\left[A_m(k)\exp(-i\omega_m(k)t)\right]. \tag{4.20}$$

Here,

$$A_m(k) = -iP_{n-1}(\omega_m(k),k)\left[\prod_{j\neq m}^{n-1}(\omega_m(k) - \omega_j(k))\right]^{-1}. \tag{4.21}$$

The amplitudes (4.21) depend on the characteristic function and its derivatives at $t = 0$ — i.e., on $A^{(q)}(t=0,k) \equiv A_0^{(q)}(k)$ $(q = 0,1,\ldots,n-1)$ — through the coefficients of the polynomial $P_{n-1}(\omega,k)$. The derivatives can be determined from the initial conditions of the form (4.14). But there is a simpler way. Substituting solution (4.20) into initial conditions (4.14), we obtain the set of linear algebraic equations that coincide with Eqs. (2.5b):

$$\sum_{m=1}^{n} A_m(k)\begin{pmatrix} 1 \\ L_2(\omega_m,k) \\ \vdots \\ L_n(\omega_m,k)\end{pmatrix} = \begin{pmatrix} b_1(k) \\ b_2(k) \\ \vdots \\ b_n(k)\end{pmatrix}. \tag{4.22}$$

Having determined the amplitudes $A_m(k)$ from Eqs. (4.22), we immediately find the explicit form of solution (4.20), without calculating the derivatives of the characteristic function at $t = 0$ (these calculations turn out to be unnecessary). It is obvious that the result of solving the initial-value problem (4.12), (4.20) for a homogeneous equation by using the Laplace transform coincides with the solution obtained by the eigenmode method, described in Secs. 1 and 2.

Now, we apply the Laplace transform method to solve inhomogeneous equation (4.13) with zero initial conditions. The zero initial conditions imply that the right-hand sides of conditions (4.14) are zero, i.e., we have $b_s(k) = 0$ for all $s = 1, 2, \ldots, n$. In turn, when the problem is well-posed mathematically, the operators $L_s(\hat{\omega}, k)$ in initial conditions (4.14) contain the derivatives of order up to $n - 1$ with respect to time t. Consequently, the relationships $b_s(k) = 0$ can be reduced to zero initial conditions for the characteristic function of the state vector and for the first $n - 1$ derivatives of this function:

$$A^{(q)}(t = 0, k) \equiv A_0^{(q)}(k) = 0, \quad q = 0, 1, \ldots, n - 1\,. \tag{4.23}$$

We switch to the Laplace transformed functions in Eq. (4.13) and take into account relationships (4.23) and formula (4.3) to obtain

$$D(\omega, k)A(\omega, k) = F(\omega, k) \rightarrow A(\omega, k) = G(\omega, k)F(\omega, k)\,, \tag{4.24}$$

where

$$F(\omega, k) = \int_0^\infty F(t, k) \exp(i\omega t) dt \tag{4.25}$$

is the Laplace transform of the right-hand side of Eq. (4.13) and the notation $G(\omega, k) = D^{-1}(\omega, k)$ is introduced. In order to calculate the inverse Laplace transform of the function $A(\omega, k)$, we use property 4 of the Laplace transform. The function $F(\omega, k)$ is the Laplace transformed function $F(t, k)$. Assume that $G(\omega, k)$, too, is the Laplace transform of a certain function $G(t, k)$. Under this assumption, the function $A(\omega, k)$ is the product of the Laplace transforms (see (4.24)). In accordance with formulas (4.6) and (4.7), this indicates that the inverse Laplace transform of the function $A(\omega, k)$ is described by the integral (convolution)

$$A(t, k) = \int_0^t G(\tau, k)F(t - \tau, k) d\tau = \int_0^t G(t - \tau, k)F(\tau, k) d\tau \tag{4.26}$$

and that the original function $G(t, k)$ is calculated from the Mellin formula,

$$G(t, k) = \frac{1}{2\pi} \int_{C(\omega)} \frac{1}{D(\omega, k)} \exp(-i\omega t) d\omega\,. \tag{4.27}$$

We can easily see that the function $G(t, k)$ is a solution to the following inhomogeneous equation with zero initial conditions (this assertion will be proved below):

$$D(\hat{\omega}, k)G(t, k) = \delta(t)\,. \tag{4.28}$$

Here, $\delta(t)$ is the delta function. The function $G(t, k)$ is called the unit point source function (the adjective "point" here means "instantaneous" because the right-hand side of Eq. (4.28) is the function $\delta(t)$), and formula (4.26) is called the Duhamel integral.

It should be stressed that expression (4.27) for the point source function can be obtained from formula (4.17) by setting $P_{n-1}(\omega, k) = 1$. This circumstance is not accidental, but rather the result of the fact that function (4.27) coincides with the solution to the initial-value problem for the homogeneous equation

$$D(\hat{\omega}, k)G(t, k) = 0 \tag{4.29}$$

with the specific initial conditions

$$\begin{aligned} \frac{d^s G(0,k)}{dt^s} &= 0, \quad s = 0, 1, 2, \ldots, n-2, \\ \frac{d^{n-1} G(0,k)}{dt^{n-1}} &= 1\,. \end{aligned} \tag{4.30}$$

In fact, applying the Laplace transformation in the time t to Eq. (4.29) and using property (4.3) and initial conditions (4.30), we readily obtain the algebraic relationship

$$D(\omega, k)G(\omega, k) = 1\,. \tag{4.31}$$

We then express the function $G(\omega, k)$ from relationship (4.31) and make the inverse Laplace transformation to arrive again at expression (4.27) for the unit point source function. Note that, in deriving relationship (4.31) from Eq. (4.29) and initial conditions (4.30), it has been assumed that the coefficient of $\hat{\omega}^n$ in the dispersion operator $D(\hat{\omega}, k)$ is unity (see also expansion (4.19)) — an assumption that can clearly be made without any loss of generality. Consequently, the solution to the initial-value problem for the homogeneous equation and particular solutions to the inhomogeneous equation are similar in structure, so the most general properties of a physical system can be analyzed by using any of the solutions obtained above, depending on what is more convenient.

Chapter 2

A Case Study of Linear Waves in Dispersive Media

5. Transverse Electromagnetic Waves in an Isotropic Dielectric

The simplest electromagnetic waves propagating along the z axis in an unbounded homogeneous medium of constant permittivity ε_0 are described by the following Maxwell's equations for the electric field strength E_x and magnetic field induction B_y (similar equations for E_y and B_x describe the same waves but with a different orientation of the polarization plane):[1]

$$\begin{aligned} \frac{\partial E_x}{\partial t} + \frac{c}{\varepsilon_0}\frac{\partial B_y}{\partial z} &= 0\,, \\ \frac{\partial B_y}{\partial t} + c\frac{\partial E_x}{\partial z} &= 0\,, \end{aligned} \tag{5.1}$$

where c is the speed of light in vacuum. The set of Eqs. (5.1) has the same structure as Eqs. (1.1) and the state vector has the form $\mathbf{\Psi}(t,z) = \{E_x, B_y\}$. We represent the solution as

$$\mathbf{\Psi}(t,z) = \{E_x(t,z), B_y(t,z)\} = \{e_x(\omega,k), b_y(\omega,k)\}\exp(-i\omega t + ikz)\,, \tag{5.2}$$

and substitute it into Eqs. (5.1) to obtain the set of linear homogeneous algebraic equations for the components $e_x(\omega,k)$ and $b_y(\omega,k)$ of the complex state vector of the electromagnetic field:

$$\begin{aligned} -i\omega e_x + i(c/\varepsilon_0)k\, b_y &= 0\,, \\ ick\, e_x - i\omega b_y &= 0\,. \end{aligned} \tag{5.3}$$

The solvability condition for Eqs. (5.3) is given by the dispersion relation

$$D(\omega,k) \equiv -\omega^2 + k^2 c_0^2 = 0\,, \tag{5.4}$$

where $c_0 = c/\sqrt{\varepsilon_0}$ is the speed of light in a dielectric.

From dispersion relation (5.4) we find the frequencies of the two eigenmodes:

$$\omega_1 = kc_0\,, \quad \omega_2 = -kc_0\,. \tag{5.5}$$

[1] If the permittivity is constant, it should necessarily be real.

According to the structure of the dispersion relation (but not according to their physical nature), spectra (5.5) should be classified as acoustic (see (1.13)).

We insert spectra (5.5) into Eqs. (5.3) and take into account dispersion relation (5.4) and representation (5.2) to arrive at the following expression for the state vector of harmonic electromagnetic waves in an anisotropic dielectric:

$$\boldsymbol{\Psi}(t,z) = \begin{Bmatrix} E_x \\ B_y \end{Bmatrix} = A_1 \begin{pmatrix} 1 \\ \sqrt{\varepsilon_0} \end{pmatrix} \exp[ik(z - c_0 t)] + A_2 \begin{pmatrix} 1 \\ -\sqrt{\varepsilon_0} \end{pmatrix} \exp[ik(z + c_0 t)]\,. \tag{5.6a}$$

In writing expression (5.6a), we accounted for the fact that the solution to a set of homogeneous equations is defined to within a constant factor. Accordingly, for the sake of convenience and brevity, we set $e_x = 1$ and calculated the amplitude b_y from Eqs. (5.3). The first term in solution (5.6a) describes a wave propagating in the positive direction of the z axis (with the phase velocity $\omega_1/k = c_0 > 0$), and the second term refers to a wave propagating in the opposite (negative) direction (with the phase velocity $\omega_2/k = -c_0 < 0$).

From spectra (5.5) we can see that the wave phase velocities in an unbounded isotropic dielectric of constant permittivity ($\varepsilon_0 = \text{const}$) are equal to $\pm c_0 = \pm c/\sqrt{\varepsilon_0}$ and thus do not depend on the wavenumber. Consequently, in accordance with formulas (1.13) and (1.15), waves in an unbounded dielectric of constant permittivity are nondispersive.

Let us now determine the complex amplitudes $A_{1,2}$ in solution (5.6a). Assume that, initially (at $t = 0$), the state vector of the system is given by the formula

$$\boldsymbol{\Psi}(0,t) = \begin{pmatrix} E_0 \\ B_0 \end{pmatrix} \exp(ikz)\,, \tag{5.7}$$

where E_0 and B_0 are the electric field strength and magnetic field induction in the initial perturbation. With solution (5.6a) and formula (5.7), Eqs. (2.5) reduce to the following set of two linear algebraic equations:

$$\begin{aligned} A_1 + A_2 &= E_0\,, \\ A_1 - A_2 &= \frac{1}{\sqrt{\varepsilon_0}} B_0\,. \end{aligned} \tag{5.8}$$

We then express the complex amplitudes $A_{1,2}$ from Eqs. (5.8) and substitute them into solution (5.6a) to write the final general solution to the problem of harmonic waves excited in an isotropic dielectric by an initial harmonic perturbation (5.7):

$$\begin{aligned} \boldsymbol{\Psi}(t,z) = \begin{Bmatrix} E_x \\ B_y \end{Bmatrix} &= \frac{1}{2}\left(E_0 + \frac{1}{\sqrt{\varepsilon_0}} B_0\right) \begin{pmatrix} 1 \\ \sqrt{\varepsilon_0} \end{pmatrix} \exp[ik(z - c_0 t)] \\ &+ \frac{1}{2}\left(E_0 - \frac{1}{\sqrt{\varepsilon_0}} B_0\right) \begin{pmatrix} 1 \\ -\sqrt{\varepsilon_0} \end{pmatrix} \exp[ik(z + c_0 t)]\,. \end{aligned} \tag{5.6b}$$

In dispersion relation (5.4), we go over from the frequency and wavenumber to operators (3.7). As a result, we obtain the following homogeneous partial differential equation for the characteristic function of the state vector of arbitrary linear electromagnetic perturbations in an isotropic dielectric:

$$\left(\frac{\partial^2}{\partial t^2} - c_0^2 \frac{\partial^2}{\partial z^2}\right) A(t, z) = 0\,. \tag{5.9}$$

This is a particular example of the general equation (3.11). The state vector of the electromagnetic field in an isotropic dielectric is expressed through the characteristic function as (see (3.10))

$$\mathbf{\Psi}(t, z) = \{E_x(t, z), B_y(t, z)\} = \left\{A(t, z), -\frac{\varepsilon_0}{c}\int \frac{\partial A}{\partial t}(t, z)dz\right\}. \tag{5.10}$$

Equation (5.9), which is known as the wave (d'Alembert) equation, describes not only the dynamics of transverse electromagnetic fields in a homogeneous isotropic dielectric but also the transverse oscillations of a string, the longitudinal oscillations of a rod, acoustic oscillations, and many other physical phenomena. It can be easily verified by direct substitution that Eq. (5.9) is satisfied by doubly differentiable functions of the arguments $z \pm c_0 t$: $f^{(+)}(z - c_0 t)$ and $f^{(-)}(z + c_0 t)$. These are two linearly independent solutions to second-order differential equation (5.9). Consequently, the general solution to this equation has the form

$$A(t, z) = f^{(+)}(z - c_0 t) + f^{(-)}(z + c_0 t)\,. \tag{5.11}$$

The specific form of the functions $f^{(+)}$ and $f^{(-)}$ is determined from the initial conditions.

In solution (5.11), the first term describes a perturbation that propagates with the velocity c_0 in the positive direction of the z axis and keeps its shape unchanged. The second term describes a perturbation that propagates with the same velocity but in the negative direction of the z axis. That is why the constant c_0 in Eq. (5.9) has the meaning of the propagation velocity of the perturbation.

A particular solution of the form (5.11) is that in which the functions $f^{(+)}$ and $f^{(-)}$ are linear combinations of sines and cosines of the arguments $z \pm c_0 t$. With the exponential representation of the trigonometric functions, this solution can be written as

$$f^{(\pm)}(t, z) = A_{1,2} \exp[ik(z \mp c_0 t)]\,, \tag{5.12}$$

where $A_{1,2}$ and k are constants. It is easy to verify by elementary manipulations that general solution (5.11) with particular solution (5.12) puts expression (5.10) into harmonic solution (5.6).

A comparison between solutions (5.12) and (5.11) shows that the velocity c_0 in the wave equation (5.9) has the meaning not only of the phase velocity of harmonic waves but also of the propagation velocity of perturbations of arbitrary shape. Moreover, as the perturbations propagate, their shape remains unchanged. These exclusive properties of the solutions to the wave equation (5.9) stem from the fact that the waves described by this equation are nondispersive.

6. Longitudinal Electrostatic Waves in a Cold Isotropic Plasma. Collisional Dissipation of Plasma Waves

Consider one-dimensional electrostatic perturbations of the electric field and of the electron velocity and density in an unbounded cold electron plasma and assume that the perturbed quantities depend only on the time t and spatial coordinate z. In the unperturbed state, the electric field is absent, the electron density is a constant equal to N_{0p} = const, and the electron velocity is zero. We denote the perturbed electric field and electron velocity and density as $E_z(t,z)$, $U_p(t,z)$, and $N_p(t,z)$, respectively. In the linear approximation, the perturbed current density in the plasma is described by the relationship

$$J_z(t,z) = e(N_{0p} + N_p)U_p \approx eN_{0p}U_p(t,z) \equiv J_p(t,z)\,. \tag{6.1}$$

From Maxwell's equations and electron hydrodynamic equations it follows that, for one-dimensional electrostatic perturbations in the plasma under consideration, Eqs. (1.1) have the form

$$\begin{aligned} \frac{\partial J_p}{\partial t} - \frac{\omega_p^2}{4\pi}E_z = 0\,, \\ \frac{\partial E_z}{\partial t} + 4\pi J_p = 0\,, \end{aligned} \tag{6.2}$$

where $\omega_p = \sqrt{4\pi e^2 N_{0p}/m}$ is the Langmuir frequency of the plasma electrons and e and m are the electron charge and mass. The state vector of the plasma reduces to the form $\mathbf{\Psi}(t,z) = \{J_p, E_z\}$. Representing the solution as

$$\mathbf{\Psi}(t,z) = \{J_p(t,z), E_z(t,z)\} = \{j_p(\omega,k), e_z(\omega,k)\}\exp(-i\omega t + ikz)\,, \tag{6.3}$$

and substituting it into Eqs. (6.2) yields the following set of one-dimensional algebraic equations for the components $j_p(\omega,k)$ and $e_z(\omega,k)$ of the complex state vector:

$$\begin{aligned} (\omega_p^2/4\pi)e_z + i\omega j_p = 0\,, \\ -i\omega e_z + 4\pi j_p = 0\,. \end{aligned} \tag{6.4}$$

The solvability condition for this set is given by the dispersion relation

$$D(\omega,k) \equiv -\omega^2 + \omega_p^2 = 0\,. \tag{6.5}$$

Note that, from the first of equations (6.4), one can determine the plasma conductivity $j_p = \sigma(\omega)e_z$ and also the longitudinal plasma permittivity $\varepsilon(\omega) = 1 + 4\pi i\sigma(\omega)/\omega = 1 - \omega_p^2/\omega^2$. We can see that dispersion relation (6.5) is equivalent to the equation $\varepsilon(\omega) = 0$.

Dispersion relation (6.5) gives the frequency spectra of the eigenmodes:

$$\omega_1 = \omega_p\,, \qquad \omega_2 = -\omega_p\,. \tag{6.6}$$

According to the structure of the dispersion relation (but not according to their physical nature), spectra (6.6) should be classified as optical (see (1.14)).

Substituting spectra (6.6) into Eqs. (6.4) and taking into account dispersion relation (6.5) and representation (6.3), we arrive at the following expression for the state vector of harmonic electrostatic perturbations in a cold electron plasma:

$$\mathbf{\Psi}(t,z) = \left\{ \begin{matrix} J_p \\ E_z \end{matrix} \right\} = A_1 \left\{ \begin{matrix} 1 \\ -i\dfrac{4\pi}{\omega_p} \end{matrix} \right\} \exp(-i\omega_p t + ikz) + A_2 \left\{ \begin{matrix} 1 \\ i\dfrac{4\pi}{\omega_p} \end{matrix} \right\} \exp(i\omega_p t + ikz)\,. \tag{6.7}$$

For convenience in writing this expression, we set $j_p = 1$ and calculated the state vector component e_z from Eqs. (6.4) with allowance for spectra (6.6). The first term in expression (6.7) describes a wave propagating in the positive direction of the z axis, and the second term refers to a wave propagating in the opposite direction.

Spectra (6.6) of the eigenfrequencies have the form of spectrum (1.14) with the constant α equal to zero and with $\beta = \pm\omega_p$. Consequently, according to the above terminology, electrostatic waves in a cold plasma are nondispersive.

Assume that, at the initial instant, the state vector of the system is given by the expression

$$\mathbf{\Psi}(0,t) = \left(\begin{matrix} J_0 \\ E_0 \end{matrix} \right\} \exp(ikz)\,, \tag{6.8}$$

where e_0 and J_0 are the electric field strength and plasma current density in the initial perturbation. With expressions (6.7) and (6.8), Eqs. (2.5) can be reduced to the following set of linear algebraic equations for the complex amplitudes:

$$\begin{aligned} A_1 + A_2 &= J_0\,, \\ A_1 - A_2 &= i(\omega_p/4\pi)E_0\,. \end{aligned} \tag{6.9}$$

We then express the complex amplitudes from Eqs. (6.9) and insert them into expression (6.7) in order to obtain a general solution to the problem of longitudinal harmonic waves generated in a cold plasma by an initial harmonic perturbation (6.8). This solution is not presented here because it is quite similar to (5.6b).

In dispersion relation (6.5), we switch from the frequency to the corresponding operator to obtain the following differential equation for the characteristic function of the state vector of linear longitudinal electrostatic perturbations in a cold plasma:

$$\left(\frac{\partial^2}{\partial t^2} + \omega_p^2 \right) A(t,z) = 0\,. \tag{6.10}$$

The state vector of such a plasma is expressed through the characteristic function as (see (3.10))

$$\mathbf{\Psi}(t,z) = \{J_p(t,z), E_z(t,z)\} = \left\{ A(t,z), \frac{4\pi}{\omega_p^2}\frac{\partial A(t,z)}{\partial t} \right\}. \tag{6.11}$$

Equation (6.10) is in fact an ordinary, rather than partial, differential equation. This equation, which is known as the harmonic oscillator equation, describes the dynamics of localized systems like a mathematical pendulum. Hence, a cold plasma with longitudinal electrostatic oscillations behaves as a system with localized parameters.

The general solution to differential equation (6.10) has the form

$$A(t, z) = f_1(z) \exp(-i\omega_p t) + f_2(z) \exp(i\omega_p t), \tag{6.12}$$

where $f_{1,2}(z)$ are arbitrary functions of the spatial coordinate z alone. Solution (6.12) describes oscillations that are harmonic in time. Moreover, the spatial structure of the oscillations does not vary with time: the oscillations do not propagate and their shape remains unchanged. This latter circumstance stems from the nondispersive nature of the oscillations. And that the oscillations do not propagate in space is attributed to the fact that Eq. (6.10), describing longitudinal electrostatic waves in a cold plasma, is an ordinary differential equation.

In a particular case, we can set

$$f_{1,2}(z) = A_{1,2} \exp(ikz), \tag{6.13}$$

where $A_{1,2}$ are constants. Substituting expression (6.13) into solution (6.12) and taking into account characteristic function (6.11) leads to expression (6.7) for harmonic electrostatic perturbations in a cold electron plasma.

In the above discussion of solution (6.7), it has been pointed out that its first term describes a wave propagating in the positive direction of the z axis, while the second term refers to a wave propagating in the opposite direction. As for solution (6.12), it implies that longitudinal plasma oscillations do not propagate at all. Hence, the actual spatial propagation of the perturbations and the propagation of harmonic waves differ generally from one another. For solutions (6.7) and (6.12), these are two radically different notions, whereas for solutions (5.6) and (5.11), these notions are identical. It is known that the spatial propagation of the wave perturbations is characterized not by the phase velocity but by another quantity, which is substantially governed by the wave dispersion (see below).

Let us now consider the damping of longitudinal waves in a plasma — an issue that we will need to deal with in addressing general problems concerning the dissipation of waves (perturbations) of an arbitrary nature in dispersive media. If plasma waves are dissipated by electron collisions (e.g., a weakly ionized plasma is dominated by collisions of electrons with neutral atoms), then the dissipation is accounted for by supplementing the electron hydrodynamic equation with the viscous friction force. In the first of equations (6.2), we switch from the current density J_p to the hydrodynamic velocity U_p (see (6.1)) and take into account the frictional force to write the following equation:

$$\frac{\partial U_p}{\partial t} = \frac{e}{m} E_z - v_{en} U_p . \tag{6.14a}$$

Here, v_{en} is the effective frequency of collisions of electrons with neutral atoms (the expression for this frequency is not specified here). For the current density $J_p = eN_{0p}U_p$, from Eq. (6.14a) we obtain

$$\frac{\partial J_p}{\partial t} = \frac{\omega_p^2}{4\pi} E_z + v_{en} J_p = 0 . \tag{6.14b}$$

Substituting solution (6.3) into Eq. (6.14b) and into the second of equations (6.2) yields the following set of equations for the components of the complex state vector:

$$\begin{aligned}(\omega_p^2/4\pi)e_z + i(\omega + i\nu_{en})j_p &= 0\,,\\ -i\omega e_z + A\pi j_p &= 0\,.\end{aligned} \tag{6.15}$$

This equation leads to the dispersion relation

$$D(\omega, k) \equiv -\omega^2 - i\nu_{en}\omega + \omega_p^2 = 0\,. \tag{6.16}$$

In contrast to (6.5), dispersion relation (6.16) is an algebraic equation with complex coefficients. Note that this dispersion relation is equivalent to the equation $\varepsilon(\omega) = 1 - \omega_p^2/[\omega(\omega + i\nu_{en})] = 0$, implying that the complex permittivity of a cold collisional plasma is zero.

From dispersion relation (6.16) we find the complex frequencies of the eigenmodes:

$$\omega_1 = \omega_p' - i\frac{\nu_{en}}{2}, \quad \omega_2 = -\omega_p' - i\frac{\nu_{en}}{2}, \quad \omega_p' = \sqrt{\omega_p^2 - \frac{\nu_{en}^2}{e}}\,. \tag{6.17}$$

Under the inequality $2\omega_p > \nu_{en}$, the damping rate of the modes with frequencies (6.17) is $\omega_{1,2}'' = -\nu_{en}/2 < 0$ (see solution (1.17)) and the real part of the frequency, $\omega_{1,2}' = \pm\omega_p'$, is nonzero.

In dispersion relation (6.16), we switch from the frequency to the corresponding operator to obtain the differential equation for the characteristic function of the state vector of damped electrostatic perturbations in a cold collisional plasma:

$$\left(\frac{\partial^2}{\partial t^2} + \nu_{en}\frac{\partial}{\partial t} + \omega_p^2\right) A(t, z) = 0\,. \tag{6.18}$$

The solution to ordinary differential equation (6.18) has the form

$$A(t, z) = \lfloor f_1(z)\exp(-i\omega_p' t) + f_2(z)\exp(i\omega_p' t)\rfloor \exp[-(\nu_{en}/2)t]\,, \tag{6.19}$$

where $f_{1,2}(z)$ are arbitrary functions of the spatial coordinate z solely. Solution (6.19) describes oscillations that are damped with time. As in the case of solution (6.12), such oscillations do not propagate in space and their shape remains undistorted. For damped oscillations, characteristic function (6.11) becomes

$$\boldsymbol{\Psi}(t, z) = \{J_p(t, z), E_z(t, z)\} = \left\{A(t, z), \frac{4\pi}{\omega_p^2}\left(\frac{\partial}{\partial t} + \nu_{en}\right) A(t, z)\right\}. \tag{6.20}$$

Under the inequality $2\omega_p \le \nu_{en}$, the real parts of frequencies (6.17) vanish, in which case the plasma oscillations become damped aperiodically. Since $|\omega_1''| < |\omega_2''|$, we can see that, on long time scales, only the first term in solution (6.19), which describes oscillations with the slowest damping rate, becomes important. Under the strong inequality

$$2\omega_p \ll \nu_{en}\,, \tag{6.21}$$

formulas (6.17) yield the following expressions for the imaginary frequencies:

$$\omega_1 = -i\omega_p \frac{\omega_p}{\nu_{en}}\,, \qquad \omega_2 = -i\nu_{en}\,. \tag{6.22}$$

The frequency ω_1 describes the collective damping of electrostatic perturbations in a highly collisional plasma (the Maxwell relaxation; see Sec. 8 below), and the frequency ω_2 describes the single-particle damping of the perturbations that are not affected by the self-consistent electric field E_z.

Under inequality (6.21), we have $|\omega_1''| \ll |\omega_2''|$, so the general expression (6.19) for the characteristic function can be written as

$$A(t, z) = f(z)\exp[-(\omega_p/\nu_{en})\omega_p t]\,, \tag{6.23}$$

where $f(z)$ is an arbitrary function. In this approximation, dispersion relation (6.16) reduces to

$$D(\omega, k) \equiv -i\nu_{en}\omega + \omega_p^2 = 0\,. \tag{6.24}$$

In this dispersion relation, we go over to the frequency and wavenumber operators to arrive at the ordinary differential equation describing aperiodic damped oscillations:

$$\left(\frac{\partial}{\partial t} + \omega_p \frac{\omega_p}{\nu_{en}}\right) A(t, z) = 0\,. \tag{6.25}$$

The general solution to Eq. (6.25) is the characteristic function (6.23).

7. Transverse Electromagnetic Waves in a Cold Isotropic Plasma. Dissipation of Transverse Waves in a Plasma

It has been shown above that a cold plasma with longitudinal waves behaves as a localized system. But a plasma with transverse electromagnetic waves behaves in a radically different manner. The simplest equations for the components of the state vector of a "transverse electromagnetic field + a cold isotropic electron plasma" system — those like Eqs. (1.1) — have the form

$$\begin{aligned}
\frac{\partial E_x}{\partial t} + c\frac{\partial B_y}{\partial z} + 4\pi J_x &= 0\,,\\
\frac{\partial B_y}{\partial t} + c\frac{\partial E_x}{\partial z} &= 0\,,\\
\frac{\partial J_x}{\partial t} + \frac{\omega_p^2}{4\pi}E_x &= 0\,.
\end{aligned} \tag{7.1}$$

The state vector is $\boldsymbol{\Psi}(t, z) = \{E_x, B_y, J_x\}$, where J_x is the density of the current induced in a plasma by the transverse electric field E_x of the wave. This current density J_x is calculated from the third of Eqs. (7.1), which is a consequence of the cold hydrodynamic equations for the plasma electrons (see also the first of Eqs. (6.2)).

Representing the solution as

$$\mathbf{\Psi}(t,z) = \{E_x, B_y, J_x\} = \{e_x(\omega,k), b_y(\omega,k), j_x(\omega,k)\}\exp(-i\omega t + ikz)\,, \tag{7.2}$$

and substituting it into Eqs. (7.1), we obtain the following set of homogeneous algebraic equations for the components $e_x(\omega,k)$, $b_y(\omega,k)$ and $j_x(\omega,k)$ of the complex state vector:

$$\begin{aligned} -i\omega\, e_x + ikc\, b_y + 4\pi j_x &= 0\,, \\ ikc\, e_x - i\omega\, b_y &= 0\,, \\ (\omega_p^2/4\pi)e_x + i\omega j_x &= 0\,. \end{aligned} \tag{7.3}$$

The solvability condition for this set is given by the dispersion relation

$$D(\omega,k) \equiv i\omega(-\omega^2 + k^2c^2 + \omega_p^2) = 0\,. \tag{7.4a}$$

One of the solutions to dispersion relation (7.4a) — specifically, $\omega = 0$ — describes a nonuniform static magnetic field in the plasma and as such has nothing to do with the wave processes in which we are interested here. Accordingly, we exclude this solution from the analysis and write the dispersion relation as

$$D(\omega,k) \equiv -\omega^2 + k^2c^2 + \omega_p^2 = 0\,. \tag{7.4b}$$

Dispersion relation (7.4b) yields the following frequency spectra of the eigenmodes:

$$\omega_1 = \sqrt{k^2c^2 + \omega_p^2}\,, \quad \omega_2 = -\sqrt{k^2c^2 + \omega_p^2}\,. \tag{7.5}$$

In what follows, waves with frequency spectra (7.5) will be considered as optical (electromagnetic). This is somewhat inconsistent with what was said above in discussing spectra (1.13) and (1.14) but better reflects the physical nature of the most important wave types in dispersive media.

In the initial-value problem, we substitute frequency spectra (7.5) into Eqs. (7.3) and take into account representation (7.2) and dispersion relations (7.4) to obtain the following formula for the state vector of an isotropic plasma with transverse harmonic electromagnetic perturbations:

$$\mathbf{\Psi}(t,z) = \{E_x, B_y, J_x\} = \sum_{m=1}^{2} A_m \left\{1, \frac{kc}{\omega_m(k)},\; i\frac{\omega_p^2}{4\pi\omega_m(k)}\right\} \exp[-i\omega_m(k)t + ikz]\,. \tag{7.6a}$$

The first term in formula (7.6a) describes a wave propagating in the positive direction of the z axis (with the phase velocity $\omega_1/k > 0$), and the second term refers to a wave propagating in the opposite direction (with the phase velocity $\omega_2/k < 0$).

Solution (7.6a) coincides qualitatively with solution (5.6). But formulas (7.5) imply that, in contrast to the case of an isotropic dielectric, the phase velocities of the transverse electromagnetic waves in an isotropic plasma depend strongly on the wavenumber k:

$$V_{ph}^{(1,2)} = \pm\frac{\sqrt{k^2c^2 + \omega_p^2}}{k} = \pm c\sqrt{1 + \frac{\omega_p^2}{k^2c^2}}\,. \tag{7.7}$$

By virtue of criterion (1.15), waves with frequencies (7.5) are dispersive. For $k \to 0$, the phase velocities (7.7) become infinite. As the wavenumber k increases, the dispersion of electromagnetic waves in an isotropic plasma becomes weaker and, at $k \to \infty$, the waves become nondispersive since $\lim_{|k|\to\infty} V_{ph}^{(1,2)} = \pm c = \text{const}$

The case at hand illustrates a situation in which the number of state vector components (three) is greater than the number of independent eigenmode branches (two). Recall that the branch $\omega = 0$ has been excluded from consideration. This does not present a difficulty, however. In fact, from formula (7.6a) we can see that the two components B_y and J_x of the state vector differ only by a constant factor and are therefore linearly independent, specifically,

$$b_y^{(m)} = \frac{kc}{\omega_m}\,, \quad j_x^{(m)} = i\frac{\omega_p^2}{4\pi\omega_m} = i\frac{\omega_p^2}{4\pi kc} b_y^{(m)} \equiv C \cdot b_y^{(m)}\,, \tag{7.8}$$

where the constant C does not depend on the number m of the branch. Consequently, any of the components B_y and J_x — say, J_x — can be omitted from the set of independent components, in which case the state vector (7.6a) can be represented as

$$\boldsymbol{\Psi}(t,z) = \begin{pmatrix} E_x(t,z) \\ B_y(t,z) \end{pmatrix} = \sum_{m=1}^{2} A_m \begin{pmatrix} 1 \\ kc/\omega_m \end{pmatrix} \exp[-i\omega_m(k)t + ikz]\,. \tag{7.6b}$$

Substituting solution (7.6b) and the initial state vector (5.7) into Eqs. (2.5), we obtain the following set of equations for determining the amplitudes A_m:

$$\begin{gathered} A_1 + A_2 = E_0\,, \\ \frac{kc}{\omega_1}A_1 + \frac{kc}{\omega_2}A_2 = B_0\,. \end{gathered} \tag{7.9}$$

By expressing the complex amplitudes $A_{1,2}$ from Eqs. (7.9) and inserting them into formulas (7.6), we can readily write out a general solution to the problem of transverse harmonic waves generated in a cold plasma by an initial harmonic perturbation (5.7). But the solution is quite obvious (see formula (5.6b)) and we do not present it here.

In dispersion relation (7.4b), we replace the frequency and wavenumber by operators (3.7) to arrive at the following partial differential equation for the characteristic function of the state vector of arbitrary linear transverse electromagnetic perturbations in an isotropic plasma:

$$\left(\frac{\partial^2}{\partial t^2} - c^2\frac{\partial^2}{\partial z^2} + \omega_p^2\right) A(t,z) = 0\,. \tag{7.10}$$

This equation, which is similar to Eq. (3.11), is known as the Klein–Gordon equation. Along with the phenomena mentioned in discussing Eq. (5.9) — those from the electrodynamics, acoustics, and elasticity theory — it also describes the motion of a relativistic quantum particle. The state vector of the transverse electromagnetic field in an isotropic plasma is expressed in terms of the characteristic function by the formula (see (3.10))

$$\boldsymbol{\Psi}(t,z) = \{E_x(t,z), B_y(t,z)\} = \left\{A(t,z), -c\int \frac{\partial A}{\partial z}(t,z)dt\right\}. \tag{7.11}$$

It can be easily verified that, for any constant value of the velocity V, arbitrary functions $f(z \pm Vt)$ are not solutions to Eq. (7.10). All dispersive systems — in particular, those described by differential equation (7.10) — have this property. The only function of the argument $z \pm Vt$ that satisfies Eq. (7.10) is that describing a harmonic wave:

$$A(t,z) = A_1 \exp\lfloor ik(z - V_{ph})t\rfloor + A_2 \exp\lfloor ik(z + V_{ph})t\rfloor, \tag{7.12}$$

where $V_{ph} = V_{ph}(k)$ is one of the phase velocities (7.7). The fundamental difference of functions (7.12) from functions (5.12) is that, in contrast to c_0, the phase velocity V_{ph} in (7.12) is independent of the wavenumber k. This circumstance, which is attributed to the dispersion of transverse electromagnetic waves in the plasma, does not allow us to construct solutions of the form $f(z \pm Vt)$ to Eq. (7.10) that would have an only constant velocity V.

The dissipation of transverse waves in a plasma can be accounted for in the same manner as was done in considering longitudinal waves. The third of Eqs. (7.1) — the equation for the transverse current density component J_x — should be written as

$$\frac{\partial J_x}{dt} - \frac{\omega_p^2}{4\pi}E_x + v_{en}J_x = 0, \tag{7.13}$$

where v_{en} is the frequency of collisions of electrons with neutral atoms (see Eq. (6.14b)). We then substitute solution (7.2) into the first two of Eqs. (7.1) and into Eq. (7.13) to obtain, instead of (7.3), the following set of algebraic equations:

$$\begin{aligned} -i\omega\, e_x + ikc\, b_y + 4\pi j_x &= 0, \\ ikc\, e_x - i\omega\, b_y &= 0, \\ (\omega_p^2/4\pi)e_x + i(\omega + iv_{en})j_x &= 0. \end{aligned} \tag{7.14}$$

The solvability condition for this set is represented by the following dispersion relation for the spectra of transverse electromagnetic waves in a collisional plasma:

$$D(\omega,k) \equiv -\omega^2 + k^2c^2 + \omega_p^2\frac{\omega}{\omega + iv_{en}} = 0. \tag{7.15}$$

It is meaningful to compare dispersion relation (7.15) with collisionless dispersion relation (7.4a): both are third-order in frequency.

Under the inequality

$$v_{en} \ll \sqrt{k^2c^2 + \omega_p^2}, \tag{7.16}$$

dispersion relation (7.15) yields the following frequency spectra of weakly damped transverse electromagnetic waves in a plasma:

$$\omega_{1,2} = \pm\sqrt{k^2c^2 + \omega_p^2} - i\frac{v_{en}}{2}\frac{\omega_p^2}{k^2c^2 + \omega_p^2}. \tag{7.17}$$

With dissipative effects taken into account, dispersion relation (7.15) has a third nonzero solution instead of the trivial root $\omega = 0$ of dispersion relation (7.4a). Assuming that inequality (7.16) is satisfied and that $|\omega| \ll kc$, we can write this third solution as

$$\omega_3 = -i v_{en} \frac{k^2c^2}{k^2c^2 + \omega_p^2} . \tag{7.18}$$

For $kc \ll \omega_p$, formula (7.18) is applicable under the condition $v_{en} kc \ll \omega_p^2$.

Solutions (7.17) and (7.18) are quite different in character. While spectra (7.17) describe weakly damped high-frequency electromagnetic waves, spectrum (7.18) describes an aperiodic process that is not in fact a wave process. By virtue of inequality (7.16), spectra (7.17) and (7.18) refer to substantially different frequency ranges. This is why the phenomena that the spectra characterize can be dealt with independently. For instance, in considering high-frequency electromagnetic waves in a collisional plasma, solution (7.18) can be excluded from the analysis. In this case, using solution (7.6b), we write the following formula for the state vector of an isotropic collisional plasma with high-frequency transverse harmonic electromagnetic perturbations in the initial-value problem:

$$\begin{aligned}\mathbf{\Psi}(t,z) &= \begin{pmatrix} E_x(t,z) \\ B_y(t,z) \end{pmatrix} \\ &= \sum_{m=1}^{2} A_m \begin{pmatrix} 1 \\ kc/\omega_m \end{pmatrix} \exp[-i\omega_m(k)t + ikz] \exp(-t(v_{en}\omega_p^2)/(2\omega_m^2)) ,\end{aligned} \tag{7.19}$$

where the eigenfrequencies $\omega_m(k)$ are given by formulas (7.5). As for spectrum (7.18), its analogue will be considered in the next section, devoted to metals. At this point, note only that, for $k^2c^2 \ll \omega_p^2$, spectrum (7.18) describes magnetic field diffusion (as well as current diffusion) in a plasma, and, for $k^2c^2 \gg \omega_p^2$, this spectrum yields the single-particle damping rate ω_2 from (6.22).

Returning again to high-frequency electromagnetic waves, we assume that inequality (7.16) is satisfied and rewrite dispersion relation (7.15) as

$$D(\omega, k) \equiv -\omega^2 + k^2c^2 + \omega_p^2 \left(1 - i\frac{v_{en}}{\omega}\right) = 0 . \tag{7.20}$$

In this dispersion relation, we go over to the frequency and wavenumber operators (3.7) and use the inverse operators (3.9) to obtain the following integrodifferential equation for the characteristic function of the state vector of transverse electromagnetic perturbations in a collisional plasma:

$$\left(\frac{\partial^2}{\partial t^2} - c^2\frac{\partial^2}{\partial z^2} + \omega_p^2\right) A(t,z) + \omega_p^2 v_{en} \int A(t',z)dt' = 0 . \tag{7.21}$$

In Eq. (7.21), the integral term arises from strong dispersion, i.e., from the strong dependence of the wave damping rate (7.17) on the wavenumber k. Under inequality (7.16), function (7.19) is the simplest harmonic solution to Eq. (7.21).

8. Electromagnetic Waves in Metals

The density of the conduction current induced by an electric field in metals is given by the formula

$$\mathbf{j} = \sigma \mathbf{E}\,, \tag{8.1}$$

where σ is the conductivity, which is assumed to be a constant scalar quantity. The conductivity of metals is so high that the displacement current $\partial \mathbf{E}/\partial t$ can be ignored in comparison with the conduction current (8.1). Let us consider electromagnetic waves that propagate in an unbounded homogeneous metal along the OZ axis and whose field has two nonzero components, $E_x(t,z)$ and $B_y(t,z)$. We ignore the displacement current and, in accordance with formula (8.1), set $j_x = \sigma E_x$ to rewrite Eqs. (7.1) as

$$\begin{gathered} \frac{\partial B_y}{\partial z} + \frac{4\pi\sigma\mu_0}{c} E_x = 0\,, \\ \frac{\partial B_y}{\partial t} + c\frac{\partial E_x}{\partial z} = 0\,. \end{gathered} \tag{8.2}$$

Here, μ_0 is the relative magnetic permeability, which is introduced phenomenologically and is assumed to be constant. Equations (8.2), which describe purely transverse electromagnetic waves, do not cover the entire scope of electromagnetic waves (perturbations) in metals. If charge density perturbations arise within a metal, then they give rise to a longitudinal electric field that should be described by the equations

$$\begin{gathered} \frac{\partial E_z}{\partial z} - \frac{4\pi}{\varepsilon_0}\rho = 0\,, \\ \frac{\partial \rho}{\partial t} + \sigma\frac{\partial E_z}{\partial z} = 0\,, \end{gathered} \tag{8.3}$$

where ε_0 is the relative permittivity of the metal. The second of Eqs. (8.3) is obviously the continuity equation ($j_z = \sigma E_z, \partial/\partial x = \partial/\partial y = 0$).

We can see that the set of Eqs. (8.2) for transverse waves and the set of Eqs. (8.3) for longitudinal waves are not coupled to one another. Let us first consider longitudinal waves. We eliminate the derivative $\partial E_z/\partial z$ in Eqs. (8.3) to obtain a single equation for the charge density perturbation:

$$\frac{\partial \rho}{\partial t} + \frac{4\pi\sigma}{\varepsilon_0}\rho = 0\,. \tag{8.4}$$

Integrating this equation, we find

$$\rho(t,z) = \rho_0(z)\exp\left(-\frac{t}{\tau}\right), \quad \tau = \frac{\varepsilon_0}{4\pi\sigma}\,, \tag{8.5}$$

where the function $\rho_0(z)$ is determined by the initial charge density perturbation and τ is the so-called Maxwell relaxation time. We can thus see that any perturbation of the charge density in a metal decreases aperiodically in amplitude with time. For metals, the relaxation time τ is extremely short. Thus, for copper, the relaxation time is about $\tau \approx 10^{-19}$ s, so a light wave with this period is an X-ray wave. Consequently, it is quite legitimate to say that, in metals, there are no longitudinal electromagnetic waves due to the charge density perturbations $\rho(t, z)$. Note that we have already dealt with an equation like Eq. (8.4) and with its aperiodic solution (8.5) in considering longitudinal waves in a highly collisional plasma — see Eq. (6.25) and formula (6.23). In a plasma, the Maxwell relaxation time τ coincides with $|\omega_1|^{-1}$, where the frequency ω_1 is defined by the first of expressions (6.22).

Let us now consider transverse waves. Substituting solution (5.2) into Eqs. (8.2) yields the following set of linear homogeneous algebraic equations for the components e_x and b_y of the complex state vector of an electromagnetic field in a metal:

$$\begin{aligned} ikb_y + \frac{4\pi\sigma\mu_0}{c} e_x &= 0\,, \\ \omega b_y - kce_x &= 0\,. \end{aligned} \tag{8.6}$$

The solvability condition for Eqs. (8.6) is represented by the dispersion relation

$$D(\omega, k) \equiv \omega + ik^2c^2 \frac{1}{4\pi\sigma\mu_0} = 0\,, \tag{8.7}$$

which gives the following expression for the only eigenfrequency:

$$\omega = -ik^2c^2 \frac{1}{4\pi\sigma\mu_0} = 0\,. \tag{8.8}$$

The state vector of a harmonic electromagnetic perturbation in a metal is given by the formula

$$\psi(t, z) = A \exp(-k^2c^2(4\pi\sigma\mu_0)^{-1}t - kz)\,, \tag{8.9}$$

where A is a constant. Since Eqs. (8.6) imply that the components e_x and b_y are proportional, the state vector (8.9) has only one component.

Solution (8.9) describes a perturbation that is harmonic in the z direction, does not propagate in space, and is damped aperiodically with time. The damping results obviously from the energy dissipation due to Joule heating. Note that, in such disciplines as electrodynamics, metal optics, and mathematical physics, electromagnetic waves in metals are traditionally considered by formulating a boundary-value problem in order to study the skin effect — the penetration of a wave into a metal. But here these waves are considered in formulating an initial-value problem.

In dispersion relation (8.7), we switch to operators (3.7) to obtain the following homogeneous partial differential equation for the characteristic function of the state vector of arbitrary linear electromagnetic perturbations in a metal:

$$\left(\frac{\partial}{\partial t} - \frac{c^2}{4\pi\sigma\mu_0}\frac{\partial^2}{\partial z^2}\right) A(t, z) = 0\,. \tag{8.10}$$

Equation (8.10) is known as the diffusion, or thermal conductivity, equation. The material substance whose diffusion is described by Eq. (8.10) is the magnetic field component β_y (or the electric field component $E_x \sim j_x$) transverse to the diffusion direction.

Let us now return to formula (7.18) for the damping rate of low-frequency perturbations of the transverse electromagnetic field in a collisional plasma. Assuming that $k^2c^2 \ll w^2p$ and passing over to the frequency and wavenumber operators in formula (7.18), we arrive at the following diffusion equation for the transverse electromagnetic field in a collisional plasma:

$$\left(\frac{\partial}{\partial t} - v_{en}\frac{c^2}{\omega_p^2}\frac{\partial^2}{\partial z^2}\right) A(t,z) = 0\,. \tag{8.11}$$

The physical relationship between Eqs. (8.10) and (8.11) can be established as follows. In the low-frequency range, the term with the time derivative in Eq. (7.13) can be ignored, so we have

$$J_x \equiv j_x = \sigma_e E_x\,, \quad \sigma_e = \frac{\omega_p^2}{4\pi v_{en}}\,, \tag{8.12}$$

where σ_e is the conductivity of an isotropic collisional electron plasma. From relationship (8.12) we can see that, to within the relative magnetic permeability μ_0, Eqs. (8.10) and (8.11) coincide with one another.

9. Electromagnetic Waves in a Waveguide with an Isotropic Dielectric

The simplest model of a waveguide is one in which a homogeneous isotropic dielectric is placed between two unbounded parallel metal planes. Let the z axis be parallel to the planes and the x axis be perpendicular to them. In such a coordinate system, we consider waves propagating along the z axis. A waveguide with two metal planes differs from an unbounded dielectric in two main respects: first, the field in the x direction has the structure of a standing wave, and, second, in addition to the components E_x and B_y, the state vector has the third component E_z. Since the electric field component E_z vanishes at the metal planes, the components of the state vector can be represented as

$$\begin{aligned} E_x &= \cos k_\perp x \cdot \tilde{E}_x(t,z)\,, \\ B_y &= \cos k_\perp x \cdot \tilde{B}_y(t,z)\,, \\ E_z &= \sin k_\perp x \cdot \tilde{E}_z(t,z)\,, \end{aligned} \tag{9.1}$$

where $k_\perp = \pi q/L$, with $q = 1, 2, \ldots$, and L is the distance between the planes. The first two of relationships (9.1) are a consequence of both the third relationship and Maxwell's equations. In turn, the third of relationships (9.1) follows from the fact that the tangential electric field components vanish at the metal planes, $E_z|_{x=0} = E_z|_{x=L} = 0$.

For the waveguide system under consideration, we reformulate Eqs. (1.1) in terms of the quantities $\tilde{E}_x$, $\tilde{B}_y$ and $\tilde{E}_z$. The equations in question derive from

Maxwell's equations and have the form (the tilde "~" from the field components is omitted)

$$\begin{aligned}\frac{\partial E_x}{\partial t} + \frac{c}{\varepsilon_0}\frac{\partial B_y}{\partial z} = 0\,,\\ \frac{\partial B_y}{\partial t} + c\frac{\partial E_x}{\partial z} - k_\perp c E_z = 0\,,\\ \frac{\partial E_z}{\partial t} + \frac{1}{\varepsilon_0}k_\perp c B_y = 0\,.\end{aligned} \tag{9.2}$$

For $L \to \infty$ (the case of an unbounded dielectric), we have $k_\perp \to 0$ and Eqs. (9.2) go over to Eqs. (5.1). In Eqs. (9.2), the state vector is $\mathbf{\Psi}(t,z) = \{E_x, B_y, E_z\}$.

We represent the solution to Eqs. (9.2) as

$$\mathbf{\Psi}(t,z) = \{E_x, B_y, E_z\} = \{e_x(\omega,k), b_y(\omega,k), e_z(\omega,k)\}\exp(-i\omega t + ikz)\,. \tag{9.3}$$

Substituting this solution into the equations yields the following set of homogeneous algebraic equations for the components $e_x(\omega,k)$, $b_y(\omega,k)$ and $e_z(\omega,k)$ of the complex state vector:

$$\begin{aligned}-i\omega\, e_x + ik(c/\varepsilon_0)b_y = 0\,,\\ ikc\, e_x - i\omega\, b_y - k_\perp c e_z = 0\,,\\ k_\perp(c/\varepsilon_0)b_y - i\omega\, e_z = 0\,.\end{aligned} \tag{9.4}$$

The solvability condition for Eqs. (9.4) is expressed by the dispersion relation

$$D(\omega,k) \equiv -\omega^2 + k^2c_0^2 + k_\perp^2 c_0^2 = 0\,, \tag{9.5}$$

from which we have excluded the trivial root $\omega = 0$.

Dispersion relation (9.5) leads to the following spectrum of the eigenmodes:

$$\omega_1 = \sqrt{k^2c_0^2 + k_\perp^2 c_0^2}\,, \quad \omega_2 = -\sqrt{k^2c_0^2 + k_\perp^2 c_0^2}\,. \tag{9.6}$$

In accordance with the terminology adopted in discussing spectra (7.5), the waves with frequency spectra (9.6) are optical.

We substitute spectra (9.6) into Eqs. (9.4) and take into account representation (9.3) and dispersion relation (9.5) to obtain the following expression for the state vector of harmonic electromagnetic perturbations in a waveguide with an isotropic dielectric filling:

$$\mathbf{\Psi}(t,z) = \{E_x, B_y, E_z\} = \sum_{m=1}^{2} A_m \left\{1,\ \frac{\varepsilon_0\omega_m}{kc},\ -i\frac{k_\perp}{k}\right\}\exp[-i\omega_m(k)t + ikz]\,. \tag{9.7a}$$

From formulas (9.6) we can see that, in a waveguide with an isotropic dielectric, the wave phase velocities depend strongly on the wavenumber (cf. formulas (7.7)):

$$V_{ph}^{(1,2)} = \pm c_0\frac{\sqrt{k^2 + k_\perp^2}}{k} = \pm c_0\sqrt{1 + \frac{k_\perp^2}{k^2}}\,. \tag{9.8}$$

According to criterion (1.15), such waves are dispersive. It should be noted, however, that the dispersion of waves with frequency spectra (9.6) is attributed not to the presence of a continuous dispersive medium but to restrictions on the possible propagation directions between two metal planes $x = 0, L$ of the waveguide.

Recall that, in dispersion relation (9.5), we have omitted the trivial root $\omega = 0$. That is why it should be expected that one of the components of the state vector (9.7a) of the system can also be omitted. In fact, the components E_x and E_z differ only by a constant factor (independent of the frequency ω_m):

$$e_z^{(m)} = -i\frac{k_\perp}{k}e_x^{(m)} . \tag{9.9}$$

Accordingly, the state vector (9.7) can be represented as

$$\boldsymbol{\Psi}(t,z) = \begin{pmatrix} E_x(t,z) \\ B_y(t,z) \end{pmatrix} = \sum_{m=1}^{2} A_m \begin{pmatrix} 1 \\ \varepsilon_0\omega_m/kc \end{pmatrix} \exp[-i\omega_m(k)t + ikz] . \tag{9.7b}$$

The set of equations for determining the complex amplitudes A_m in solution (9.7b) can be obtained by substituting it, along with the initial state vector (5.7), into Eqs. (2.5):

$$\begin{aligned} A_1 + A_2 &= E_0 , \\ \frac{\varepsilon_0\omega_1}{kc}A_1 + \frac{\varepsilon_0\omega_2}{kc}A_2 &= B_0 . \end{aligned} \tag{9.10}$$

And finally, inserting the solutions to Eqs. (9.10) into solution (9.7b), we can readily find a general solution to the problem of harmonic waves generated by an initial harmonic perturbation (5.7) in a waveguide with an isotropic dielectric filling. This general solution is quite obvious (see expression (5.6a)) and we do not present it here.

Like Eq. (7.10), Eq. (3.11) for the characteristic function of the state vector of arbitrary linear electromagnetic perturbations in a waveguide filled with an isotropic dielectric is the Klein–Gordon equation:

$$\left(\frac{\partial^2}{\partial t^2} - c^2\frac{\partial^2}{\partial z^2} + k_\perp^2 c_0^2\right) A(t,z) = 0 . \tag{9.11}$$

This equation follows from dispersion relation (9.5). As for the state vector of the electromagnetic field in the waveguide under analysis, it is given by the formula

$$\boldsymbol{\Psi}(t,z) = \{E_x(t,z), B_y(t,z)\} = \left\{A(t,z), -\frac{\varepsilon_0}{c}\int\frac{\partial A}{\partial t}(t,z)dz\right\} . \tag{9.12}$$

10. Longitudinal Waves in a Hot Isotropic Plasma. Electron Diffusion in a Plasma

In Sec. 6, we have considered one-dimensional longitudinal electrostatic waves in an unbounded cold electron plasma. We have established that such waves are nondispersive and do not propagate in space. The spatial propagation of electrostatic plasma waves can occur in a bounded plasma or in a plasma with thermal electron motion. The effects stemming from the finite spatial dimensions of the system will be thoroughly investigated later. Now, we take into account only the thermal motion of plasma electrons. In this way, Eq. (6.14a) with a zero frequency is to be supplemented with the pressure force of the electron gas:

$$\frac{\partial U_p}{\partial t} = \frac{e}{m} E_z - \frac{1}{N_{0p} m} \frac{\partial P_e}{\partial z} . \tag{10.1a}$$

Here, for a nondegenerate electron plasma with a Maxwellian electron velocity distribution, the electron pressure P_e is given by the formula

$$P_e = P_0 + p(t, z) = (N_{0p} + N_p(t, z))\kappa_B T_e , \tag{10.2}$$

where P_0 is the equilibrium electron pressure, $p = N_p \kappa_B T_e$ is the perturbed pressure of the electron gas, N_p is the perturbed electron density, T_e is the electron temperature, and κ_B is Boltzmann's constant. Assuming for simplicity that the electron temperature is constant, we rewrite Eq. (10.1a) as

$$\frac{\partial J_p}{\partial t} = \frac{\omega_p^2}{4\pi} E_z - V_{Te}^2 e \frac{\partial N_p}{\partial z} , \tag{10.1b}$$

where $V_{Te} = \sqrt{\kappa_B T_e / m}$ is the thermal velocity of the plasma electrons.

Replacing the first of Eqs. (6.2) by Eq. (10.1b) and taking into account the continuity equation, we write the following set of equations for the state vector $\mathbf{\Psi}(t, z) = \{J_p, E_z, \rho_p\}$ of electrostatic perturbations in a hot electron plasma:

$$\begin{aligned} \frac{\partial J_p}{\partial t} - \frac{\omega_p^2}{4\pi} E_z + V_{Te}^2 \frac{\partial \rho_p}{\partial z} &= 0 , \\ \frac{\partial E_z}{\partial t} + 4\pi J_p &= 0 , \\ \frac{\partial \rho_p}{\partial t} + \frac{\partial J_p}{\partial z} &= 0 , \end{aligned} \tag{10.3}$$

where $\rho_p = e N_p$ is the perturbed charge density of the plasma electrons.

Representing the solution as

$$\mathbf{\Psi}(t, z) = \{J_p, E_z, \rho_p\} = \{J_p(\omega, k), e_z(\omega, k), r_p(\omega, k)\} \exp(-i\omega t + ikz) , \tag{10.4}$$

and substituting it into Eqs. (10.3) yields the following set of homogeneous algebraic equations for the components of the complex state vector:

$$\begin{aligned} (\omega_p^2/4\pi) e_z + i\omega j_p - i V_{Te}^2 k r_p &= 0 , \\ -i\omega\, e_z + 4\pi j_p &= 0 , \\ k j_p - \omega\, r_p &= 0 . \end{aligned} \tag{10.5}$$

The solvability condition for Eqs. (5.3) is given by the dispersion relation

$$D(\omega, k) \equiv -\omega^2 + k^2 V_{Te}^2 + \omega_p^2 = 0 . \tag{10.6}$$

In writing dispersion relation (10.6), we have discarded the trivial root $\omega = 0$, as is the practice in similar situations. This root describes the Debye screening of a static longitudinal field in a hot plasma — an issue that is not considered in detail in the present monograph.

Dispersion relation (10.6) yields the eigenfrequency spectra

$$\omega_1 = \sqrt{k^2 V_{Te}^2 + \omega_p^2}, \quad \omega_2 = -\sqrt{k^2 V_{Te}^2 + \omega_p^2}, \tag{10.7}$$

and Eqs. (10.5) lead to the following expression for the state vector of harmonic electromagnetic perturbations in an unbounded hot plasma:

$$\boldsymbol{\Psi}(t,z) = \begin{pmatrix} J_p(t,z) \\ E_z(t,z) \end{pmatrix} = \sum_{m=1}^{2} A_m \begin{pmatrix} 1 \\ -i4\pi/\omega_m \end{pmatrix} \exp[-i\omega_m(k)t + ikz]. \tag{10.8}$$

It should be noted that, according to the analysis of longitudinal waves in a hot plasma on the basis of the Vlasov kinetic equation, the quantity V_{Te}^2 in formulas (10.7) should be replaced with $3V_{Te}^2$. In what follows, this replacement will be taken into account. In addition, spectra (10.7) are valid under the inequality $k^2 V_{Te}^2 < \omega_p^2$. More precisely, under the opposite inequality, longitudinal waves in a hot plasma are strongly damped, $|\omega'| \sim |\omega''|$, $\omega'' < 0$. The same damping, known as Landau damping, occurs also for $k^2 V_{Te}^2 < \omega_p^2$, in which case, however, the damping rate is exponentially slow, $\omega'' \sim -\exp[-\omega_p^2/(k^2 V_{Te}^2)]$. Note that transverse waves in an isotropic plasma are not subject to Landau damping at all, so, in this case, we deal only with collisional damping, which has been considered in Sec. 7. According to the structure of their spectra, waves with frequencies (10.7) should be considered as optical.

Since we have discarded the trivial root $\omega = 0$ in dispersion relation (10.6), we must also exclude the component ρ_p from the state vector (10.4), because the components ρ_p and E_z differ only by a constant factor (independent of the frequency ω_m):

$$r_p^{(m)} = i\frac{k}{4\pi} e_z^{(m)}. \tag{10.9}$$

Formulas (10.7) and (10.8) have the same structure as formulas (7.5) and (7.6b). Consequently, further analysis of electrostatic waves in an unbounded hot electron plasma will largely repeat the previous analysis. In particular, dispersion relation (10.6) implies that an equation for the characteristic function of the state vector of such waves is the Klein–Gordon equation:

$$\left(\frac{\partial^2}{\partial t^2} - 3V_{Te}^2 \frac{\partial^2}{\partial z^2} + \omega_p^2\right) A(t,z) = 0. \tag{10.10}$$

The state vector of a hot plasma with electrostatic (longitudinal) plasma waves is given by the formula (see also (6.11))

$$\boldsymbol{\Psi}(t,z) = \{J_p(t,z), E_z(t,z)\} = \{A(t,z), -4\pi \int A(t,z)dt\}. \tag{10.11}$$

Let us now consider the effect of collisions on longitudinal waves in a hot electron plasma. With collisions taken into account, we must rewrite the first of Eqs. (10.3) as (see (6.14b))

$$\frac{\partial J_p}{\partial t} - \frac{\omega_p^2}{4\pi} E_z + V_{Te}^2 \frac{\partial \rho_p}{\partial z} + v_{en} J_p = 0\,, \tag{10.12}$$

while keeping the remaining two of the equations unchanged. In this case, the general solution to Eqs. (10.4) is the same and dispersion relation (10.6) transforms to

$$D(\omega, k) \equiv -\omega(\omega + i v_{en}) + k^2 V_{Te}^2 + \omega_p^2 = 0\,. \tag{10.13}$$

For low collision frequencies, $v_{en} \ll \omega_p$, dispersion relation (10.13) yields the frequency spectra

$$\omega_{1,2} = \pm\sqrt{k^2 V_{Te}^2 + \omega_p^2} - i\frac{v_{en}}{2}\,, \tag{10.14}$$

and the state vector is again given by formula (10.8).

In the opposite case of high collision frequencies, $v_{en} \gg \omega_p$, dispersion relation (10.13) yields the spectra

$$\omega_1 = -i v_{en}^{-1}(k^2 V_{Te}^2 + \omega_p^2), \quad \omega_2 = -i v_{en}\,, \tag{10.15}$$

which determine the damping rates of aperiodic longitudinal perturbations in a plasma. For $V_{Te} \to 0$, frequency spectra (10.15) go over to (6.22). In the limit $v_{en} \gg \omega_p$, the most slowly damped perturbation is that with the frequency ω_1; consequently, on long time scales ($v_{en} t > 1$), the characteristic function is described by the formula

$$A(t, z) = A_1 \exp(-v_{en}^{-1}(k^2 V_{Te}^2 + \omega_p^2) + ikz)\,. \tag{10.16}$$

In the first of expressions (10.15), we switch to operators (3.7) to obtain the following differential equation for the characteristic function of the longitudinal perturbations in a hot, highly collisional plasma:

$$\left(\frac{\partial}{\partial t} - \frac{V_{Te}^2}{v_{en}}\frac{\partial^2}{\partial z^2} + \frac{\omega_p^2}{v_{en}}\right) A(t, z) = 0\,. \tag{10.17}$$

Since, for high collision frequencies, the electron density perturbation ρ_p, as well as the quantities E_z and J_p, is proportional to the characteristic function, Eq. (10.17) describes the equalization of the density perturbations. The term proportional to ω_p^2 describes density equalization by the electric field (Maxwell relaxation); this term also refers to a cold plasma (see Eq. (6.25)). The term proportional to V_{Te}^2 in Eq. (10.17) accounts for a free (monopolar) diffusion of the plasma electrons. The free electron diffusion coefficient can be written as

$$D_e = \frac{V_{Te}^2}{v_{en}} = V_{Te} l_{en}\,, \quad l_{en} = \tau_{en} V_{Te}\,, \quad \tau_{en} = \frac{1}{v_{en}}\,, \tag{10.18}$$

where l_{en} is the electron mean free path and τ_{en} is the mean time between collisions of electrons with neutral atoms. If we ignore the electric field (i.e., Maxwell relaxation), then we can rewrite Eq. (10.17) as a "classical" diffusion equation:

$$\frac{\partial N_p}{\partial t} = D_e \frac{\partial^2 N_p}{\partial z^2}\,, \tag{10.19}$$

where N_p is the perturbed density of the plasma electrons.

11. Longitudinal Waves in an Isotropic Degenerate Plasma. Waves in a Quantum Plasma

At a zero absolute temperature, the electrons in a degenerate plasma are in the lowest energy states. The result is that, in the gas approximation, the electron distribution over the momenta $\mathbf{p}$ is described by the Fermi function normalized to the density:

$$f_e(\mathbf{p}) = \begin{cases} 2(2\pi\hbar)^{-3}, & |\mathbf{p}| \le p_F \\ 0, & |\mathbf{p}| > p_F \end{cases}, \tag{11.1}$$

where $p_f = (3\pi^2)^{1/3}\hbar N_e^{1/3}$ is the Fermi momentum, with N_e being the electron density. Using distribution function (11.1) and applying the methods of kinetic gas theory, one can easily obtain the following relationship for the electron pressure in the plasma:

$$P_e = \frac{1}{5}(3\pi^2)^{2/3}\frac{\hbar^2}{m}N_e^{5/3} = \frac{1}{5}\frac{p_F^2}{m}N_e\,. \tag{11.2}$$

This relationship should be regarded as the equation of state for a degenerate electron plasma, in particular, for an electron gas in metals.

To first order in the perturbed electron density N_p, we set $N_e = N_{0p} + N_p(t,z)$ to convert (11.2) to the relationship (cf. (10.2))

$$P_e = P_0 + p(t,z) = \frac{1}{5}\frac{p_{0F}^2}{m}N_{0p} + \frac{1}{3}\frac{p_{0F}^2}{m}N_p(t,z)\,, \tag{11.3}$$

where $p_{0F} = (3\pi^2)^{1/3}\hbar N_{0p}^{1/3}$ is the unperturbed Fermi momentum. Substituting relationship (11.3) into Eq. (10.1), we then obtain the following dynamic equation for a degenerate electron plasma at a zero absolute temperature:

$$\frac{\partial J_p}{\partial t} - \frac{\omega_p^2}{4\pi}E_z + \frac{1}{3}V_{0F}^2\frac{\partial \rho_p}{\partial z} = 0\,. \tag{11.4}$$

Here, $V_{0F} = p_{0F}/m$ is the Fermi velocity and the perturbations of the current density and electron charge, J_p and ρ_p, are the same as those in the first of Eqs. (10.3). The second and third of Eqs. (10.3) are valid for a degenerate plasma, too. Consequently, relationships (10.6)–(10.11) with the replacement of V_{Te}^2 by $V_{0F}^2/3$ are also valid for longitudinal waves in a degenerate electron plasma. In particular, the spectra of longitudinal waves in a degenerate plasma are given by the formulas

$$\omega_1 = \sqrt{\omega_p^2 + \theta k^2 V_{0F}^2}\,, \quad \omega_2 = -\sqrt{\omega_p^2 + \theta k^2 V_{0F}^2}\,, \tag{11.5}$$

with $\theta = 1/3$. According to the structure of their spectra, waves with frequencies (11.5) should be classified as optical.

Note that spectra (11.5) have been derived using the hydrodynamic model. But with a more rigorous model based on the kinetic equation, it can be shown that these spectra are valid only for $k^2V_{0F}^2 \ll \omega_p^2$, with $\theta = 3/5$. For $k^2V_{0F}^2 \sim \omega_p^2$, spectra (11.5) become more complicated, while remaining valid qualitatively. A more important fact is, however, that, in a degenerate plasma, longitudinal waves are not subject to collisionless Landau damping and, consequently, exist for $k^2V_{0F}^2 > \omega_p^2$ as well (in a nondegenerate plasma, for which V_{Te} is an analogue of V_{0F}, weakly damped longitudinal waves with spectra (10.7) exist only for $k^2V_{Te}^2 < \omega_p^2$). In particular, from the kinetic equation it can be inferred that, in the short-wavelength limit $k^2V_{0F}^2 \gg \omega_p^2$, the frequency spectrum of longitudinal waves in a degenerate plasma is given by the formula

$$\omega = \pm kV_{0F}(1 + 2\exp(-2 - 2k^2V_{0F}^2/3\omega_p^2)\,. \tag{11.6}$$

According to the structure of their spectrum, waves with frequencies (11.6) should be treated as acoustic (see 1.13)). In the theory of metals, waves with frequency spectrum (11.6) are known as zero sound.

The degeneration that leads to distribution (11.1) is a quantum effect resulting from the Pauli exclusion principle for particles with a half-integer spin, such as electrons. In the short-wavelength limit, quantum effects arising from the wave properties of the electrons can also become important. Let us analyze how these effects influence longitudinal waves in plasma.

The Schrödinger equation for the wave function $\psi(t,z)$ of an electron executing one-dimensional motion has the form

$$i\hbar\frac{\partial\psi}{\partial t} = \left[-\frac{\hbar^2}{2m}\frac{\partial^2}{\partial z^2} + e\varphi\right]\psi\,, \tag{11.7}$$

where $\varphi(t,z)$ is a scalar potential. In order to go over from the Schrödinger equation to the quantum hydrodynamic description of a plasma, we represent the wave function as

$$\psi(t,z) = a(t,z)\exp\left[\frac{i}{\hbar}S(t,z)\right] \tag{11.8}$$

and use the definitions of the current and charge densities from quantum mechanics,

$$\begin{aligned} \rho_e &= eN_e = e|\psi|^2 = ea^2\,, \\ j_{ze} &= eN_eU_{ze} = \frac{ie\hbar}{2m}\left(\psi\frac{\partial\psi^*}{\partial z} - \psi^*\frac{\partial\psi}{\partial z}\right) = \frac{e}{m}a^2\frac{\partial S}{\partial z}\,. \end{aligned} \tag{11.9}$$

As a result, from Schrödinger equation (11.7) we obtain the set of equations

$$\begin{aligned} &\frac{\partial N_e}{\partial t} + \frac{\partial}{\partial z}(N_eU_{ze}) = 0\,, \\ &\frac{\partial U_{ze}}{\partial t} + U_{ze}\frac{\partial U_{ze}}{\partial z} = -\frac{e}{m}\frac{\partial\varphi}{\partial z} + \frac{\hbar^2}{4m^2}\frac{\partial}{\partial z}\left[\frac{1}{N_e}\left(\frac{\partial^2 N_e}{\partial z^2} - \frac{1}{2N_e}\left(\frac{\partial N_e}{\partial z}\right)^2\right)\right], \end{aligned} \tag{11.10}$$

the first of which coincides with the continuity equation and the second has the same structure as the Euler hydrodynamic equation. This is why Eqs. (11.10) are called the quantum hydrodynamic equations for a cold electron plasma. Moreover, from definitions (11.9) we can see that classical expressions for the current and charge densities remain also valid in quantum hydrodynamics. Equations (11.10) differ from the corresponding classical equations in that the Euler equation contains the so-called "quantum force" arising from the Heisenberg uncertainty principle.

To first order in the perturbations N_p and U_e, we as usual set $N_e = N_{0p} + N_p$, $\rho_p = eN_p$, and $J_p = eN_{0p}U_e$. As a result, from Eqs. (11.10) we obtain the following set of equations:

$$\frac{\partial J_p}{\partial t} - \frac{\omega_p^2}{4\pi}E_z - \frac{\hbar^2}{4m^2}\frac{\partial^3 \rho_p}{\partial z^3} = 0\,,$$
$$\frac{\partial E_z}{\partial t} + 4\pi J_p = 0\,, \tag{11.11}$$
$$\frac{\partial \rho_p}{\partial t} + \frac{\partial J_p}{\partial z} = 0\,.$$

In writing these equations, we have switched from the scalar potential φ to the longitudinal electric field component E_z and have used the equation for E_z from Eqs. (10.3). Inserting solution (10.4) into Eqs. (11.11), we obtain a dispersion relation for the frequency spectra of longitudinal waves in a quantum cold electron plasma:

$$D(\omega, k) \equiv -\omega^2 + \omega_p^2 + \frac{\hbar^2}{4m^2}k^4\,. \tag{11.12}$$

If we take into account the electron gas degeneration as well, then we must add the electron pressure force to the first of Eqs. (11.11),

$$\frac{\partial J_p}{\partial t} - \frac{\omega_p^2}{4\pi}E_z - \frac{\hbar^2}{4m^2}\frac{\partial^3 \rho_p}{\partial z^3} + \theta V_{0F}^2\frac{\partial \rho_p}{\partial z} = 0\,, \tag{11.13}$$

in which case dispersion relation (11.12) changes accordingly,

$$D(\omega, k) \equiv -\omega^2 + \omega_p^2 + \theta k^2 V_{0F}^2 + \frac{\hbar^2}{4m^2}k^4\,. \tag{11.14}$$

This dispersion relation incorporates the quantum effects associated with both degeneration of the electron gas and its wave properties (the Heisenberg uncertainty principle).

From dispersion relation (11.14) we find the frequency spectra of the eigenmodes:

$$\omega_{1,2} = \pm\sqrt{\omega_p^2 + \theta k^2 V_{0F}^2 + (\hbar^2/4m^2)k^4}\,. \tag{11.15}$$

The state vector of harmonic electrostatic perturbations in a quantum degenerate plasma is obviously described by formula (10.8).

In dispersion relation (11.14), we switch to the frequency and wavenumber operators to arrive at the following equation for the characteristic function of the state vector of electrostatic perturbations in the plasma under consideration:

$$\left(\frac{\partial^2}{\partial t^2} - \theta V_{0F}^2\frac{\partial^2}{\partial z^2} + \frac{\hbar^2}{4m^2}\frac{\partial^4}{\partial z^4} + \omega_p^2\right)A(t, z) = 0\,. \tag{11.16}$$

Setting $N_{0P} = 0$ reduces this equation to

$$\left(\frac{\partial^2}{\partial t^2} + \frac{\hbar^2}{4m^2}\frac{\partial^4}{\partial z^4}\right) A(t,z) = 0 \rightarrow \left(\frac{\partial}{\partial t} \pm i\frac{\hbar}{2m}\frac{\partial^2}{\partial z^2}\right) A(t,z) = 0\,, \tag{11.17}$$

which is the Schrödinger equation for a free particle (see Eq. (11.7) with $\varphi = 0$).

12. Ion Acoustic Waves in a Nonisothermal Plasma. Ambipolar Diffusion

Up to this point, we considered waves in a plasma without allowance for ion motion. It should be noted, however, that, in a nonisothermal plasma in which the electron temperature T_e is much higher than the ion temperature (the latter may even be set equal to zero), there is an interesting type of waves resulting from ion motion — ion acoustic waves. Now we are going to consider these waves in the one-dimensional potential approximation. Let the electron and ion densities in an unperturbed state be equal to N_{0p} and N_{0i}. By virtue of plasma quasineutrality, we have $N_{0p} = ZN_{0i}$, where Z is the ion charge number. We denote the perturbations of the electric field, electron density, ion velocity, and ion density by $E_z(t,z)$, $N_p(t,z)$, $U_i(t,z)$, and $N_i(t,z)$, respectively. The linearized differential equations (1.1) describing the main properties of ion acoustic waves have the form

$$\begin{gathered}
\frac{\partial N_i}{\partial t} + N_{0i}\frac{\partial U_i}{\partial z} = 0\,, \\
\frac{\partial U_i}{\partial t} - \frac{e_i}{M}E_z = 0\,, \\
\frac{\partial E_z}{\partial z} - 4\pi e N_p - 4\pi e_i N_i = 0\,, \\
\frac{1}{N_{0p}}\frac{\partial P_e}{\partial z} - eE_z = 0\,.
\end{gathered} \tag{12.1}$$

Here, $e_i = -Ze$ and M are the charge and mass of an ion and $P_e = (N_{0p}+N_p)\kappa_B T_e$ is the electron gas pressure. The first two of Eqs. (12.1) are the cold hydrodynamic equations for the ion plasma component, the third is Poisson's equation, and the fourth is the equilibrium equation for a hot electron gas in an electric field. This fourth equation implies that electron thermal motion is the fastest among all other motions characteristic of the waves under consideration, so the electrons have enough time to relax to their equilibrium temperature, $T_e = \text{const}$. Accordingly, the fourth of Eqs. (12.1) can be written as

$$\frac{\partial N_p}{\partial z} - V_{Te}^{-2}\frac{eN_{0p}}{m}E_z = 0\,, \tag{12.2}$$

where $V_{Te} = \sqrt{\kappa_B T_e/m}$ is the electron thermal velocity.

Substituting the solution of the form

$$\mathbf{\Psi}(t,z) = \{N_i, U_i, E_z, N_p\} = \{n_i(\omega,k), u_i(\omega,k), e_z(\omega,k), n_p(\omega,k)\}\exp(-i\omega t + ikz)\,, \tag{12.3}$$

into Eqs. (12.1) and (12.2), we obtain the following set of homogeneous algebraic equations for the components of the complex state vector:

$$\begin{aligned} \omega n_i - N_{0i} k u_i &= 0\,, \\ i\omega u_i - Z(e/M) e_z &= 0\,, \\ 4\pi e Z n_i + ik e_z - 4\pi e n_p &= 0\,, \\ V_{Te}^{-2}(eN_{0p}/m) e_z - ik n_p &= 0\,. \end{aligned} \tag{12.4}$$

The solvability condition for Eqs. (12.4) is represented by the dispersion relation

$$D(\omega, k) \equiv \omega^2 (k^2 + r_{D_e}^{-2}) - k^2 \omega_i^2 = 0\,, \tag{12.5}$$

where $\omega_i = \sqrt{4\pi Z^2 e^2 N_{0i}/M}$ is the ion Langmuir frequency and $r_{D_e} = V_{Te}/\omega_p$ is the electron Debye radius.

From dispersion relation (12.5) we find expressions for the frequencies of the ion acoustic waves:

$$\omega_{1,2} = \pm \omega_i \frac{k r_{D_e}}{\sqrt{1 + k^2 r_{D_e}^2}}\,, \qquad V_{ph}^{(1,2)} = \pm \frac{\omega_i r_{D_e}}{\sqrt{1 + k^2 r_{D_e}^2}}\,. \tag{12.6}$$

Spectra (12.6) can be expressed in terms of the so-called ion acoustic speed by the relationship

$$V_S = \omega_i r_{D_e} = \sqrt{Z \frac{\kappa T_e}{M}}\,, \tag{12.7}$$

which gave the name of the waves at hand. In what follows, the waves with frequency spectra (12.6) will be classified as acoustic (or sound) waves. Although this classification is somewhat inconsistent with what was said above in discussing spectra (1.13) and (1.14), it better reflects the physical nature of the most important wave types in dispersive media.

Since there are only two eigenfrequencies in spectrum (12.7), two components of the state vector (12.3) are redundant. In fact, from Eqs. (12.4) we can see that the quantities e_z and n_p are proportional to n_i and, moreover, the proportionality coefficients are independent of the number of the eigenfrequency. Consequently, the state vector that unambiguously describes the plasma under analysis can be written as

$$\Psi(t, z) = \begin{pmatrix} N_i(t, z) \\ U_i(t, z) \end{pmatrix} = \sum_{m=1}^{2} A_m \begin{pmatrix} 1 \\ \omega_m / k N_{0i} \end{pmatrix} \exp[-i\omega_m(k)t + ikz]\,. \tag{12.8}$$

In the long-wavelength range $k r_{D_e} \ll 1$, waves with frequency spectra (12.6) are nondispersive because $V_{ph}^{(1,2)} = \pm V_S =$ const. In the short-wavelength range $k r_{D_e} \gg 1$, these waves are nondispersive, too, because $\omega_{1,2} = \pm \omega_i =$ const. A detailed description of waves of this type will be given later. Since the maximum velocity of ion acoustic waves is V_S, the applicability condition for Eq. (12.2) reduces to the inequality $V_S \ll V_{Te}$ or $\sqrt{Zm/M} \ll 1$. Recall that, along with this inequality, which is certainly satisfied, the existence of ion acoustic waves requires that the degree to which the plasma is nonisothermal should be sufficiently high.

In dispersion relation (12.5), we replace the frequency and wavenumber with operators (3.7) to obtain the following partial differential equation for the characteristic function of the state vector of arbitrary ion acoustic perturbations in an isotropic nonisothermal plasma:

$$\left(\frac{\partial^2}{\partial t^2} - V_S^2\frac{\partial^2}{\partial z^2} - r_{D_e}^2\frac{\partial^2}{\partial t^2}\frac{\partial^2}{\partial z^2}\right)A(t,z) = 0\,. \tag{12.9}$$

This equation, which is similar to Eq. (3.11) and is known as the linear Boussinesq equation or the Love equation, will be thoroughly investigated below. The state vector of an isotropic nonisothermal plasma is expressed in terms of the characteristic function by the formula

$$\mathbf{\Psi}(t,z) = \{N_i(t,z), U_i(t,z)\} = \left\{A(t,z), -\frac{1}{N_{0i}}\int\frac{\partial A}{\partial t}(t,z)dz\right\}. \tag{12.10}$$

Let us now take into account collisions in the theory of ion acoustic waves in a nonisothermal plasma. In the low-frequency range (i.e., for frequencies of $\omega \sim \omega_i$ or lower), a weakly ionized plasma is dominated by collisions of ions with neutral atoms, whereas electron collisions are far less important and can be ignored. When ion collisions are taken into account, the second of Eqs. (12.1) becomes

$$\frac{\partial U_i}{\partial t} - \frac{e_i}{M}E_z + v_{in}U_i = 0\,, \tag{12.11}$$

while the dispersion relation is derived by the same procedure as above. Omitting the details of the derivation, we immediately write out the final result:

$$D(\omega,k) \equiv \omega(\omega + iv_{in})(k^2 + r_{D_e}^{-2}) - k^2\omega_i^2 = 0\,. \tag{12.12}$$

For low frequencies of collisions of ions with neutrals, $v_{in} \ll \omega_i$, the frequencies $\omega_{1,2}$ of the ion acoustic waves are again described by formulas (12.6) but with nonzero imaginary parts, $\omega''_{1,2} = -v_{in}/2$. In the opposite limit of high ion collision frequencies, $v_{in} \gg \omega_i$, dispersion relation (12.12) yields the spectra

$$\omega_1 = -i\frac{k^2 r_{D_e}^2}{1 + k^2 r_{D_e}^2}\frac{\omega_i^2}{v_{in}}\,, \qquad \omega_2 = -iv_{in}\,. \tag{12.13}$$

For $k^2 r_{D_e}^2 \gg 1$, the frequency ω_1 describes the Maxwell relaxation of electrostatic perturbations in a nonisothermal collisional plasma (and thereby is an analogue of the frequency ω_1 from expressions (6.22)). For $k^2 r_{D_e}^2 \ll 1$, the spectrum ω_1 describes a certain diffusion process that will be considered later. The frequency ω_2 describes the single-particle damping (see the frequency ω_2 in (6.22)).

For $v_{in} \gg \omega_i$, we have $|\omega_1| \ll |\omega_2|$. Consequently, on long time scales ($v_{in}t > 1$), the characteristic function is given by the expression

$$A(t,z) = A_1\exp(-|\omega_1|t + ikz)\,. \tag{12.14}$$

In the first of formulas (12.13), we go over to the frequency and wavenumber operators to obtain the following differential equation for the characteristic function of ion acoustic perturbations in a nonisothermal, highly collisional plasma:

$$\left(\frac{\partial}{\partial t} - r_{D_e}^2 \frac{\omega_i^2}{v_{in}} \frac{\partial^2}{\partial z^2} - r_{D_e}^2 \frac{\partial}{\partial t} \frac{\partial^2}{\partial z^2}\right) A(t,z) = 0\,. \tag{12.15}$$

If, in the first of formulas (12.13), we switch to the operators and assume that $k^2 r_{D_e}^2 \ll 1$, then we arrive at the diffusion equation

$$\left(\frac{\partial}{\partial t} - D_A \frac{\partial^2}{\partial z^2}\right) A(t,z) = 0\,, \quad D_A = r_{D_e}^2 \frac{\omega_i^2}{v_{in}}\,, \tag{12.16}$$

where D_A is the diffusion coefficient. Let us clarify what sort of diffusion is described by Eq. (12.16). In formula (10.18), we have introduced the free electron diffusion coefficient. By analogy, we can also introduce the free ion diffusion coefficient

$$D_i = \frac{V_{Ti}^2}{v_{in}}\,, \tag{12.17}$$

where $V_{Ti} = \sqrt{\kappa_B T_i / M}$ is the ion thermal velocity and T_i is the ion temperature. Since the diffusion coefficients (10.18) and (12.17) differ from one another, the electrons and ions will not diffuse independently: the electrons are lighter than the ions (and are hotter in a nonisothermal plasma) and as such diffuse faster. This rapid diffusion of electrons gives rise to a space charge and the associated electric field, which in turn prevents the electrons from escaping from the space charge region and expels the ions away from there. As the steady state is established, the electron and ion fluxes become the same. The diffusion resulting from that self-consistent process came to be called ambipolar diffusion. The ambipolar diffusion coefficient is known to be given by the formula

$$D_A = \frac{(T_e + T_i) D_e D_i}{T_i D_e + D_e D_i}\,. \tag{12.18}$$

The difference in mass between electrons and ions is so large that, even for $T_e \gg T_i$, it is necessary to set $T_e D_i \ll T_i D_e$. Accordingly, formula (12.18) simplifies to

$$D_A = \left(1 + \frac{T_e}{T_i}\right) D_i\,. \tag{12.19}$$

For a nonisothermal plasma such that $T_e \gg T_i$, formula (12.19) leads to the expression for the diffusion coefficient D_A from (12.16). Hence, in the case at hand, we deal with ambipolar diffusion in a nonisothermal plasma.

13. Electromagnetic Waves in a Waveguide with an Anisotropic Plasma in a Strong External Magnetic Field

Here, we consider a waveguide system which was described in Sec. 9 and in which the space between two parallel metal plates is occupied by a highly anisotropic plasma rather than by an isotropic dielectric. Let the plasma anisotropy be created by a strong external magnetic field directed parallel to the z axis, which in turn is assumed to be parallel to the metal plates. We will be interested in the waves propagating along this axis. In a strong external magnetic field, the electric field component E_z induces a plasma current with the density $J_z \equiv J_p$ (see the first of Eqs. (6.2)). The onset of the remaining components of the plasma current is prevented by the external magnetic field. With allowance for the third of relationships (9.1), we can write

$$J_p = \sin k_\perp x \cdot \tilde{J}_p(t, z)\,. \tag{13.1}$$

Substituting relationships (9.1) and (13.1) into Maxwell's equations and into the cold hydrodynamic equations for the plasma electrons, we obtain the following equations for the quantities $\tilde{E}_x$, $\tilde{B}_y$, $\tilde{J}_p$ and $\tilde{E}_z$ (from which the tilde "$\sim$" will be omitted for brevity):

$$\begin{aligned} \frac{\partial E_x}{\partial t} + c\frac{\partial B_y}{\partial z} &= 0\,, \\ \frac{\partial B_y}{\partial t} + c\frac{\partial E_x}{\partial z} - k_\perp c E_z &= 0\,, \\ \frac{\partial J_p}{\partial t} - \frac{\omega_p^2}{4\pi} E_z &= 0\,, \\ \frac{\partial E_z}{\partial t} + 4\pi J_p + k_\perp c B_y &= 0\,. \end{aligned} \tag{13.2}$$

The state vector of the system described by Eqs. (13.2) has the form $\mathbf{\Psi}(t, z) = \{E_x, B_y, J_p, E_z\}$.

We represent the solution as

$$\mathbf{\Psi}(t, z) = \{e_x(\omega, k), b_y(\omega, k), j_p(\omega, k), e_z(\omega, k)\} \exp(-i\omega t + ikz)\,, \tag{13.3}$$

and insert it into Eqs. (13.2) to arrive at the following set of homogeneous algebraic equations for the components e_x, b_y, j_p and e_z of the complex state vector:

$$\begin{aligned} -i\omega e_x + ikcb_y &= 0\,, \\ ikce_x - i\omega b_y - k_\perp c e_z &= 0\,, \\ -i\omega j_p - (\omega_p^2/4\pi) e_z &= 0\,, \\ k_\perp c b_y + 4\pi j_p - i\omega e_z &= 0\,. \end{aligned} \tag{13.4}$$

The solvability condition for this set is given by the dispersion relation

$$D(\omega, k) \equiv -(k_\perp^2 c^2 \omega^2 + (k^2c^2 - \omega^2)(\omega^2 - \omega_p^2)) = 0\,. \tag{13.5}$$

The roots of fourth-order biquadratic equation (13.5) can be written as

$$\begin{gathered} \omega_1 = \Omega_1\,, \quad \omega_2 = -\Omega_1\,, \quad \omega_3 = \Omega_2\,, \quad \omega_4 = -\Omega_2\,; \\ \Omega_{1,2}^2 = \frac{1}{2}\left\{(k^2 + k_\perp^2)c^2 + \omega_p^2 \pm \sqrt{[(k^2 + k_\perp^2)c^2 + \omega_p^2]^2 - 4k^2c^2\omega_p^2}\right\}, \end{gathered} \tag{13.6}$$

where $\Omega_{1,2}$ are auxiliary quantities having the dimension of frequency. Using (13.6), (13.4), and (13.3) and again setting $e_x = 1$ yields the following expression for the state vector of harmonic electromagnetic perturbations in a waveguide filled with an anisotropic plasma:

$$\begin{aligned}\boldsymbol{\Psi}(t,z) &= \{E_x, B_y, J_p, E_z\} \\ &= \sum_{m=1}^{4} A_m \left\{1, \frac{\omega_m}{kc}, \frac{1}{4\pi}\omega_m \frac{k_\perp}{k}\frac{\omega_p^2}{\omega_m^2-\omega_p^2}, -i\frac{k_\perp}{k}\frac{\omega_m^2}{\omega_m^2-\omega_p^2}\right\} \\ &\quad \times \exp[-i\omega_m(k)t + ikz]\,. \end{aligned} \tag{13.7}$$

According to the numbering of the frequencies and the order of the subscripts in the quantities $\Omega_{1,2}^2$ in relationships (13.6), one can easily understand the meaning of each of the terms in solution (13.7). The first two terms ($m = 1, 2$) describe the electromagnetic waveguide modes that propagate in opposite directions along the z axis and are modified by the presence of plasma. For a waveguide without a plasma ($\omega_p \to 0$), these terms go over to solution (9.7) (with $\varepsilon_0 = 1$). The last two terms ($m = 3, 4$) in solution (13.7) describe purely plasma waves: for $\omega_p \to 0$, the frequencies $\omega_{3,4}$ vanish.

The dispersion of waves with frequencies (13.6) is rather complicated. Electromagnetic waves with the frequencies $\omega_{1,2}$ (according to the above classification, these are optical waves) have qualitatively the same dispersion as waves with frequencies (7.5) and (9.6). The situation with plasma waves with the frequencies $\omega_{3,4}$ is radically different. For instance, the smaller the wavenumber k, the weaker the dispersion of waves with small k values,

$$V_{ph}^{(3,4)} = \pm c\left(1 + \frac{k_\perp^2 c^2}{\omega_p^2}\right)^{-1/2} = \text{const}\,, \quad \text{for } k \to \infty\,. \tag{13.8}$$

For large wavenumbers, the dispersion of plasma waves is also weak (as that in criterion (1.15) with $\alpha = 0$ and $\beta = \pm\omega_p$) because

$$V_{ph}^{(3,4)} = \pm\frac{\omega_p}{k}\,, \quad \text{for } |k| \to \infty\,. \tag{13.9}$$

In the range of intermediate wavenumbers, the dispersion of plasma waves with the frequencies $\omega_{3,4}(k)$ is strong. We will classify waves with the frequencies $\omega_{3,4}(k)$ as acoustic, although their dispersion differs from that of waves with frequencies (12.6).

Assume that the initial state vector of an electromagnetic field and anisotropic plasma in the waveguide has the form

$$\boldsymbol{\Psi}(0,z) = \begin{pmatrix} E_{x0} \\ B_{y0} \\ J_{p0} \\ E_{z0} \end{pmatrix} \exp(ikz)\,. \tag{13.10}$$

Introducing the notation $\varepsilon_1 = 1-\omega_p^2/\Omega_1^2$ and $\varepsilon_2 = 1-\omega_p^2/\Omega_2^2$ (see spectra (13.6)) and using expression (13.7) for the state vector, we rewrite Eqs. (2.5) for the unknown complex amplitudes $A_{1,2,3,4}$ as

$$A_1 \begin{pmatrix} 1 \\ \dfrac{\Omega_1}{kc} \\ \dfrac{1}{4\pi}\dfrac{k_\perp \omega_p^2}{k\Omega_1\varepsilon_1} \\ -i\dfrac{k_\perp}{k\varepsilon_1} \end{pmatrix} + A_2 \begin{pmatrix} 1 \\ -\dfrac{\Omega_1}{kc} \\ -\dfrac{1}{4\pi}\dfrac{k_\perp \omega_p^2}{k\Omega_1\varepsilon_1} \\ -i\dfrac{k_\perp}{k\varepsilon_1} \end{pmatrix} + A_3 \begin{pmatrix} 1 \\ \dfrac{\Omega_2}{kc} \\ \dfrac{1}{4\pi}\dfrac{k_\perp \omega_p^2}{k\Omega_2\varepsilon_2} \\ -i\dfrac{k_\perp}{k\varepsilon_2} \end{pmatrix} + A_4 \begin{pmatrix} 1 \\ -\dfrac{\Omega_2}{kc} \\ -\dfrac{1}{4\pi}\dfrac{k_\perp \omega_p^2}{k\Omega_2\varepsilon_2} \\ -i\dfrac{k_\perp}{k\varepsilon_2} \end{pmatrix}$$

$$= \begin{pmatrix} E_{x0} \\ B_{y0} \\ J_{p0} \\ E_{z0} \end{pmatrix} \tag{13.11}$$

Finding the amplitudes $A_{1,2,3,4}$ from Eqs. (13.11) and inserting them into expression (13.7) leads to a general solution to the initial-value problem of the excitation of harmonic electromagnetic waves in a waveguide with an anisotropic plasma. This solution is not presented here because it is fairly involved and its derivation is quite obvious.

Dispersion relation (13.5) for electromagnetic waves in a waveguide with an anisotropic plasma filling can be written as

$$D(\omega, k) \equiv (k^2c^2 - \omega^2)(\omega^2 - \omega_p^2) + k_\perp^2 c^2 \omega^2 = 0\,. \tag{13.12}$$

In this dispersion relation, we go over to operators (3.7) to obtain the fourth-order differential equation for the characteristic function of the state vector:

$$\left[\left(\frac{\partial^2}{\partial t^2} - c^2\frac{\partial^2}{\partial z^2}\right)\left(\frac{\partial^2}{\partial t^2} + \omega_p^2\right) + k_\perp^2 c^2 \frac{\partial^2}{\partial t^2}\right] A(t,z) = 0\,. \tag{13.13}$$

The state vector of the electromagnetic field and anisotropic plasma is given by the formula

$$\begin{aligned} \mathbf{\Psi}(t,z) &= \{e_x(t,z), B_y(t,z), J_p(t,z), E_z(t,z)\} \\ &= \Bigg\{ A(t,z), -\frac{1}{c}\int \frac{\partial A}{\partial t}(t,z)dz, -\frac{\omega_p^2}{4\pi}\frac{1}{k_\perp c^2} \\ &\qquad \iint \hat{\mathrm{D}}A(t,z)dzdt, -\frac{1}{k_\perp c^2}\int \hat{\mathrm{D}}A(t,z)dz \Bigg\}, \end{aligned} \tag{13.14}$$

where $\hat{\mathrm{D}} = (\partial^2/\partial t^2 - c^2\partial^2/\partial z^2)$ is the d'Alambertian (the wave operator). Equation (13.13) and formula (13.14) will be further investigated below.

In the potential approximation (i.e., in the formal limit $c \to \infty$), dispersion relation (13.12) reduces to

$$D(\omega, k) \equiv \omega^2(k^2 + k_\perp^2) - k^2\omega_p^2 = 0\,. \tag{13.15}$$

This dispersion relation determines two frequencies of electrostatic waves in a fully magnetized plasma in the waveguide (in the limit $c \to \infty$, the remaining two frequencies of the electromagnetic waves become infinite):

$$\omega_1 = \omega_p \frac{k}{\sqrt{k^2 + k_\perp^2}}, \quad \omega_2 = -\omega_p \frac{k}{\sqrt{k^2 + k_\perp^2}}. \tag{13.16}$$

Dispersion relation (13.15) and solutions (13.16) have exactly the same structure as (12.5) and (12.6).

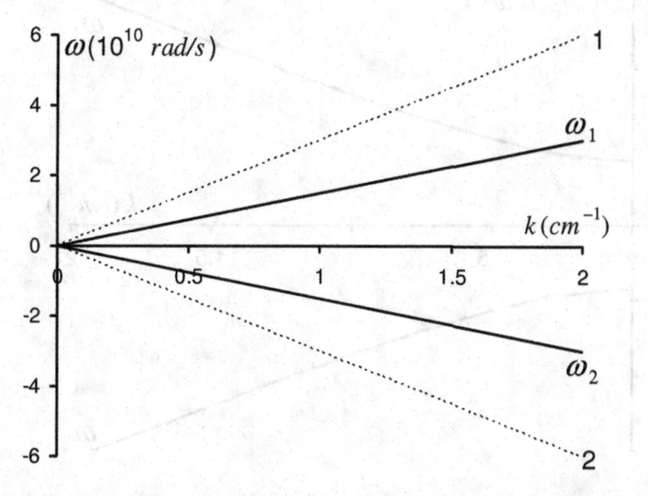

Fig. 3. Frequencies $\omega_{1,2}$ of electromagnetic waves in an isotropic dielectric with $\varepsilon_0 = 4$. The dotted curves *1* and *2* show the frequencies of electromagnetic waves in vacuum.

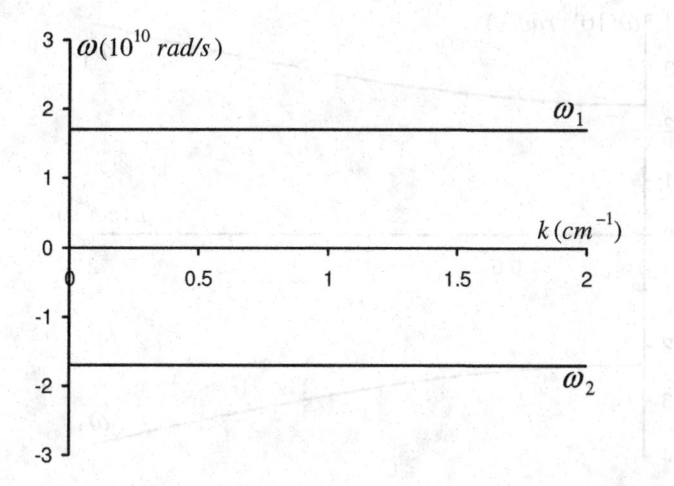

Fig. 4. Frequencies $\omega_{1,2}$ of longitudinal electrostatic waves in a plasma with the unperturbed electron density $N_{0p} = 10^{11}$ cm^{-3}.

In the potential approximation at hand, Eq. (13.13) goes over to the Love equation (see (12.9))

$$\left(\frac{\partial^2}{\partial t^2} - \frac{\omega_p^2}{k_\perp^2}\frac{\partial^2}{\partial z^2} - k_\perp^{-2}\frac{\partial^2}{\partial t^2}\frac{\partial^2}{\partial z^2}\right) A(t,z) = 0\,. \tag{13.17}$$

It is obvious that this same equation can also be obtained from dispersion relation (13.15) in which the frequency and wavenumber are replaced with the corresponding

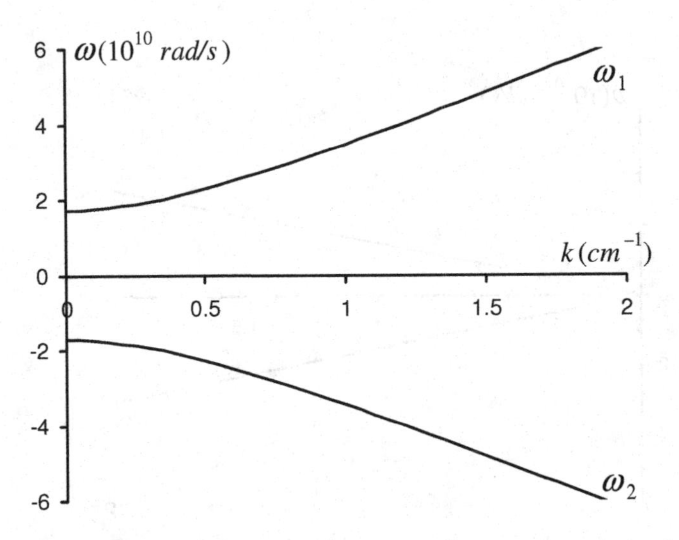

Fig. 5. Dispersion curves $\omega_{1,2}$ of transverse electromagnetic waves in an isotropic plasma with the unperturbed electron density $N_{0p} = 10^{11}$ cm^{-3}.

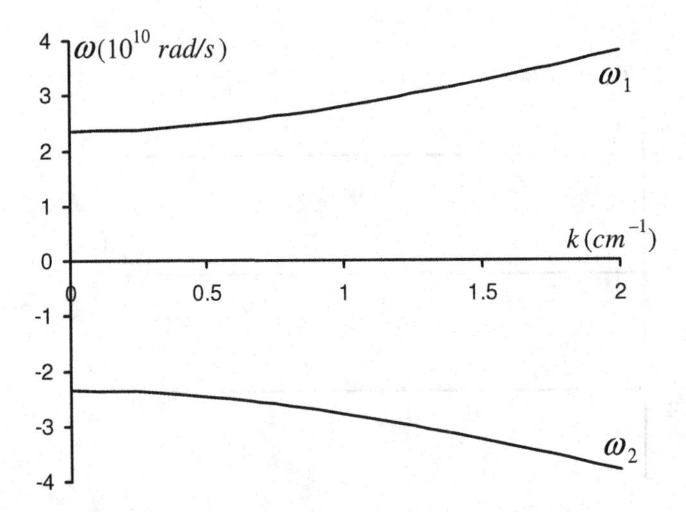

Fig. 6. Dispersion curves $\omega_{1,2}$ of transverse electromagnetic waves in a waveguide filled with an isotropic dielectric with $\varepsilon_0 = 4$. The distance between the metal planes is $L = 2$ cm.

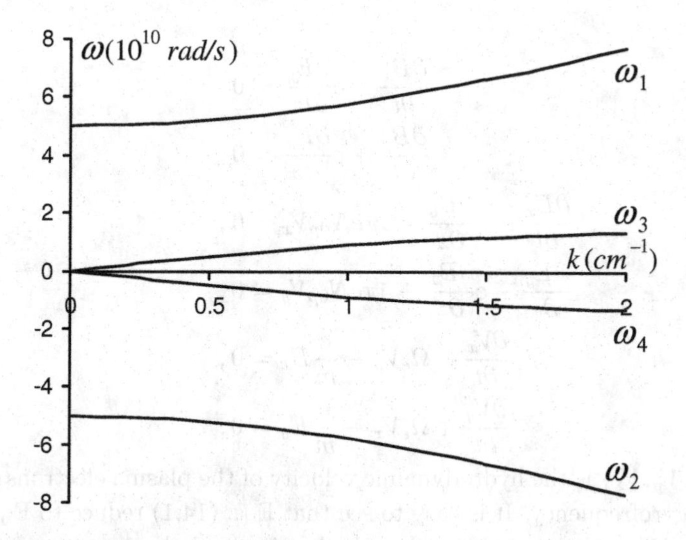

Fig. 7. Dispersion curves $\omega_{1,2}$ of electromagnetic waves and dispersion curves $\omega_{3,4}$ of plasma waves in a waveguide filled with an anisotropic plasma with the density $N_{0p} = 10^{11}$ cm^{-3}. The distance between the metal planes is $L = 2$ cm.

operators. It is easy to show that, in the potential approximation, the state vector (13.14) reduces to (6.11).

The frequency spectra of the eigenmodes can conveniently be represented graphically in the (k, ω) plane. The solutions $\omega = \omega_m(k)$ to dispersion relation (1.8) in this plane are called dispersion curves. Figures 3–7 show the dispersion curves described by formulas (5.5), (6.6), (7.5), (9.6), and (13.6). The numbering of the wave branches in the figures corresponds to that adopted in the text. The dispersion curves are plotted for systems with specific parameters, whose values are indicated in figure captions.

14. Electromagnetic Waves Propagating in a Magnetized Electron Plasma along a Magnetic Field

Let us consider the propagation of high-frequency electromagnetic waves in an unbounded homogeneous plasma in a uniform external magnetic field with the induction $\mathbf{B}_0 = \{0, 0, B_0\}$. Under the assumption $\omega \gg \omega_i$, Ω_i (where Ω_i is the ion gyrofrequency), the plasma ions can be treated as immobile. For waves propagating exactly along the external magnetic field, the complete set of Maxwell's equations and of the cold hydrodynamic equations for the plasma electrons splits into two independent subsets:

$$\frac{\partial N_p}{\partial t} + N_{0p}\frac{\partial V_z}{\partial z} = 0\,, \qquad \frac{\partial V_z}{\partial t} - \frac{e}{m}E_z = 0\,, \qquad \frac{\partial E_z}{\partial z} - 4\pi e N_p = 0\,, \tag{14.1}$$

and

$$\begin{gathered}
\frac{\partial B_x}{\partial t} - c\frac{\partial E_y}{\partial z} = 0\,, \\
\frac{\partial B_y}{\partial t} + c\frac{\partial E_x}{\partial z} = 0\,, \\
\frac{\partial E_x}{\partial t} + c\frac{\partial B_y}{\partial z} + 4\pi e N_{0p} V_x = 0\,, \\
\frac{\partial E_y}{\partial t} - c\frac{\partial B_x}{\partial z} + 4\pi e N_{0p} V_y = 0\,, \\
\frac{\partial V_x}{\partial t} - \Omega_e V_y - \frac{e}{m} E_x = 0\,, \\
\frac{\partial V_y}{\partial t} + \Omega_e V_x - \frac{e}{m} E_y = 0\,.
\end{gathered} \tag{14.2}$$

Here, $\mathbf{V} = \{V_x, V_y, V_z\}$ is the hydrodynamic velocity of the plasma electrons and Ω_e is the electron gyrofrequency. It is easy to see that Eqs. (14.1) reduce to Eqs. (6.2) (with $J_p = eN_{0p}V_z$) and thereby describes the longitudinal electrostatic waves in a plasma that have been considered above. This is why we will be interested only in Eqs. (14.2).

We introduce the complex functions

$$E = E_x \pm iE_y\,, \quad B = B_y \mp iB_x\,, \quad V = V_x \pm iV_y \tag{14.3}$$

to write Eqs. (14.2) as

$$\begin{gathered}
\frac{\partial E}{\partial t} + c\frac{\partial B}{\partial z} + 4\pi e N_{0p} V = 0\,, \\
\frac{\partial B}{\partial t} + c\frac{\partial E}{\partial z} = 0\,, \\
\frac{\partial V}{\partial t} \pm i\Omega_e V - \frac{e}{m} E = 0\,.
\end{gathered} \tag{14.4}$$

In these terms, the state vector of the system, $\psi(t,z) = \{E(t,z), B(t,z), V(t,z)\}$, is complex. Let this vector be represented by the formula

$$\mathbf{\Psi}(t,z) = \{\varepsilon(\omega,k), b(\omega,k), v(\omega,k)\}\exp(-i\omega t + ikz)\,, \tag{14.5}$$

in which, in contrast to formula (13.3) and other similar formulas, both the imaginary part and the real part have physical meaning. If representation (14.5) is valid, then, according to relationships (14.3), the solutions to Eqs. (14.4) in which we are interested here describe circularly polarized waves.

With representation (14.5), Eqs. (14.4) can be reduced to the following set of linear algebraic equations for the amplitudes $\varepsilon(\omega,k)$, $b(\omega,k)$ and $v(\omega,k)$:

$$\begin{gathered}
-i\omega\varepsilon + ikcb + 4\pi e N_{0p} v = 0\,, \\
-i\omega b + ikc\varepsilon = 0\,, \\
-i(\omega \mp \Omega_e)v - (e/m)\varepsilon = 0\,.
\end{gathered} \tag{14.6}$$

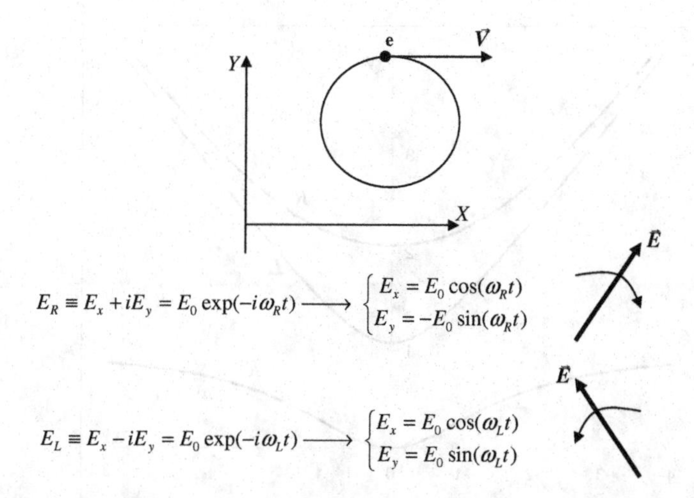

Y
e
V
X
E
$E_R \equiv E_x + iE_y = E_0 \exp(-i\omega_R t)$
$E_x = E_0 \cos(\omega_R t)$
$E_y = -E_0 \sin(\omega_R t)$
E
$E_L \equiv E_x - iE_y = E_0 \exp(-i\omega_L t)$
$E_x = E_0 \cos(\omega_L t)$
$E_y = E_0 \sin(\omega_L t)$

Fig. 8. Diagram illustrating the coordinate system (the Z axis points towards the reader, the magnetic field $\mathbf{B}_0$ points away from the reader, and $\Omega_e > 0$), the direction of gyration of an electron in the field $\mathbf{B}_0$, the directions of rotation of the vector $\mathbf{E}$ in an ordinary electromagnetic wave (with right circular polarization), E_R, and in an extraordinary electromagnetic wave (with left circular polarization), E_L.

The solvability condition for Eqs. (14.6) is a dispersion relation of third order in the frequency ω,

$$D(\omega, k) = \begin{vmatrix} -i\omega & ikc & 4\pi e N_{0p} \\ ikc & -i\omega & 0 \\ e/m & 0 & i(\omega \mp \Omega_e) \end{vmatrix}$$

$$= -i[(\omega^2 - k^2c^2)(\omega \mp \Omega_e) - \omega\omega_p^2] = 0\,, \tag{14.7}$$

in which the signs are arranged in the same manner as in relationships (14.3). When the sign of the frequency is changed, the plus sign in Eq. (14.7) is to be replaced with the minus sign, and vice versa. That is why, for definiteness, we set $\omega > 0$. In this case, from relationships (14.3) and formula (14.5) one can see that the upper sign corresponds to a right-polarized wave and the lower sign, to a left-polarized wave. The right-polarized wave is called an ordinary electromagnetic wave, because the vector $\mathbf{E}$ in this wave rotates in the direction of electron gyration in an external magnetic field. The left-polarized wave is called an extraordinary electromagnetic wave. Figure 8 illustrates the orientations of the coordinate system and of the external magnetic field, as well as the direction of gyration of an electron along its Larmor orbit and the direction of rotation of the vector $\mathbf{E}$ in both of the waves (in the figure, E_0 is a certain constant).

In terms of the dimensionless quantities

$$\frac{\omega}{\Omega_e} = y\,, \quad \frac{kc}{\Omega_e} = x\,, \quad \frac{\omega_p}{\Omega_e} = \alpha\,, \tag{14.8}$$

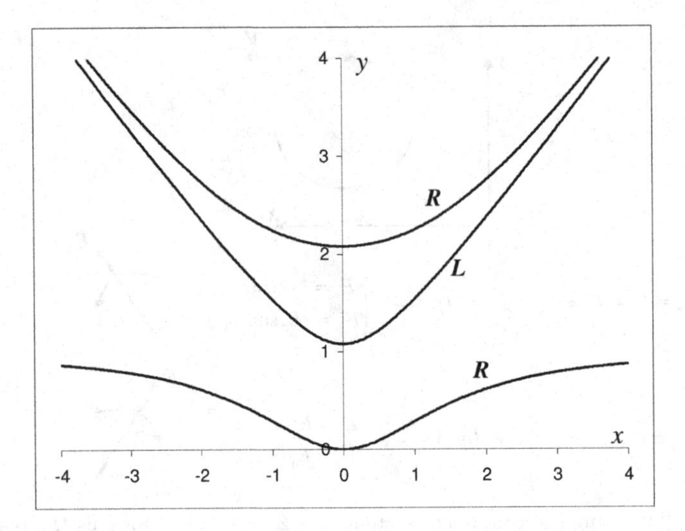

Fig. 9. Dispersion curves of ordinary waves (**R**) and an extraordinary wave (**L**) in a high-density plasma ($\alpha = \omega_p/\Omega_e = 1.5$).

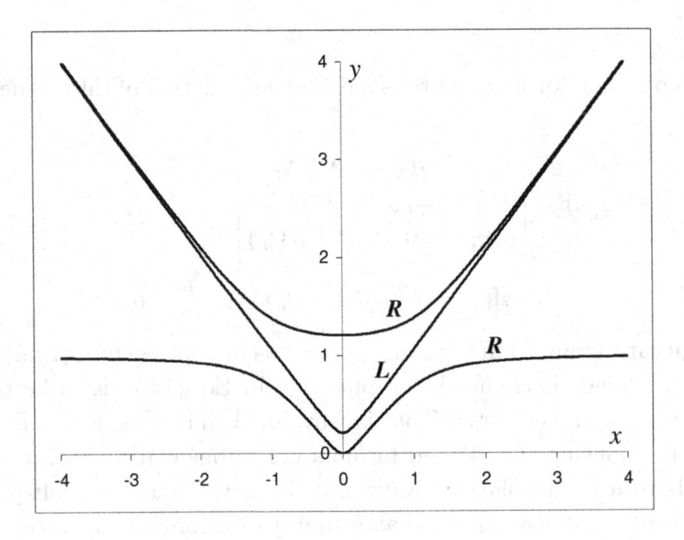

Fig. 10. Dispersion curves of ordinary waves (**R**) and an extraordinary wave (**L**) in a low-density plasma ($\alpha = \omega_p/\Omega_e = 0.5$).

dispersion relation (14.7) reads

$$(y^2 - x^2)(y \mp 1) - \alpha^2 y = 0\,, \quad y \geq 0\,, \quad x \in (-\infty, +\infty)\,. \tag{14.9}$$

Figures 9 and 10 show dimensionless dispersion curves of ordinary waves and of an extraordinary wave for two cases, $\omega_p > \Omega_e$ and $\omega_p < \Omega_e$, respectively. It is evident that the high-frequency ordinary wave and the extraordinary wave are both optical. In the short-wavelength range, these waves become nondispersive and their phase velocities approach $\pm e$. For $k \to 0$ (i.e., in the long-wavelength range), the frequencies of the waves are given by the formulas

$$\omega_R(0) = \sqrt{\omega_p^2 + \frac{\Omega_e^2}{4}} + \frac{\Omega_e}{2}\,,$$
$$\omega_L(0) = \sqrt{\omega_p^2 + \frac{\Omega_e^2}{4}} - \frac{\Omega_e}{2}\,. \tag{14.10}$$

The frequency $\omega_R(0)$ of the ordinary wave is always lower than the electron gyrofrequency Ω_e. For the extraordinary wave, we have $\omega_L(0) < \Omega_e$ for $\omega_p < \sqrt{2}\Omega_e$ and $\omega_L(0) \geq \Omega_e$ for $\omega_p \geq \sqrt{2}\Omega_e$. The higher the electron Langmuir frequency ω_p, the higher the two frequencies (14.10).

In its dispersion law, a low-frequency ordinary wave with a large wavenumber resembles an acoustic wave (see Fig. 7, curves $\omega_{3,4}$). For $k \to \infty$, its frequency approaches Ω_e from below and its dispersion becomes weaker. But in the long-wavelength range, $k \to 0$, the dispersion law of a low-frequency ordinary wave is unusual: for $\omega \ll \Omega_e$, dispersion relation (14.7) with the upper sign yields

$$\omega = k^2 c^2 \frac{\Omega_e}{\omega_p^2}\,. \tag{14.11}$$

Formula (14.11) for the frequency is valid for wavenumbers in the range $k^2c^2 \ll \omega_p^2$ ($x^2 \ll \alpha^2$). A low-frequency ordinary wave with the wavenumber from this range is called a helicon (see Fig. 11).

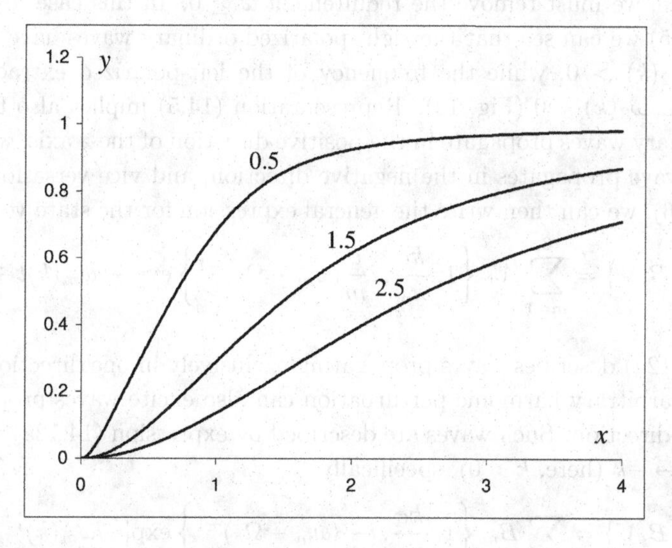

Fig. 11. Dispersion curves of a low-frequency ordinary wave for $\alpha = \omega_p/\Omega_e = 0.5$; 1.5 and 2.5. In the region $x^2 \ll \alpha^2$, the wave is a helicon.

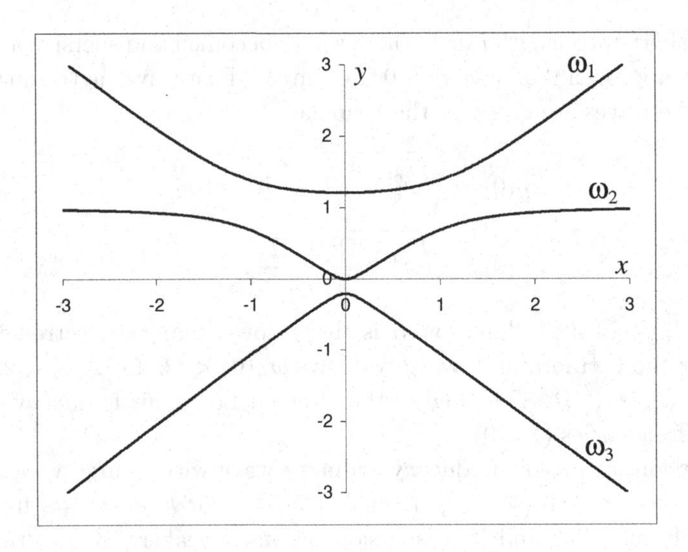

Fig. 12. Illustration of the structure of solution (14.12): $\omega_{1,2}$ are for ordinary waves and ω_3 is for an extraordinary wave.

We denote the solutions to dispersion relation (14.7) by $\omega_m(k)$, with $m = 1, 2, 3$, to find a general expression for the state vector of harmonic electromagnetic perturbations propagating in the plasma along the external magnetic field. The state vector cannot be represented as a simple sum over the branches of the eigenmodes because the waves have different polarizations corresponding to two signs in (14.3), (14.6), and (14.7). Let us choose one of the signs — say, the upper sign. Having made this choice, we must remove the requirement $\omega \geq 0$. In this case, from representation (14.5) we can see that the right-polarized ordinary waves have positive frequencies, $\omega_{1,2}(k) > 0$, while the frequency of the left-polarized extraordinary wave is negative, $\omega_3(k) < 0$ (Fig. 12). Representation (14.5) implies also that, for $k > 0$, the ordinary waves propagate in the positive direction of the z axis, while the extraordinary wave propagates in the negative direction, and vice versa for $k < 0$. Using Eqs. (14.6), we can then write the general expression for the state vector as

$$\mathbf{\Psi}(t, z) = \{E, B, V\} = \sum_{m=1}^{3} A_m \left\{1, \frac{kc}{\omega_m}, i\frac{e}{m}(\omega_m - \Omega_e)^{-1}\right\} \exp[-i\omega_m(k)t + ikz] . \tag{14.12a}$$

Expression (14.12a) describes waves propagating exclusively in one direction of the z axis. But an arbitrary harmonic perturbation can also excite waves propagating in the opposite direction. Such waves are described by expression (14.12a) with the replacement $k \to -k$ (here, $k > 0$), specifically,

$$\mathbf{\Psi}(t, z) = \{E, B, V\} = \sum_{m=1}^{3} B_m \left\{1, \frac{kc}{\omega_m}, i\frac{e}{m}(\omega_m - \Omega_e)^{-1}\right\} \exp[-i\omega_m(k)t - ikz] . \tag{14.12b}$$

Hence, in order to find a complete solution to the initial-value problem, it is necessary to determine six constants. We do not write out this solution because it is very involved.

In dispersion relation (14.7), we switch to the operators and obtain the following equation for the characteristic function of the state vector of circularly polarized perturbations in a homogeneous magnetized plasma:

$$\left[(\hat{D}+\omega_p^2)\frac{\partial}{\partial t} \pm i\Omega_e\hat{D}\right] A(t,z) = 0\,, \tag{14.13}$$

where $\hat{D} = (\partial^2/\partial t^2 - c^2\partial^2/\partial z^2)$ is the d'Alambertian and the signs correspond to those in relationships (14.3). A distinctive feature of differential equation (14.13) is that it contains the imaginary unit, which arises from the circular polarization of the perturbations under consideration. In this case, the characteristic function is necessarily complex. Setting

$$A(t,z) = A_x(t,z) \pm iA_y(t,z)\,, \tag{14.14}$$

we transform complex equation (14.13) to the following set of two real equations:

$$\begin{aligned}(\hat{D}+\omega_p^2)\frac{\partial A_x}{\partial t} - \Omega_e\hat{D}A_y = 0\,,\\ (\hat{D}+\omega_p^2)\frac{\partial A_y}{\partial t} + \Omega_e\hat{D}A_x = 0\,.\end{aligned} \tag{14.15}$$

In contrast to Eq. (14.13), Eqs. (14.5) do not contain two different signs. In order to convince ourselves that these equations do nevertheless describe perturbations having circular polarizations of opposite sign, we substitute the solutions of the form

$$\begin{aligned}A_x = A_{x0}\cos(\omega t - kz)\,,\\ A_y = A_{y0}\sin(\omega t - kz)\end{aligned} \tag{14.16}$$

into these equations and obtain the set of equations

$$\begin{aligned}\omega((\omega^2-k^2c^2)-\omega_p^2)A_{x0} + \Omega_e(\omega^2-k^2c^2)A_{y0} = 0\,,\\ \omega((\omega^2-k^2c^2)-\omega_p^2)A_{y0} + \Omega_e(\omega^2-k^2c^2)A_{x0} = 0\,,\end{aligned} \tag{14.17}$$

which yields the dispersion relation

$$\omega^2((\omega^2-k^2c^2)-\omega_p^2)^2 = \Omega_e^2(\omega^2-k^2c^2)^2\,. \tag{14.18a}$$

Extracting the square root gives

$$\omega((\omega^2-k^2c^2)-\omega_p^2) = \pm\Omega_e(\omega^2-k^2c^2)\,. \tag{14.18b}$$

It is easy to see that this dispersion relation coincides with (14.7). From Eqs. (14.17) and dispersion relations (14.18) we also obtain the relationships $A_{x0}/A_{y0} = \mp 1 \to A_x = A_0\cos(\omega t - kx)$, $A_y = \mp A_0\sin(\omega t - kz) \to A = A_x \pm iA_y = A_0\exp(-i\omega t + ikz)$, which show that the characteristic function has the same structure as that of the state vector (14.5).

To conclude this section, we present the differential equation for the characteristic function of a helicon. Going over to the operators in formula (14.11) yields the Schrödinger equation for a free particle,

$$\left(i\omega_p \frac{\partial}{\partial t} + c^2 \frac{\Omega_e}{\omega_p} \frac{\partial^2}{\partial z^2}\right) A(t, z) = 0\,. \tag{14.19}$$

15. Electrostatic Waves Propagating in a Magnetized Electron Plasma at an Angle to a Magnetic Field

In the potential approximation ($c \gg |\omega/k|$), dispersion relation (14.7) has the single solution $\omega = \Omega_e$, describing an electron cyclotron wave. This elementary wave, which corresponds to a free rotation of an electron in a plasma with a zero self-consistent field, is obviously unimportant for our analysis. The situation with perturbations propagating in a plasma at an angle to the external magnetic field is far more interesting. In the general, nonpotential case, investigation of such perturbations involves laborious and lengthy manipulations. This is why we restrict ourselves here to the potential approximation.

As in Sec. 13, we consider a plasma waveguide that is formed by metal planes perpendicular to the x axis and is placed in an external magnetic field parallel to the z axis. Note that, in such a waveguide, waves propagate only in a strictly fixed direction, namely, along the waveguide axis (in the case at hand, along the guiding metal planes, i.e., parallel to the OZ axis). The waveguide waves are a superposition of waves propagating at an angle to the waveguide axis: a standing wave forms in the direction perpendicular to the axis and a running wave forms in the direction parallel to the axis (see Fig. 13). In a waveguide filled with a magnetized plasma, there necessarily exist waves propagating at an angle to the external magnetic field. Of course, this remark applies equally to the waveguide system considered in Sec. 13.

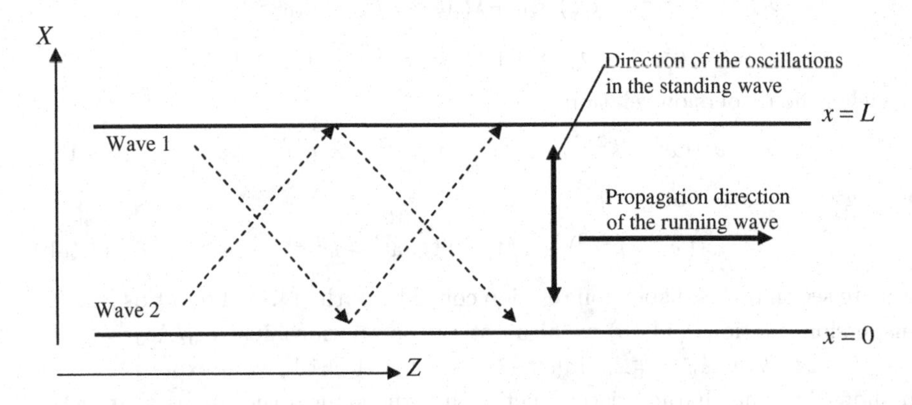

Fig. 13. Schematic explanation of the formation of a wave in a waveguide.

Without allowance for ion motion, the set of the field equations and of the cold hydrodynamic equations for potential perturbations in a plasma waveguide has the form

$$
\begin{aligned}
\frac{\partial V_x}{\partial t} - \Omega_e V_y &= \frac{e}{m} E_x \, , \\
\frac{\partial V_y}{\partial t} + \Omega_e V_x &= 0 \, , \\
\frac{\partial V_z}{\partial t} - \frac{e}{m} E_z &= 0 \, , \\
\frac{\partial N_p}{\partial t} + N_{0p} \frac{\partial V_z}{\partial z} - k_\perp N_{0p} V_x &= 0 \, , \\
\frac{\partial E_z}{\partial z} - k_\perp E_x - 4\pi e N_p &= 0 \, , \\
\frac{\partial E_x}{\partial z} - k_\perp E_z &= 0 \, .
\end{aligned}
\tag{15.1}
$$

Equations (15.1) are written for the functions $\tilde{F}(t, z)$ determined by the relationships

$$
\begin{aligned}
E_z(t, z, x) &= \tilde{E}_z(t, z) \sin(k_\perp x) \, , \\
E_x(t, z, x) &= \tilde{E}_x(t, z) \cos(k_\perp x) \, , \\
V_z(t, z, x) &= \tilde{V}_z(t, z) \sin(k_\perp x) \, , \\
V_{x,y}(t, z, x) &= \tilde{V}_{x,y}(t, z) \cos(k_\perp x) \, , \\
N_p(t, z, x) &= \tilde{N}_p(t, z) \sin(k_\perp x) \, ,
\end{aligned}
\tag{15.2}
$$

which account for the boundary conditions at the metal planes (see (9.1) and (13.1)). In writing Eqs. (15.1), we have omitted the tilde "$\sim$" for brevity. Here, we will not follow the scheme of description that was adopted in Secs. 5–14. We are now more interested in waves with dispersion laws differing qualitatively from those investigated above, as well as in new differential equations for the characteristic functions. That is why we immediately write out the dispersion relation that can be obtained from Eqs. (15.1) after simple manipulations:

$$
D(\omega, k) \equiv k^2(\omega^2 - \omega_p^2)(\omega^2 - \Omega_e^2) + k_\perp^2 \omega^2 (\omega^2 - \Omega_e^2 - \omega_p^2) = 0 \, . \tag{15.3}
$$

Note that, for $\Omega_e \to \infty$, dispersion relation (15.3) yields spectra (13.6). Figure 14 shows the dispersion curves obtained for $\Omega_e > \omega_p$ from the solutions to dispersion relation (15.3). For $\Omega_e < \omega_p$, the dispersion curves are identical to those displayed in Fig. 14. Curve 1 is of most interest: it is for the first time that we are dealing with the case in which the wave frequency decreases as the wavenumber increases (for $k > 0$ and $\omega > 0$, we have $d\omega/dk < 0$). Such waves are called waves with an anomalous dispersion. The anomalous dispersion is a conventional phenomenon in the region where the waves are absorbed or are unstable. But in the case at hand, we are considering waves in a stable nondissipative system.

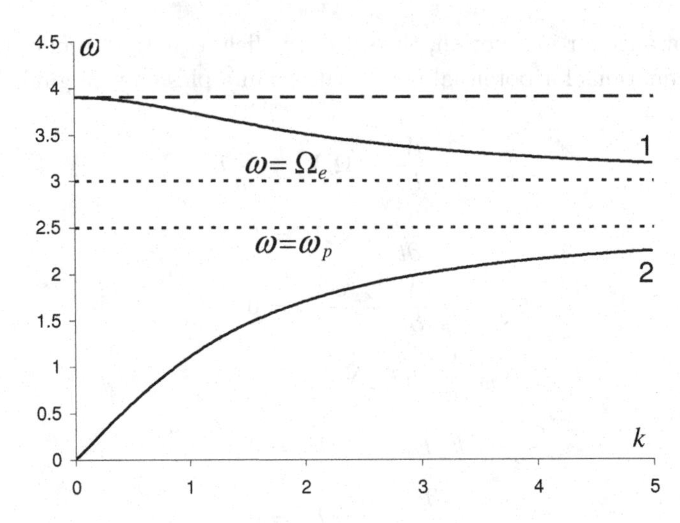

Fig. 14. Dispersion curves of electrostatic waves propagating in a plasma waveguide along the external magnetic field for $k_\perp = 1.5$.

For $k \to 0$, dispersion relation (15.3) have the roots

$$\omega^2 = 0\,, \quad \omega^2 = \omega_p^2 + \Omega_e^2\,. \tag{15.4}$$

The dispersion curves of waves with anomalous dispersion originate from the points $(k = 0,\, \omega = \pm\Omega_g)$, where $\Omega_g = \sqrt{\omega_p^2 + \Omega_e^2}$ is the so-called upper hybrid frequency. In the short-wavelength limit such that $k^2 \gg k_\perp^2$, dispersion relation (15.3) has the following solutions ($\omega_p \neq \Omega_e$):

$$\omega^2 = \omega_p^2 + \frac{k_\perp^2}{k^2}\frac{\omega_p^2\Omega_e^2}{\omega_p^2 - \Omega_e^2} \equiv \omega_1^2\,, \tag{15.5a}$$

$$\omega^2 = \Omega_e^2 + \frac{k_\perp^2}{k^2}\frac{\omega_p^2\Omega_e^2}{\Omega_e^2 - \Omega_p^2} = \omega_2^2\,. \tag{15.5b}$$

For $\omega_p > \Omega_e$, formula (15.5a) leads to an anomalous dispersion law and formula (15.5b), to a normal dispersion law, and we have $\omega_1^2 > \omega_2^2$. For $\omega_p < \Omega_e$, the situation is opposite.

Why the dispersion of one of the waves is anomalous can be explained as follows. For $k = 0$ — when the wave vector $\mathbf{k} = \{k_\perp, 0, k\}$ and electric field $\mathbf{E} = \{E_x, 0, E_z\}$ are perpendicular to the external magnetic field $\mathbf{B}_0 = \{0, 0, B_0\}$ — the frequency is equal to the upper hybrid frequency Ω_g (or is zero, see solutions (15.4)). For $k \to \infty$ — when the vectors $\mathbf{k}$, $\mathbf{E}$ and $\mathbf{B}_0$ are parallel — the wave frequencies are equal to Ω_e and ω_p and thus are lower than Ω_g. Hence, as the electric field $\mathbf{E}$ is turned to become parallel or antiparallel to the magnetic field $\mathbf{B}_0$, the frequency of one of the waves decreases.

To simplify matters, instead of the fairly complicated dispersion relation (15.3), we use relationships (15.5). Going over to the frequency and wavenumber operators in these relationships, we obtain the following differential equation for the characteristic function:

$$\left[\frac{\partial^2}{\partial z^2}\left(\frac{\partial^2}{\partial t^2}+\Omega_0^2\right)-k_\perp^2\Omega_0^2\theta\right]A(t,z)=0\,. \tag{15.6}$$

Here, $\Omega_0=\omega_p$ or $\Omega_0=\Omega_e$ and

$$\theta=\begin{cases}\Omega_e^2(\Omega_0^2-\Omega_e^2)^{-1}, & \Omega_0=\omega_p,\\ \omega_p^2(\Omega_0^2-\omega_p^2)^{-1}, & \Omega_0=\omega_e.\end{cases} \tag{15.7}$$

Let us clarify whether among the equations we have considered above there are equations analogous to Eq. (15.6). If we rewrite the Love equation (13.17) as

$$\left[\frac{\partial^2}{\partial z^2}\left(\frac{\partial^2}{\partial t^2}+\Omega_0^2\right)-k_\perp^2\frac{\partial^2}{\partial t^2}\right]A(t,z)=0\,, \tag{15.8}$$

then we can see that it goes over to Eq. (15.6) under the equality

$$\frac{\partial^2}{\partial t^2}=\Omega_0^2\theta\,, \tag{15.9}$$

which is to be understood in the operator sense. For $\theta<0$ — a case in which the dispersion of waves with frequencies (15.5) is normal — we have

$$\frac{\partial}{\partial t}=\pm i\Omega_0\sqrt{|\theta|}\,. \tag{15.10}$$

In this case, Eq. (15.6), which is in fact equivalent to the Love equation, describes acoustic waves with a normal dispersion law (see the lower dispersion curve in Fig. 14). For $\theta>0$ — a case in which the dispersion of waves with frequencies (15.5) is anomalous — equality (15.9) yields

$$\frac{\partial}{\partial t}=\pm\Omega_0\sqrt{|\theta|}\,. \tag{15.11}$$

In this case, Eq. (15.6) does not reduce to Eq. (15.8). Let us explain that, for $\partial/\partial t=i\alpha$ and when the constant α is real, the operator $\partial/\partial t$ characterizes an oscillatory process. When the constant α is imaginary, this same operator describes an aperiodic process.

Dispersion relation (15.3) is valid not only for potential waves in a magnetized plasma waveguide but also for waves in an unbounded system. In the latter case, both of the components of the wave vector $\mathbf{k}=\{k_\perp,0,k\}$ can be arbitrary. Let us consider the solutions to dispersion relation (15.3) and the wave dispersion that is described by these solutions as a function of $k_\perp$ for a fixed wave vector component k. Recall that $k_\perp$ is the wave vector component transverse to the external magnetic field. For $k_\perp=0$, dispersion relation (15.3) gives two frequencies,

$$\omega^2=\omega_p^2,\quad \omega^2=\Omega_e^2\,. \tag{15.12}$$

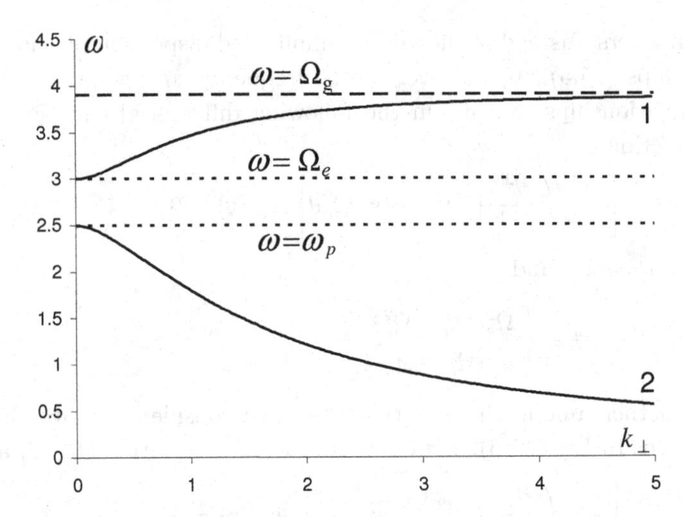

Fig. 15. Dispersion curves of electrostatic waves propagating in a plasma waveguide across the external magnetic field for $k = 1.5$.

Under the inequality $k_\perp^2 \gg k^2$, dispersion relation (15.3) has the solutions

$$\omega^2 = \Omega_g^2 - \frac{k^2}{k_\perp^2}\frac{\omega_p^2\Omega_e^2}{\Omega_g^2} \equiv \omega_1^2\,, \tag{15.13a}$$

$$\omega^2 = \frac{k^2}{k_\perp^2}\frac{\omega_p^2\Omega_e^2}{\Omega_g^2} \equiv \omega_2^2\,. \tag{15.13b}$$

Formula (15.13a) describes a wave with a normal dispersion law and formula (15.13b) leads to an anomalous dispersion law, and we have $\omega_1^2 > \omega_2^2$ (Fig. 15).

Switching to the frequency and wavenumber operators, from formulas (15.13) we obtain the following differential equations for the characteristic function:

$$\left[\frac{\partial^2}{\partial x^2}\left(\frac{\partial^2}{\partial t^2} + \Omega_g^2\right) + k^2\Omega_g^2\vartheta\right] A(t,z) = 0\,, \tag{15.14a}$$

$$\left[\frac{\partial^2}{\partial x^2}\frac{\partial^2}{\partial t^2} - k^2\Omega_g^2\vartheta\right] A(t,z) = 0\,. \tag{15.14b}$$

where $\vartheta = \omega_p^2\Omega_e^2/\Omega_g^4$. Equation (15.14a) has the form of Eq. (15.6) with $\theta < 0$, and Eq. (15.14b), although different from Eq. (15.6), also reduces to the latter with $\theta > 0$.

16. Magnetohydrodynamic Waves in a Conducting Fluid

Unlike in Secs. 13–15, here we consider waves in a medium in which charged and neutral particles interact so strongly that the medium can be described in the conducting fluid approximation. It is known that waves in a conducting medium, in particular, in a highly collisional plasma, are described by the magnetohydrodynamic (MHD) equations. Let us consider MHD waves in the presence of a uniform external magnetic field $\mathbf{B}_0$ directed along the OZ axis. We assume that, in an equilibrium state, the hydrodynamic velocity of the medium is zero and that the density and pressure of the medium are ρ_0 and P_0, respectively. For the moment, we restrict ourselves to waves propagating exactly along the external magnetic field. We will work with the following set of linearized MHD equations:

$$\frac{\partial \rho}{\partial t} + \rho_0 \frac{\partial V_z}{\partial z} = 0\,, \qquad \frac{\partial V_z}{\partial t} + \frac{U_0^2}{\rho_0}\frac{\partial \rho}{\partial z} = 0\,, \tag{16.1a}$$

$$\frac{\partial B_x}{\partial t} - B_0 \frac{\partial V_x}{\partial z} = 0\,, \qquad \frac{\partial V_x}{\partial t} - \frac{B_0}{4\pi\rho_0}\frac{\partial B_x}{\partial z} = 0\,, \tag{16.1b}$$

$$\frac{\partial B_y}{\partial t} - B_0 \frac{\partial V_y}{\partial z} = 0\,, \qquad \frac{\partial V_y}{\partial t} - \frac{B_0}{4\pi\rho_0}\frac{\partial B_y}{\partial z} = 0\,. \tag{16.1c}$$

Here, ρ, (V_x, V_y, V_z), and (B_x, B_y) are perturbations of the density, hydrodynamic velocity, and magnetic field induction, respectively, and $U_0 = \sqrt{\partial P_0/\partial \rho_0}$ is the speed of sound in a conducting fluid.

It is easy to see that Eqs. (16.1) split into three independent subsets of equations. Subset (16.1a) describes longitudinal acoustic waves propagating in a conducting fluid along the external magnetic field. Representing the solution as

$$\psi(t,z) = \{\rho, V_z\} = \{r(\omega,k), v_z(\omega,k)\}\exp(-i\omega t + ikz) \tag{16.2}$$

and substituting it into Eqs. (16.1a), we obtain the following set of homogeneous algebraic equations for the components $r(\omega,k)$ and $v_z(\omega,k)$ of the complex state vector:

$$\begin{aligned} -i\omega r + ik\rho_0 v_z &= 0\,, \\ -i\omega v_z + ikU_0^2/\rho_0 r &= 0\,. \end{aligned} \tag{16.3}$$

The solvability condition for Eqs. (16.3) is given by the simplest dispersion relation for acoustic waves:

$$D(\omega,k) \equiv -\omega^2 + k^2U_0^2 = 0\,. \tag{16.4}$$

We insert the frequencies of the acoustic waves, $\omega_{1,2} = \pm kU_0$, into Eqs. (16.3) and take into account representation (16.2) and dispersion relation (16.4). As a result, we arrive at the following formula for the state vector of longitudinal harmonic acoustic perturbations propagating in a conducting fluid along the external magnetic field:

$$\psi(t,z) = \left\{\begin{matrix}\rho \\ V_z\end{matrix}\right\} = A_1\left\{\begin{matrix}1 \\ U_0/\rho_0\end{matrix}\right\}\exp(ik(z-U_0t) + A_2\left\{\begin{matrix}1 \\ -U_0/\rho_0\end{matrix}\right\}\exp(ik(z+U_0t)\,. \tag{16.5}$$

The first term in this formula describes a wave propagating in the positive direction of the OZ axis, and the second term describes a wave propagating in the opposite direction. By virtue of criterion (1.15), the acoustic waves in question are nondispersive.

Switching to the frequency and wavenumber operators in dispersion relation (16.4) leads to the d'Alembert wave equation (5.9) with the speed of sound $U_0 = \sqrt{\partial P_0/\partial \rho_0}$ in place of the speed of light c_0. The state vector of acoustic perturbations propagating in a conducting fluid along the external magnetic field is expressed in terms of the characteristic function by the formula

$$\psi(t,z) = \{\rho(t,z), V_z(t,z)\} = \left\{A(t,z), -\frac{1}{\rho_0}\int \frac{\partial A}{\partial t}(t,z)dz\right\}, \tag{16.6}$$

which is quite similar to formula (5.10).

In the case at hand, namely, that with perturbations propagating along the external magnetic field $\mathbf{B}_0$, subsets of equations (16.1b) and (16.1c) are identical: they describe transverse waves with two different polarizations, (B_x, V_x) and (B_y, V_y). Note that subset (16.1a) describes a longitudinal wave, because the perturbed velocity vector in it, $\mathbf{V} = \{0, 0, V_z\}$, is collinear with its propagation direction. All the three subsets of equations (16.1) have the same mathematical structure. Accordingly, the waves described, e.g., by subset (16.1b), can be considered in a manner analogous to that provided by dispersion relation (16.4) and formulas (16.5) and (16.6). Thus, the dispersion relation has the form

$$D(\omega, k) \equiv -\omega^2 + k^2 V_A^2 = 0\,, \tag{16.7}$$

where

$$V_A = \sqrt{\frac{B_0^2}{4\pi\rho_0}} \tag{16.8}$$

is the so-called Alfvén speed. The frequencies are given by the relationships $\omega_{1,2} = \pm k V_A$, and the state vector is represented by the formulas

$$\psi(t,z) = \left\{\begin{matrix} B_x \\ V_x \end{matrix}\right\} = A_1 \left\{\begin{matrix} 1 \\ -V_A/B_0 \end{matrix}\right\} \exp(ik(z - V_A t))$$

$$+ A_2 \left\{\begin{matrix} 1 \\ V_A/B_0 \end{matrix}\right\} \exp(ik(z + V_A t))\,, \tag{16.9}$$

$$\psi(t,z) = \{B_x(t,z), V_x(t,z)\} = \left\{A(t,z), \frac{1}{B_0}\int \frac{\partial A}{\partial t}(t,z)dz\right\}. \tag{16.10}$$

Transverse waves described by subsets of equations (16.1b) and (16.1c), dispersion relation (16.7), and formulas (16.8)–(16.10) are called Alfvén MHD waves in a conducting fluid. Except for their transverse polarization, such waves have the same properties as conventional acoustic waves. Even the propagation velocity (16.8) has the same form as the speed of sound — "pressure divided by density raised to a power 1/2" — the only difference being that the role of the hydrodynamic pressure in

Alfvén waves is played by the magnetic field pressure. But this elementary analysis applies solely to Alfvén waves propagating exactly along the external magnetic field. It is therefore more interesting to consider MHD waves propagating in a conducting fluid at an angle to the vector $\mathbf{B}_0$.

As in Secs. 9, 13, and 15, we assume that the conducting medium fills a waveguide between two parallel planes oriented along the external magnetic field. The OX axis points perpendicular to the planes (and the OZ axis is again directed parallel to $\mathbf{B}_0$). The linearized set of MHD equations with $\partial/\partial z \neq 0$, $\partial/\partial x = 0$, and $\partial/\partial y = 0$ has the form

$$\begin{aligned}
&\frac{\partial \rho}{\partial t} + \rho_0 \frac{\partial V_z}{\partial z} = -\rho_0 \frac{\partial V_x}{\partial x}\,,\\
&\frac{\partial V_z}{\partial t} + \frac{U_0^2}{\rho_0}\frac{\partial \rho}{\partial z} = 0\,,\\
&\frac{\partial B_x}{\partial t} - B_0 \frac{\partial V_x}{\partial z} = 0\,,\\
&\frac{\partial V_x}{\partial t} - \frac{B_0}{4\pi\rho_0}\frac{\partial B_x}{\partial z} = -\frac{U_0^2}{\rho_0}\frac{\partial \rho}{\partial x} - \frac{B_0}{4\pi\rho_0}\frac{\partial B_z}{\partial x}\,,\\
&\frac{\partial B_z}{\partial t} - B_0 \frac{\partial V_x}{\partial x} = 0\,.
\end{aligned} \tag{16.11}$$

The equations for (B_y, V_y) are Eqs. (16.1c). Consequently, the simplest transverse Alfvén wave with the components (B_y, V_y) in the waveguide system at hand is the same as above and thereby does not require additional investigation. As for subsets (16.1a) and (16.1b), they turn out to be coupled to one another, as can be seen from Eqs. (16.11). This circumstance implies that, for $\partial/\partial x \neq 0$, we deal with the interaction between longitudinal acoustic and transverse Alfvén waves.

Taking into account the condition at the waveguide walls ($V_x|_{x=0} = V_x|_{x=L} = 0$) and proceeding from the structure of Eqs. (16.11), we write the following relationship:

$$\begin{aligned}
&V_x = \sin k_\perp x \cdot \tilde{V}_x(t,z)\,, \quad B_x = \sin k_\perp x \cdot \tilde{B}_x(t,z)\,,\\
&V_z = \cos k_\perp x \cdot \tilde{V}_z(t,z)\,, \quad B_z = \cos k_\perp x \cdot \tilde{B}_z(t,z)\,,\\
&\rho = \cos k_\perp x \cdot \tilde{\rho}(t,z)\,.
\end{aligned} \tag{16.12}$$

Substituting these relationships into Eqs. (16.11) and cancelling the common trigonometric factor, we arrive at the following set of equations, which is similar to Eqs. (1.1) and describes MHD waves propagating in a conducting fluid at an angle to the external magnetic field $\mathbf{B}_0$ (here, the tilde "$\sim$" is omitted for brevity):

$$\begin{aligned}
\frac{\partial \rho}{\partial t} + \rho_0 \frac{\partial V_z}{\partial z} &= -\rho_0 k_\perp V_x\,, \\
\frac{\partial V_z}{\partial t} + \frac{U_0^2}{\rho_0}\frac{\partial \rho}{\partial z} &= 0\,, \\
\frac{\partial B_x}{\partial t} - B_0 \frac{\partial V_x}{\partial z} &= 0\,, \\
\frac{\partial V_x}{\partial t} - \frac{B_0}{4\pi\rho_0}\frac{\partial B_x}{\partial z} &= k_\perp \frac{U_0^2}{\rho_0}\rho + k_\perp \frac{B_0}{4\pi\rho_0} B_z\,, \\
\frac{\partial B_z}{\partial t} + k_\perp B_0 V_x &= 0\,, \quad \left(\frac{\partial B_z}{\partial z} + k_\perp B_x = 0\right).
\end{aligned} \tag{16.13}$$

We have supplemented these equations with the equation $\nabla\cdot\mathbf{B} = 0$ (see the equation in parentheses), which was not presented in the initial set of equations (16.11). The magnetic field component B_z can be excluded from the state vector components because it is proportional to B_x.

We represent the solution as

$$\begin{aligned}
\psi(t,z) &= \{\rho, V_z, B_x, V_x\} \\
&= \{r(\omega,k), v_z(\omega,k), b_x(\omega,k), v_x(\omega,k)\}\exp(-i\omega t + ikz)
\end{aligned} \tag{16.14}$$

and substitute it into Eqs. (16.13) to obtain the following set of homogeneous algebraic equations for the components r, v_z, b_x and v_x of the complex state vector:

$$\begin{aligned}
-i\omega r + ik\rho_0 v_z + k_\perp \rho_0 v_x &= 0\,, \\
-i\omega v_z + ik\frac{U_0^2}{\rho_0} r &= 0\,, \\
-i\omega b_x - ikB_0 v_x &= 0\,, \\
-i\omega v_x - \frac{i}{k}(k^2 + k_\perp^2)\frac{B_0}{4\pi\rho_0} b_x - k_\perp \frac{U_0^2}{\rho_0} r &= 0\,.
\end{aligned} \tag{16.15}$$

The solvability condition for this set of equations is expressed by the dispersion relation

$$D(\omega,k) \equiv (\omega^2 - (k^2 + k_\perp^2)V_A^2)(\omega^2 - k^2 U_0^2) - k_\perp^2 U_0^2 \omega^2 = 0\,. \tag{16.16}$$

We do not write out here exact solutions to biquadratic equation (16.16) because they are poorly informative. Instead, let us restrict ourselves to considering dispersion curves in particular cases. In the short-wavelength limit such that $k^2 \gg k_\perp^2$, dispersion relation (16.16) yields the following expressions for the frequencies squared:

$$\omega^2 = k^2 V_A^2 \left(1 + \frac{k_\perp^2}{k^2}\frac{U_0^2}{V_A^2 - U_0^2}\right), \quad \omega^2 = k^2 U_0^2 \left(1 + \frac{k_\perp^2}{k^2}\frac{U_0^2}{U_0^2 - V_A^2}\right), \tag{16.17}$$

In the opposite, long-wavelength, limit, $k^2 \ll k_\perp^2$, the squares of the frequencies are given by the formulas

$$\omega^2 = k_\perp^2(U_0^2 + V_A^2)\,, \quad \omega^2 = k^2 \frac{U_0^2 V_A^2}{U_0^2 + V_A^2}\,. \tag{16.18}$$

Formulas (16.17) and (16.18) should be supplemented with the relationship

$$\omega^2 = k^2 V_A^2\,, \tag{16.19}$$

which determines the frequency spectra of a conventional Alfvén wave that has the polarization (B_y, V_y) and is described by Eqs. (16.1c). Relationship (16.19) is valid over the entire range of values of the wavenumber k.

In our opinion, the most convenient classification of MHD waves with frequencies (16.17)–(16.19) is as follows. For $V_A > U_0$, the first frequencies in formulas (16.17) and (16.18) determine the short- and long-wavelength ranges of the spectra of oblique Alfvén MHD waves and the second frequencies in this formulas give the short- and long-wavelength spectra of oblique magnetosonic waves. For $V_A < U_0$, the classification of short-wavelength waves (see formulas (16.17)) should be kept the same as that for $V_A > U_0$. But for long-wavelength waves, it seems worthwhile to propose another classification: the first of formulas (16.18) determines the frequencies of oblique magnetosonic waves and the second of the formulas, the frequencies of oblique Alfvén waves.[2] Both these types of waves originate from perturbations of the longitudinal and transverse velocities of the fluid under the action of the forces of "longitudinal" hydrodynamic and "transverse" magnetic pressures. Alfvén waves are dominated by the magnetic pressure forces and magnetosonic waves, by the hydrodynamic pressure forces. Formula (16.19) describes the frequency spectra of conventional Alfvén waves in which the hydrodynamic pressure is unperturbed. A particular case of equal pressures, $V_A = U_0$, will be examined separately.

We introduce the dimensionless variables

$$y = \frac{\omega}{k_\perp U_0}\,, \quad x = \frac{k}{k_\perp}\,, \quad \alpha = \frac{V_A}{U_0}\,, \tag{16.20}$$

in terms of which to rewrite dispersion relation (16.16) as

$$D(y,x) \equiv (y^2 - \alpha^2(1+x^2))(y^2 - x^2) - y^2 = 0\,. \tag{16.21}$$

Curves 1 and 2 in Figs. 16 and 17 are the dimensionless dispersion curves $y(x)$ of dispersion relation (16.21). The light straight line in Fig. 16 shows the dispersion curve $y = \alpha x$ of a conventional Alfvén wave with frequency spectrum (16.19).

We denote the solutions to dispersion relation (16.16) by $\omega_m(k)$. Using representation (16.14) and Eqs. (16.15) and setting $r(\omega, k) = 1$, we find the following

[2] The term "oblique" implies that the waves propagate at an angle to the external magnetic field. In a waveguide, an oblique wave is formed by several waves propagating at different angles (see Fig. 13).

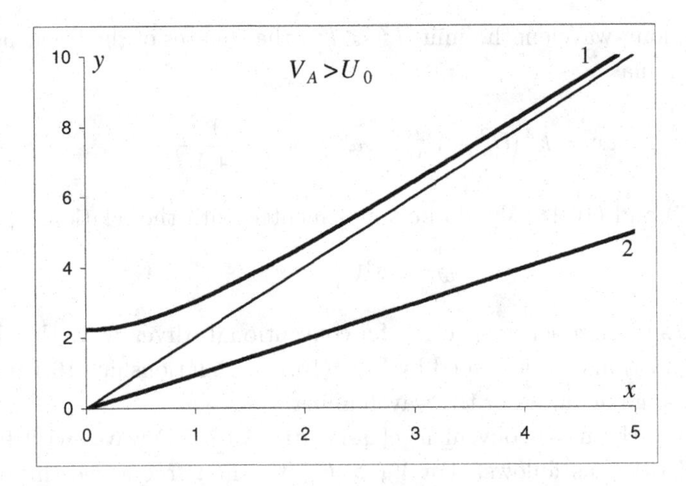

Fig. 16. Dimensionless dispersion curves of oblique MHD waves in a conducting fluid in the $(x = k/k_\perp,\ y = \omega/k_\perp U_0)$ plane for $V_A > U_0$: *1* Alfvén wave and *2* magnetosonic wave.

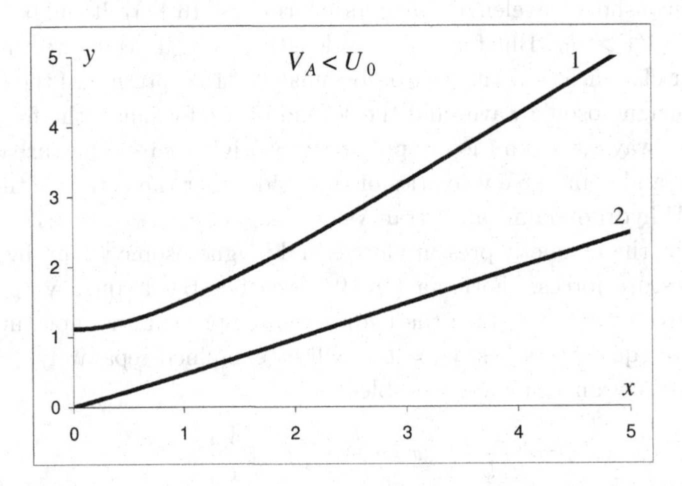

Fig. 17. Dimensionless dispersion curves of oblique MHD waves in a conducting fluid in the $(x = k/k_\perp,\ y = \omega/k_\perp U_0)$ plane for $V_A < U_0$: *1* magnetosonic wave and *2* Alfvén wave.

expression for the state vector of harmonic MHD perturbations in a conducting fluid:

$$\begin{aligned} \psi(t,z) &= \{\rho, V_z, B_x, V_x\} \\ &= \sum_{m=1}^{4} A_m \left\{1, \frac{kU_0^2}{\omega_m \rho_0}, -\frac{ikB_0(\omega_m^2 - k^2U_0^2)}{k_\perp \omega_m^2 \rho_0}, \frac{i(\omega_m^2 - k^2U_0^2)}{k_\perp \omega_m \rho_0}\right\} \\ &\quad \times \exp[-i\omega_m(k)t + ikz]\,. \end{aligned} \tag{16.22}$$

Formula (16.22) is very similar in structure to expression (13.7) for the state vector of electromagnetic perturbations in a waveguide with an anisotropic plasma filling in a strong external magnetic field. This similarity stems from the structural similarity between dispersion relations (13.12) and (16.16). Thus, switching to the frequency and wavenumber operators in dispersion relation (16.16), we obtain the following general differential equation describing MHD perturbations in a conducting fluid:

$$\left[\left(\frac{\partial^2}{\partial t^2} - U_0^2\frac{\partial^2}{\partial z^2}\right)\left(\frac{\partial^2}{\partial t^2} - V_A^2\frac{\partial^2}{\partial z^2} + k_\perp^2 V_A^2\right) + k_\perp^2 U_0^2 \frac{\partial^2}{\partial t^2}\right] A(t,z) = 0\,. \qquad (16.23)$$

Although more complicated, Eq. (16.23) is similar to differential equation (13.13). Moreover, with the thermal motion of plasma electrons taken into account, the structures of the equations become the same, as can be seen from a comparison between differential equations (6.10) and (10.10), describing longitudinal waves in a plasma.

17. Acoustic Waves in Crystals

In this section, we consider acoustic waves in crystalline media, both nonconducting (dielectric crystals) and conducting (piezosemiconductors). We restrict ourselves to the long-wavelength limit, assuming that the length of acoustic waves is much greater than the period of the crystal lattice. This assumption enables us to describe the lattice dynamics in terms of the displacement vector $\mathbf{U}(t,\mathbf{r})$ independent of the lattice period, i.e., to treat a crystal as a continuous medium. If the crystal conductivity is ignored, then the lattice dynamics is described by the following equation from elasticity theory:

$$\rho\frac{\partial^2 U_i}{\partial t^2} - \lambda_{imlj}\frac{\partial^2 U_j}{\partial r_m \partial r_l} = 0\,. \qquad (17.1)$$

Here, ρ is the crystal density and λ_{imlj} is the elastic modulus tensor, which describes the force arising from the crystal lattice deformation. Since Eq. (17.1) is valid only for small lattice deformations, the density ρ and tensor λ_{imlj} are assumed to be constant. We also explain that the subscripts i, j, etc. stand for the spatial coordinates (x, y, z) and that, in Eq. (17.1), summation over the repeated indices is implied.

As in the previous sections, we consider waves propagating along the OZ axis, i.e., we assume that $\mathbf{U} = \mathbf{U}(t,z)$. In this case, the spatial differential operator in Eq. (17.1) reduces to

$$\lambda_{imlj}\frac{\partial}{\partial r_m}\frac{\partial}{\partial r_l} = \lambda_{izzj}\frac{\partial^2}{\partial z^2}\,. \qquad (17.2)$$

The nonzero elements of the tensor λ_{imlj} are determined by the spatial symmetry of the crystal. For instance, for crystals of cubic symmetry — the case to which our analysis will be restricted here — the nonzero tensor elements are

$$\begin{aligned} \lambda_{xzzx} &= \lambda_{yzzy} = \lambda_{\perp}\,, \\ \lambda_{zzzz} &= \lambda_{||}\,, \end{aligned} \tag{17.3}$$

where $\lambda_{\perp}$ and $\lambda_{||}$ are the transverse and longitudinal elastic moduli of the lattice. The remaining elements of the tensor λ_{izzj} are zero. It is, of course, assumed that the OZ axis is directed along one of the crystal axes.

From what has been said in this section, we can see that, in a crystal of cubic symmetry, there are two types of lattice displacements: across and along the propagation direction of small acoustic perturbations. Accordingly, the state vector has two components:

$$\psi(t,z) = \{U_{\perp}(t,z), U_{||}(t,z)\}\,, \tag{17.4}$$

where $U_{||}(t,z) = U_z(t,z)$ and $U_{\perp}(t,z) = U_x(t,z)$ (or $U_y(t,z)$). With the nonzero tensor elements (17.3), Eq. (17.1) splits into two independent equations,

$$\frac{\partial^2 U_{\perp}}{\partial t^2} - C_{\perp}^2 \frac{\partial^2 U_{\perp}}{\partial z^2} = 0\,, \qquad \frac{\partial^2 U_{||}}{\partial t^2} - C_{||}^2 \frac{\partial^2 U_{||}}{\partial z^2} = 0\,. \tag{17.5}$$

The quantities

$$C_{\perp} = \sqrt{\frac{\lambda_{\perp}}{\rho}}\,, \quad C_{||} = \sqrt{\frac{\lambda_{||}}{\rho}} \tag{17.6}$$

are known, respectively, as the transverse and longitudinal speeds of sound in a crystal of cubic symmetry. Equations (17.5) yield the following spectra of transverse and longitudinal acoustic waves in a cubic crystal:

$$\omega = \pm k C_{\perp}\,, \quad \omega = \pm k C_{||}\,. \tag{17.7}$$

It is from the linear spectrum (17.7) of acoustic waves in crystals that the term "acoustic spectrum" was derived. Since such waves are quite similar to magnetosonic waves described by dispersion relations (16.4) and (16.7), there is no need to investigate them further. They are interesting in that they provide a fairly simple study of some new effects — that is just what we are going to do now.

Up to this point, we have ignored the presence of charge carriers and considered a purely dielectric crystal. Let us now turn to crystalline media with free charge carriers. Such media can be exemplified by metals and semiconductors. Of course, the charge carriers interact with ions and atoms located at the lattice points. Consequently, the displacements $\mathbf{U}(t,\mathbf{r})$ of the lattice points causes the motion of charge carriers, and vice versa. This coupling between the charge carriers and the lattice is most pronounced in the so-called piezosemiconductors (e.g., in quartz). In the general case, the force exerted on a lattice by an applied electric field $\mathbf{E}(t,\mathbf{r})$ can be represented as

$$F_i = \beta_{ijm} \frac{\partial E_m}{\partial r_j}\,. \tag{17.8}$$

Accordingly, in a piezosemiconductor, the lattice deformation under the action of the force (17.8) induces the current, which, with allowance for the Onsager symmetry principle, is written as

$$\delta j_i = -\beta_{iml}\frac{\partial^2 U_m}{\partial t \partial r_l}\,. \tag{17.9}$$

This current should be added to the current of charge carriers in the electric field **E**.

Finally, for a piezosemiconductor, we have the equation of motion of the lattice (see Eq. (17.1)),

$$\rho\frac{\partial^2 U_i}{\partial t^2} = \lambda_{imlj}\frac{\partial^2 U_j}{\partial r_m \partial r_l} + \beta_{ijm}\frac{\partial E_m}{\partial r_j}\,; \tag{17.10}$$

the equations of motion of the free charge carriers (presented in the previous sections for different models of the medium); and Maxwell's equations (in which it is necessary to take into account both the current of charge carriers and the piezocurrent (17.9)).

Let us make some simplifying assumptions. Specifically, we consider a crystal with hexagonal lattice symmetry and assume that the OZ axis is aligned with the symmetry axis of the crystal (i.e., with the wave propagation direction). Such a crystal differs from a cubic one in that it has no centre of symmetry. Consequently, all what has been said about the tensor λ_{imlj} remains valid. As for the tensor β_{ijm}, its nonzero components are $\beta_{xxz} = \beta_{yyz} = \beta_{xzx} = \beta_{yzy} = \beta_1$, $\beta_{zxx} = \beta_{zyy} = \beta_2$, $\beta_{zzz} = \beta_3$. We describe the charge carriers in the cold dissipative electron plasma model (with v_{eff} being a certain effective collision frequency) under the assumption that the electric field is purely longitudinal. We supplement the last of equations (6.2) with the piezocurrent (17.9) and use the equations of electron motion (6.14) and the lattice dynamic equation (17.10) to write the following set of equations:

$$\begin{aligned}
&\frac{\partial E_z}{\partial t} + 4\pi J_p - 4\pi\beta_3\frac{\partial^2 U_{||}}{\partial t \partial z} = 0\,,\\
&\frac{\partial J_p}{\partial t} - \frac{\omega_p^2}{4\pi}E_z + v_{eff}J_p = 0\,,\\
&\rho\frac{\partial^2 U_{||}}{\partial t^2} - \lambda_{||}\frac{\partial^2 U_{||}}{\partial z^2} - \beta_3\frac{\partial E_z}{\partial z} = 0\,.
\end{aligned} \tag{17.11}$$

Representing the solution as

$$\begin{aligned}
\mathbf{\Psi}(t,z) &= \{J_p(t,z), E_z(t,z), U_{||}(t,z)\}\\
&= \{j_p(\omega,k), e_z(\omega,k), u_{||}(\omega,k)\}\exp(-i\omega t + ikz)\,,
\end{aligned} \tag{17.12}$$

and substituting it into Eqs. (17.11) yields the following set of homogeneous algebraic equations for the components $j_p(\omega,k)$, $e_z(\omega,k)$ and $u_{||}(\omega,k)$ of the complex state vector:

$$\begin{aligned}
&i\omega e_z - 4\pi j_p + 4\pi k\omega\beta_3 u_{||} = 0\,,\\
&(\omega_p^2/4\pi)e_z + i(\omega + iv_{eff})j_p = 0\,,\\
&(\rho\omega^2 - k^2\lambda_{||})u_{||} + ik\beta_3 e_z = 0\,.
\end{aligned} \tag{17.13}$$

The solvability condition for this set is given by the dispersion relation

$$D(\omega, k) \equiv \left(1 - \frac{\omega_p^2}{\omega(\omega + i v_{eff}}\right)(\omega^2 - k^2 C_{\|}^2) - \frac{4\pi\beta_3^2}{\rho} k^2 = 0 . \tag{17.14}$$

This dispersion relation describes a new effect, which will be analyzed later. Now, we immediately present the final result for longitudinal acoustic waves. In an actual piezosemiconductor, the frequencies (17.7) of acoustic waves satisfy the inequalities

$$|\omega| \ll v_{eff} \ll \omega_p . \tag{17.15}$$

Consequently, for acoustic waves, dispersion relation (17.14) can be substantially simplified to the form

$$D(\omega, k) \equiv -\omega^2 - i \frac{4\pi\beta_3^2 v_{eff} k^2}{\rho \omega_p^2} \omega + k^2 C_{\|}^2 = 0 . \tag{17.16}$$

This dispersion relation coincides in structure with dispersion relation (6.16), which determines the frequency spectra of damped longitudinal waves in a collisional electron plasma. But in contrast to (6.16), dispersion relation (17.16) determines the spectra of acoustic waves in a piezosemiconductor that are damped with time due to collisional energy dissipation in the electron subsystem. With inequalities (17.15), the damping rate of acoustic waves is described by the formula

$$\omega'' = -\frac{2\pi\beta_3^2 v_{eff} k^2}{\rho \omega_p^2} , \tag{17.17}$$

and the real part of the frequency is given by the second of formulas (17.7). This type of damping is known as the plasma damping of acoustic waves in piezosemiconductors. The plasma damping, as well as other related effects, will be considered below in discussing the interaction between linear waves of different nature.

18. Longitudinal Electrostatic Waves in a One-Dimensional Electron Beam

Let us consider an unbounded electron beam moving along the z axis. Let N_{0b} and U_{0b} be the unperturbed density and velocity of the electrons and let $N_b(t, z)$ and $N_b(t, z)$ be small perturbations of these beam parameters. The perturbations in the beam produce a longitudinal electric field $E_z(t, z)$, which in turn influences the beam electrons. This process is described by the following set of field and hydrodynamic equations:

$$\begin{gathered} \frac{\partial J_{b1}}{\partial t} + U_{0b} \frac{\partial J_{b1}}{\partial z} - \frac{\omega_v^2}{4\pi} E_z = 0 , \\ \frac{\partial J_{b2}}{\partial t} + U_{0b} \frac{\partial J_{b2}}{\partial z} + U_{0b} \frac{\partial J_{b1}}{\partial z} = 0 \\ \frac{\partial E_z}{\partial t} + 4\pi J_{b1} + 4\pi J_{b2} = 0 . \end{gathered} \tag{18.1}$$

Here, $\omega_b = \sqrt{4\pi e^2 N_{0b}/M}$ is the Langmuir frequency of the beam electrons, $J_{b1} = eN_{0b}U_b$ is a quantity having the dimension of current and proportional to the perturbed electron velocity, and $J_{b2} = eU_{0b}N_b$ is a quantity having the dimension of current and proportional to the perturbed electron density. It is obvious that, to first order in the perturbations, the quantity $J_b = J_{b1} + J_{b2}$ describes the total perturbed beam current. Equations (18.1) have the form of general equations (1.1). Their state vector is $\mathbf{\Psi}(t,z) = \{J_{b1}, J_{b2}, E_z\}$. For $U_{0b} = 0$, the second of Eqs. (18.1) is unnecessary and the remaining two equations reduce to Eqs. (6.2) for longitudinal waves in a cold plasma.

Representing the solution as

$$\mathbf{\Psi}(t,z) = \{J_{b1}, J_{b2}, E_z\} = \{j_1(\omega,k), j_2(\omega,k), e_z(\omega,k)\} \exp(-i\omega t + ikz)\,, \quad (18.2)$$

and substituting it into Eqs. (18.1) leads to the following set of homogeneous algebraic equations for the components j_1, j_2 and e_z of the complex state vector:

$$\begin{aligned} -i(\omega - kU_{0b})j_1 - (\omega_b^2/4\pi)e_z &= 0\,, \\ -i(\omega - kU_{0b})j_2 + ikU_{0b}j_1 &= 0\,, \\ -i\omega e_z + 4\pi j_1 + 4\pi j_2 &= 0\,. \end{aligned} \quad (18.3)$$

The solvability condition for Eqs. (18.3) is given by the dispersion relation

$$D(\omega,k) \equiv -(\omega - kU_{0b})^2 + \omega_b^2 = 0\,, \quad (18.4)$$

in which we have discarded the trivial root $\omega = 0$.

From dispersion relation (18.4) we find the frequencies of the eigenmodes,

$$\omega_1 = kU_{0b} + \omega_b\,, \quad \omega_2 = kU_{0b} - \omega_b\,. \quad (18.5)$$

From expressions (18.5), Eqs. (18.3), and representation (18.2) we in turn find the following formula for the state vector of a one-dimensional electron beam with harmonic electrostatic perturbations in the initial-value problem:

$$\begin{aligned} \mathbf{\Psi}(t,z) &= \{J_{b1}, J_{b2}, E_z\} \\ &= \left(A_1\left\{1, \frac{kU_{0b}}{\omega_b}, -i\frac{4\pi}{\omega_b}\right\}\exp(-i\omega_b t)\right. \\ &\quad \left. + A_2\left\{1, -\frac{kU_{0b}}{\omega_b}, i\frac{4\pi}{\omega_b}\right\}\exp(+i\omega_b t)\right)\exp[ik(z - U_{0b}t)]\,. \end{aligned} \quad (18.6)$$

The first term in this formula describes the so-called fast beam charge density wave, and the second term, the so-called slow beam charge density wave. The phase

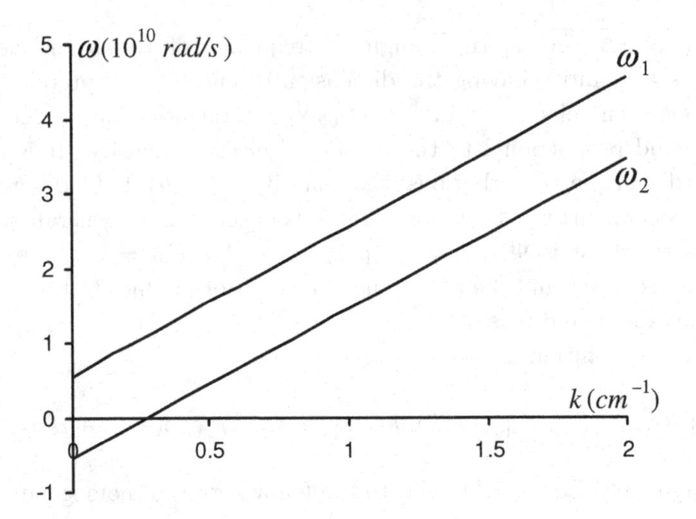

Fig. 18. Dispersion curves of longitudinal waves in a one-dimensional electron beam with $U_{0b} = 10^{10}$ cm/s and $N_{0b} = 10^{10}$ cm^{-3}.

velocities of these waves are given by the formulas

$$V_{ph}^{(1,2)} = U_{0b} \pm \frac{\omega_b}{k} . \tag{18.7}$$

For $k > \omega_b/U_{0b}$, both of the charge density waves propagate in the same direction — that in which the beam moves. For $0 < k < \omega_b/U_{0b}$, the waves propagate in opposite directions: the fast wave again propagates in the direction of the beam motion and the slow wave propagates towards the beam. For $k = \omega_b/U_{0b}$, the velocity of the slow space charge density wave vanishes. In the range of large wavenumbers, the phase velocities of the beam charge density waves are close to the unperturbed beam velocity U_{0b}. Eigenfrequency spectra (18.5) of the beam waves have the form of spectrum (1.14); consequently, in accordance with the above classification, the waves in question are nondispersive. It is also easy to see that, for $U_{0b} = 0$, solution (18.6) goes over to solution (6.7), to within the notation. The dispersion curves of the waves with frequencies (18.5) are illustrated in Fig. 18. Having calculated the dispersion curves, we can readily investigate the wave dispersion. Thus, the closer the curve is to a straight line (not necessarily passing through the origin of the coordinates), the weaker is the wave dispersion.

In dispersion relation (18.4), we have discarded the trivial root $\omega = 0$. Accordingly, in the expression for the state vector, we should omit the "redundant" component. From formula (18.6) we can see that the state vector components E_z and J_{b2} are proportional to one another because, for any m value, the corresponding components of the complex state vector satisfy the relationship

$$j_2^{(m)} = i\frac{kU_{0b}}{4\pi} e_z^{(m)} . \tag{18.8}$$

We eliminate the component J_{b2} and represent the state vector (18.6) as

$$\mathbf{\Psi}(t,z) = \left\{ A_1 \begin{pmatrix} 1 \\ -i4\pi/\omega_b \end{pmatrix} \exp(-i\omega_b t) + A_2 \begin{pmatrix} 1 \\ i4\pi/\omega_b \end{pmatrix} \exp(+i\omega_b t) \right\} \exp[ik(z - U_{0b}t)] . \tag{18.9}$$

We also assume that, at the initial instant, the state vector is given by the formula

$$\mathbf{\Psi}(0,z) = \begin{pmatrix} J_0 \\ E_0 \end{pmatrix} \exp(ikz) . \tag{18.10}$$

Substituting formulas (18.9) and (18.10) into Eqs. (2.5) yields the following set of equations for determining the unknown complex amplitudes of the beam charge density waves:

$$\begin{aligned} A_1 + A_2 &= J_0 , \\ A_1 - A_2 &= i(\omega_b/4\pi)E_0 . \end{aligned} \tag{18.11}$$

By finding the complex amplitudes $A_{1,2}$ from Eqs. (18.11) and inserting them into formula (18.9), one can obtain a general solution to the problem of the excitation of beam waves by an initial harmonic perturbation (18.10).

In dispersion relation (18.4), we go over to operators (3.7). As a result, we obtain the following differential equation for the characteristic function of the state vector of the electron beam:

$$\left[\left(\frac{\partial}{\partial t} + U_{0b} \frac{\partial}{\partial z} \right)^2 + \omega_b^2 \right] A(t,z) = 0 . \tag{18.12}$$

The state vector of the beam is calculated from the formula

$$\mathbf{\Psi}(t,z) = \{J_{b1}, E_z\} = \{A(t,z), (4\pi/\omega_b^2)\hat{L}A(t,z)\} , \tag{18.13}$$

where $\hat{L} = \partial/\partial t + U_{0b}\partial/\partial z$ is the translation operator.

We can easily verify that the general solution to differential equation (18.12) has the form (it is expedient to compare it with solution (6.12))

$$A(t,z) = f_1(z - U_{0b}t) \exp(-i\omega_b t) + f_2(z - U_{0b}t) \exp(i\omega_b t) , \tag{18.14}$$

where $f_{1,2}$ are arbitrary functions. Solution (18.14) describes waves that are harmonic in time and propagate in space with the unperturbed beam velocity U_{0b} without changing their shape, as a single wave packet.

In a particular case, we can set (see (6.13))

$$f_{1,2}(\xi) = A_{1,2} \exp(ik\xi) , \tag{18.15}$$

where $A_{1,2}$ are constants and $\xi = z - U_{0b}t$. Substituting (18.15) into (18.14) leads to the following expression for the characteristic function of the state vector of longitudinal harmonic electrostatic waves in an electron beam:

$$A(t,z) = A_1 \exp(-ikU_{0b}t - i\omega_b t + ikz) + A_2 \exp(-ikU_{0b}t + i\omega_b t + ikz)\,. \quad (18.16)$$

With allowance for formula (18.13), from expression (18.16) it is also easy to obtain expression (18.9) for the harmonic state vector of the beam.

Notably, general solution (18.14) implies that arbitrary beam perturbations propagate in space with the velocity U_{0b}. As for the harmonic waves described by characteristic function (18.16), their phase velocities can be arbitrary, depending on the k value. Recall that it is not the phase velocity that characterizes the spatial propagation of the wave perturbations.

19. Beam Instability in a Plasma

In the examples from Secs. 5–18, the solutions to the dispersion relations are either real frequencies ω or complex frequencies, but with $\mathrm{Im}\,\omega < 0$. This implies that the corresponding physical systems are in a stable equilibrium state. Let us consider an example of a nonequilibrium system consisting of two interpenetrating unbounded objects — an electron beam and a cold electron plasma. Electrostatic perturbations that propagate along the beam (i.e., along the z axis) in such a beam–plasma system are described by equations like Eqs. (18.1). It is only necessary to supplement the equation for E_z with the plasma current density J_p, satisfying the first of Eqs. (6.2). As a result, we have the following set of equations:

$$\begin{aligned}
\frac{\partial J_{b1}}{\partial t} + U_{0b}\frac{\partial J_{b1}}{\partial z} - \frac{\omega_b^2}{4\pi}E_z &= 0\,,\\
\frac{\partial J_{b2}}{\partial t} + U_{0b}\frac{\partial J_{b2}}{\partial z} + U_{0b}\frac{\partial J_{b1}}{\partial z} &= 0\,,\\
\frac{\partial J_p}{\partial t} - \frac{\omega_p^2}{4\pi}E_z &= 0\,,\\
\frac{\partial E_z}{\partial t} + 4\pi J_{b1} + 4\pi J_{b2} + 4\pi J_p &= 0\,,
\end{aligned} \quad (19.1)$$

where J_{b1} and J_{b2} are the partial beam current densities that have been introduced in the previous section. The state vector of Eqs. (19.1) is $\mathbf{\Psi}(t,z) = \{J_{b1}, J_{b2}, J_p, E_z\}$.

Substituting the solution of the form

$$\mathbf{\Psi}(t,z) = \{j_{b1}(\omega,k), j_{b2}(\omega,k), j_p(\omega,k), e_z(\omega,k)\}\exp(-i\omega t + ikz) \quad (19.2)$$

into Eqs. (19.1), we obtain the following set of homogeneous algebraic equations for the components j_{b1}, j_{b2}, j_p and e_z of the complex state vector:

$$\begin{aligned}-i(\omega - kU_{0b})j_{b1} - (\omega_b^2/4\pi)e_z &= 0\,,\\ -i(\omega - kU_{0b})j_{b2} + ikU_{0b}j_{b1} &= 0\,,\\ -i\omega j_p - (\omega_b^2/4\pi)e_z &= 0\,,\\ -i\omega e_z + 4\pi j_{b1} + 4\pi j_{b2} + 4\pi j_p &= 0\,.\end{aligned} \tag{19.3}$$

The condition for homogeneous equations (19.3) to have a nontrivial solution is expressed by the dispersion relation

$$D(\omega, k) \equiv (\omega^2 - \omega_p^2)(\omega - kU_{0b})^2 - \omega^2\omega_b^2 = 0\,. \tag{19.4}$$

Since this dispersion relation is fourth-order in ω, it is difficult to represent its exact solution in an analytic form. To simplify matters, we assume that the following inequality is satisfied:

$$\omega_b^2 \ll \omega_p^2\,. \tag{19.5}$$

With this inequality, we can readily find approximate solutions to dispersion relation (19.4).

Figure 19 illustrates the dispersion curves $\omega_m(k)$ of relation (19.4). The asymptotes of the curves are the straight lines $\omega = \pm\omega_p$ and $\omega = kU_{0b} \pm \omega_b$ (see Figs. 4, 18). We can see that, for wavenumbers in the range between symbols *1* and *2*, corresponding to each k value there are only two real values of the frequency. Consequently, in this wavenumber range, the remaining two frequencies are complex, indicating that there is a beam instability. For wavenumbers in the ranges to the left of symbol *1* and to the right of symbol *2*, corresponding to each k value are four real values of the frequency, so there is no instability in these two ranges of the

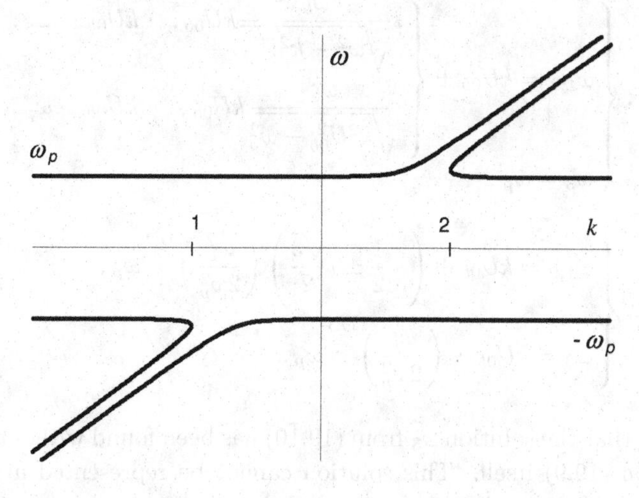

Fig. 19. Dispersion curves of a beam–plasma system.

wavenumber k. The boundaries of the instability region in terms of the wavenumber k will be determined later.

Let us now solve dispersion relation (19.4). For definiteness, we restrict ourselves to the range $k > 0$. Under inequality (19.5), the contribution of the beam to the dispersion relation in the (ω, k) plane is important only in the region where $\omega \approx kU_{0b}$. Over the rest of the (ω, k) plane, the beam contribution can be ignored, so we immediately find one of the four solutions to the dispersion relation ($k > 0$):

$$\omega_4 = -\omega_p \,. \tag{19.6}$$

The remaining three solutions can be sought for in the form

$$\omega = kU_{0b} + \delta\omega \,, \tag{19.7}$$

where $\delta\omega$ is the frequency correction, which, by virtue of inequality (19.5), satisfies the conditions

$$|\delta\omega| \ll kU_{0b} \,, \quad \omega_p \,. \tag{19.8}$$

We take into account conditions (19.8) to see that representation (19.7) converts dispersion relation (19.4) to the form

$$\left(1 - \frac{\omega_p^2}{k^2 U_{0b}^2}\right) + 2\left(\frac{\omega_p}{kU_{0b}}\right)^3 \frac{\delta\omega}{\omega_p} - \frac{\omega_b^2}{\delta\omega^2} = 0 \,. \tag{19.9}$$

For $kU_{0b} \neq \omega_p$, the second term on the left-hand side of dispersion relation (19.9) can be ignored, by virtue of conditions (19.8). Conversely, for $kU_{0b} = \omega_p$, this second term has to be taken into account. That is why there are two fundamentally different solutions to dispersion relation (19.9), as well as to dispersion relation (19.4):

$$\text{I.} \quad kU_{0b} \neq \omega_p : \begin{cases} \omega_{1,2} = kU_{0b} \pm \begin{cases} i\dfrac{\omega_b}{\sqrt{\omega_p^2 - k^2 U_{0b}^2}} kU_{0b} \,, & kU_{0b} < \omega_p \,, \\[2ex] \dfrac{\omega_b}{\sqrt{k^2 U_{0b}^2 - \omega_p^2}} kU_{0b} \,, & kU_{0b} > \omega_p \,, \end{cases} \\[2ex] \omega_3 = \omega_p \,, \end{cases} \tag{19.10}$$

$$\text{II.} \quad kU_{0b} = \omega_p : \begin{cases} \omega_{1,2} = kU_{0b} + \left(-\dfrac{1}{2} \pm i\dfrac{\sqrt{3}}{2}\right)\left(\dfrac{\omega_b^2}{2\omega_p^2}\right)^{1/3} \omega_p \,, \\[2ex] \omega_3 = kU_{0b} + \left(\dfrac{\omega_b^2}{2\omega_p^2}\right)^{1/3} \omega_p \,. \end{cases} \tag{19.11}$$

It is worth noting that the solution ω_3 from (19.10) has been found without invoking dispersion relation (19.9) itself. This solution cannot be represented in the form (19.7) with $\delta\omega$ satisfying conditions (19.8), and, in the (ω, k) plane, it lies within the region where the beam contribution to the dispersion relation is small. It is

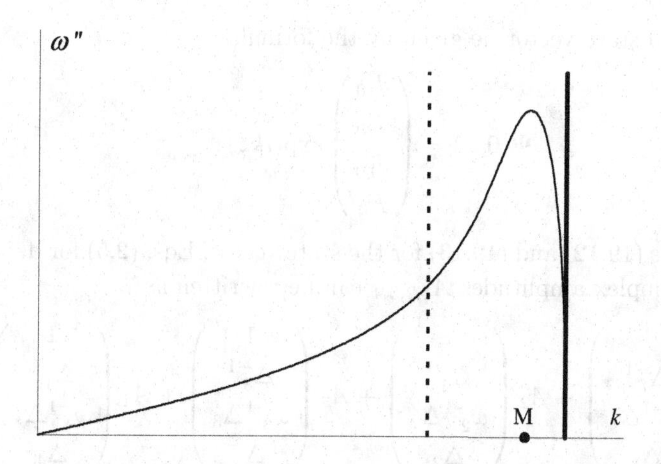

Fig. 20. Wavenumber dependence of the growth rate of the beam instability in a plasma.

this latter circumstance that leads to the solution at hand (see formula (19.6) and Fig. 19).

A representative wavenumber dependence of the growth rate of the beam instability, $\omega_1''(k) = \operatorname{Im} \omega_1(k)$, is depicted in Fig. 20. The heavy vertical line in the figure shows the right boundary of the instability region. The dotted vertical line is the boundary between solutions (19.10) and (19.11) to dispersion relation (19.9). This boundary is fairly conditional because it is determined only approximately, as well as solutions (19.10) and (19.11) themselves. The maximum growth rate $(\omega_1'')_{\max} = \sqrt{3}/2(\omega_b^2/2\omega_p^2)^{1/3}\omega_p$ follows from solution (19.11); it is reached at $k = \omega_p/U_{0b}$ (see Fig. 20, point M).

From Eqs. (19.3) we find the following expressions for the complex state eigenvectors (again setting $j_{b1} = 1$ for convenience):

$$\{j_{b1}, j_{b2}, j_p, e_z\}_{(m)} = \left\{1, kU_{0b}\Delta_m^{-1}, \omega_m^{-1}\frac{\omega_p^2}{\omega_b^2}\Delta_m, -i\frac{4\pi}{\omega_b^2}\Delta_m\right\}, \tag{19.12}$$

where $\Delta_m = \omega_m - kU_{0b}$. And finally, substituting eigenfrequencies (19.10) or (19.11) and complex state eigenvectors (19.12) into solution (1.11), we obtain the following expression for the state vector of harmonic electrostatic perturbations in the beam–plasma system under consideration:

$$\boldsymbol{\Psi}(t,z) = \begin{pmatrix} J_{b1} \\ J_{b2} \\ J_p \\ E_z \end{pmatrix} = \sum_{m=1}^{4} A_m \begin{pmatrix} j_{b1} \\ j_{b2} \\ j_p \\ e_z \end{pmatrix}_{(m)} \exp[-i\omega_m(k)t + ikz]\,. \tag{19.13}$$

Let the initial state vector be given by the formula

$$\mathbf{\Psi}(0,z) = \begin{pmatrix} J_{10} \\ J_{20} \\ J_{p0} \\ E_0 \end{pmatrix} \exp(ikz)\,. \tag{19.14}$$

With expressions (19.12) and (19.13) for the state vector, Eqs. (2.5) for determining the unknown complex amplitudes $A_{1,2,3,4}$ can be rewritten as

$$A_1 \begin{pmatrix} 1 \\ \Delta_1^{-1} \\ \omega_1^{-1}\Delta_1 \\ \Delta_1 \end{pmatrix} + A_2 \begin{pmatrix} 1 \\ \Delta_2^{-1} \\ \omega_2^{-1}\Delta_2 \\ \Delta_2 \end{pmatrix} + A_3 \begin{pmatrix} 1 \\ \Delta_3^{-1} \\ \omega_3^{-1}\Delta_3 \\ \Delta_3 \end{pmatrix} + A_4 \begin{pmatrix} 1 \\ \Delta_4^{-1} \\ \omega_4^{-1}\Delta_4 \\ \Delta_4 \end{pmatrix}$$

$$= \begin{pmatrix} J_{10} \\ (kU_{0b})^{-1}J_{20} \\ (\omega_b^2/\omega_p^2)J_{p0} \\ i(\omega_p^2/4\pi)E_0 \end{pmatrix}, \tag{19.15}$$

where the eigenfrequencies ω_m are described by formulas (19.10) or (19.11).

Since expressions (19.10) and (19.11) for the eigenfrequencies of the beam–plasma system are fairly complicated, we restrict ourselves to solving Eqs. (19.15) with particular initial conditions such that $J_{10} \neq 0$, $J_{20} = 0$, $J_0 = 0$ and $E_0 = 0$. These conditions imply that initial perturbation arises from an electron beam premodulated only in the velocity and that there are no initial perturbations in the plasma.

We begin with the case $kU_{0b} \neq \omega_p$, in which the eigenfrequencies are given by formulas (19.10) and (19.6). Introducing the notation

$$\Delta = \begin{cases} i\dfrac{\omega_b}{\sqrt{\omega_p^2 - k^2U_{0b}^2}}kU_{0b}\,, & kU_{0b} < \omega_p\,, \\[2ex] \dfrac{\omega_b}{\sqrt{k^2U_{0b}^2 - \omega_p^2}}kU_{0b}\,, & kU_{0b} > \omega_p \end{cases} \tag{19.16}$$

and taking into account the fact that, under inequality (19.5) and for $kU_{0b} \neq \omega_p$, the second term on the left-hand side of dispersion relation (19.9) can be ignored, we convert Eqs. (19.15) into the form

$$\begin{aligned} A_1 + A_2 + A_3 + A_4 &= J_{10}\,, \\ A_1 - A_2 + \Delta(\omega_p - kU_{0b})^{-1}A_3 - \Delta(\omega_p + kU_{0b})^{-1}A_4 &= 0\,, \\ A_1 - A_2 + \Delta^{-1}\frac{kU_{0b}}{\omega_p}(\omega_p - kU_{0b})A_3 + \Delta^{-1}\frac{kU_{0b}}{\omega_p}(\omega_p + kU_{0b})A_4 &= 0\,, \\ A_1 - A_2 + \Delta^{-1}(\omega_p - kU_{0b})A_3 - \Delta^{-1}(\omega_p + kU_{0b})A_4 &= 0\,. \end{aligned} \tag{19.17}$$

Equations (19.17) should be solved by using two small parameters,

$$|\Delta/\omega_p| \ll 1\,, \quad |\Delta/kU_{0b}| \ll 1\,. \tag{19.18}$$

That these parameters are indeed small follows from inequality (19.5). We can easily see that the second, third, and fourth of Eqs. (19.17) constitute a subset of linear homogeneous algebraic equations for the quantities $(A_1 - A_2)$, A_3 and A_4. For $kU_{0b} \neq \omega_p$, the determinant of this subset is nonzero. Consequently, these three linear homogeneous algebraic equations have only a trivial solution, $A_1 - A_2 = 0$ and $A_3 = A_4 = 0$. From this trivial solution, as well as from the first of Eqs. (19.17), we find the following expressions for the complex amplitudes:

$$A_1 = A_2 = J_{10}/2\,, \quad A_3 = A_4 = 0\,. \tag{19.19}$$

The sought-for solution to the initial-value problem can be obtained by substituting expressions (19.19) and (19.10) into the general expression (19.13) for the state vector of the system. We present here only the state vector components J_{b1} and J_{b2}:

$$\begin{aligned} J_{b1}(t,z) &= J_{10}\exp[ik(z - U_{0b}t)]\begin{cases} \mathrm{ch}(|\Delta|t), & kU_{0b} < \omega_p\,, \\ \cos(|\Delta|t), & kU_{0b} > \omega_p\,. \end{cases} \\ J_{b2}(t,z) &= J_{10}\frac{kU_{0b}}{\Delta}\exp[ik(z - U_{0b}t)]\begin{cases} \mathrm{sh}(|\Delta|t), & kU_{0b} < \omega_p\,, \\ \sin(|\Delta|t), & kU_{0b} > \omega_p\,. \end{cases} \end{aligned} \tag{19.20}$$

Expressions (19.19) and (19.20) imply that, to within second-order terms in the small parameters (19.18), a modulated electron beam does not excite plasma waves with the frequencies $\omega = \omega_{3,4}$ in the case $kU_{0b} \neq \omega_p$. The only waves that are excited are beam eigenmodes with frequencies of $\omega \approx kU_{0b}$. The result is that the plasma is polarized and thereby exerts an inverse effect on the evolution of perturbations in the beam. In particular, for $kU_{0b} < \omega_p$, the plasma polarization gives rise to instability (an instability due to "negative permittivity").

Let us now consider the case $kU_{0b} = \omega_p$, in which the eigenfrequencies are described by expressions (19.6) and (19.11) and the instability growth rate is maximum. We introduce the notation

$$\Delta_0 = \left(\frac{\omega_b^2}{2\omega_p^2}\right)^{1/3}\omega_p \tag{19.21}$$

and take into account the fact that, for $kU_{0b} = \omega_p$, dispersion relation can be represented as

$$\delta\omega^3 = (\omega - kU_{0b})^3 = \Delta_0^3\,. \tag{19.22}$$

We also write out the cubic roots of unity in the order that was adopted in solutions (19.11):

$$\delta_1 = \frac{-1 + i\sqrt{3}}{2} = \exp\left(i\frac{2\pi}{3}\right), \quad \delta_2 = \frac{-1 + i\sqrt{3}}{2} = \exp\left(i\frac{4\pi}{3}\right), \quad \delta_3 = 1\,. \tag{19.23}$$

With allowance for dispersion relation (19.22) and solution (19.6), the set of Eqs. (19.15) with particular initial conditions $J_{10} \neq 0$, $J_{20} = 0$, $J_0 = 0$ and $E_0 = 0$ becomes

$$\begin{aligned} A_1 + A_2 + A_3 + A_4 &= J_{10}\,, \\ A_1\delta_1^* + A_2\delta_2^* + A_3\delta_3^* - (\Delta_0/2\omega_p)A_4 &= 0\,, \\ A_1\delta_1^2 + A_2\delta_2^2 + A_3\delta_3^2 - (4\omega_p^2/\Delta_0^2)A_4 &= 0\,, \\ A_1\delta_1 + A_2\delta_2 + A_3\delta_3 - (2\omega_p/\Delta_0)A_4 &= 0\,, \end{aligned} \tag{19.24a}$$

where the asterisk "*" stands for the complex conjugate. The third of Eqs. (19.24a) has been derived by using the fourth equation and the inequality $\Delta_0/\omega_p = (\omega_p^2/2\omega_p^2)^{1/3} \ll 1$, which is consistent with inequality (19.5).

Taking into account the relationships $\delta_j^3 = 1$ and $\delta_j\delta_j^* = 1$, where the quantities δ_j are the cubic roots of unity (19.23), we rewrite the second of Eqs. (19.24a) as

$$A_1\delta_1^2 + A_2\delta_2^2 + A_3\delta_3^2 - (\Delta_0/2\omega_p)A_4 = 0\,. \tag{19.24b}$$

This equation and the third of Eqs. (19.24a) yield $A_4 = 0$. This indicates that a modulated electron beam does not excite a counterpropagating plasma wave (for $\omega_4 = -\omega_p$, the phase velocity $V_{ph}^{(4)} = \omega_4/k = -\omega_p/k = -U_{0b} < 0$ is directed opposite to the beam). As a result, Eqs. (19.24a) can be substantially simplified to become

$$\begin{aligned} A_1 + A_2 + A_3 &= J_{10}\,, \\ A_1\delta_1 + A_2\delta_2 + A_3\delta_3 &= 0\,, \\ A_1\delta_1^2 + A_2\delta_2^2 + A_3\delta_3^2 &= 0\,. \end{aligned} \tag{19.25}$$

With allowance for the cubic roots of unity (19.23), the solution to Eqs. (19.25) has the form

$$A_1 = A_2 = A_3 = \frac{1}{3}J_{10}\,. \tag{19.26}$$

For $kU_{0b} = \omega_p$, the final solution to the initial-value problem is obtained by substituting formulas (19.26) and (19.11) into the general expression (19.13) for the state vector of the system. We write out expressions for the state vector components J_{b1} and J_{b2}:

$$\begin{aligned} J_{b1}(t,z) &= \frac{1}{3}J_{10}\exp[ik(z - U_{0b}t)]\sum_{m=1}^{3}\exp(-i\delta_m\Delta_0 t)\,, \\ J_{b2}(t,z) &= \frac{1}{3}J_{10}\frac{kU_{0b}}{\Delta_0}\exp[ik(z - U_{0b}t)]\sum_{m=1}^{3}\delta_m^*\exp(-i\delta_m\Delta_0 t)\,. \end{aligned} \tag{19.27}$$

From these expressions we can see that the initial beam perturbation has been distributed in equal proportions (with the proportionality coefficient 1/3) among all the three eigenmodes with frequencies (19.11). It is incorrect to assert that one of the three waves is a plasma wave and the remaining two are beam eigenmodes: each of the waves is a perturbation of the whole system. Formulas (19.27) describe how the beam modulation grows in the course of the instability due to the Cherenkov resonance ($kU_{0b} = \omega_p$) interaction between the beam and the plasma eigenmode.

With operators (3.7), dispersion relation (19.4) leads to the following differential equation for the characteristic function of the state vector of the interacting electron beam and plasma:

$$\left[\left(\frac{\partial^2}{\partial t^2} + \omega_p^2\right)\left(\frac{\partial}{\partial t} + U_{0b}\frac{\partial}{\partial z}\right)^2 + \omega_b^2 \frac{\partial^2}{\partial t^2}\right] A(t,z) = 0\,. \tag{19.28}$$

The state vector of the beam–plasma system has the form

$$\Psi(t,z) = \{J_{b1}, J_{b2}, J_p, E_z\} = \left\{A, -U_{0b}\hat{L}^{-1}\frac{\partial A}{\partial z}, \frac{\omega_p^2}{\omega_b^2}\int \hat{L}A dt, \frac{4\pi}{\omega_b^2}\hat{L}A\right\}, \tag{19.29}$$

where $\hat{L} = \partial/\partial t + U_{0b}\partial/\partial z$ is the translation operator. The first and last components of the state vector (19.29) coincide with the corresponding components of the beam state vector (18.13). We make the replacement

$$A'(t,z) = -\omega_b^{-2}\hat{L}^{-1}A(t,z) \tag{19.30}$$

to introduce the new characteristic function of the state vector, $A'(t,z)$, which, too, satisfies Eq. (19.28). Switching to function (19.30) in expression (19.29) yields the following formula for the state vector of the beam–plasma system:

$$\Psi(t,z) = \{J_{b1}, J_{b2}, J_p, E_z\} = \left\{\omega_b^2\hat{L}A', -\omega_b^2 U_{0b}\frac{\partial A'}{\partial z}, \omega_p^2\int \hat{L}^2 A' dt, 4\pi\hat{L}^2 A'\right\}. \tag{19.31}$$

Formula (19.31) will be used below to give the mathematical formulation of additional conditions to Eq. (19.28).

20. Instability of a Current-Carrying Plasma

Here, we consider one-dimensional electrostatic perturbations in an unbounded homogeneous plasma in which cold electrons move relative to the ions with the velocity $\mathbf{U}_0 = \{0, 0, U_{0e}\}$. The cold hydrodynamic equations for the electron and ion plasma components and the field equations are similar to Eqs. (19.1), specifically:

$$\begin{aligned}
\frac{\partial J_{e1}}{\partial t} + U_{0e}\frac{\partial J_{e1}}{\partial z} - \frac{\omega_e^2}{4\pi}E_z &= 0\,,\\
\frac{\partial J_{e2}}{\partial t} + U_{0e}\frac{\partial J_{e2}}{\partial z} + U_{0e}\frac{\partial J_{e1}}{\partial z} &= 0\,,\\
\frac{\partial J_i}{\partial t} - \frac{\omega_i^2}{4\pi}E_z &= 0\,,\\
\frac{\partial E_z}{\partial t} + 4\pi J_{e1} + 4\pi J_{e2} + 4\pi J_i &= 0\,.
\end{aligned} \tag{20.1}$$

Here, ω_e and ω_i are the electron and ion gyrofrequencies; $J_{e1} = eN_{0e}U_e$ and $J_{e2} = eU_{0e}N_e$ are the partial electron current densities (see Eqs. (18.1) and the comments on them); $J_i = eN_{0i}U_i$ is the ion current density; N_{0e} and N_{0i} are the unperturbed electron and ion densities; and U_e, U_i and N_e are perturbations of the corresponding quantities. The state vector of Eqs. (20.1) is $\mathbf{\Psi}(t,z) = \{J_{e1}, J_{e2}, J_i, E_z\}$.

We substitute the solution of the form

$$\mathbf{\Psi}(t,z) = \{j_{e1}(\omega,k), J_{e2}(\omega,k), j_i(\omega,k), e_z(\omega,k)\}\exp(-i\omega t + ikz)\,, \tag{20.2}$$

into Eqs. (20.1) to obtain the following homogeneous algebraic set of equations:

$$\begin{aligned} -i(\omega - kU_{0e})j_{e1} - (\omega_e^2/4\pi)e_z &= 0\,, \\ -i(\omega - kU_{0e})j_{e2} + ikU_{0e}j_{e1} &= 0\,, \\ -i\omega j_i - (\omega_i^2/4\pi)e_z &= 0\,, \\ -i\omega e_z + 4\pi j_{e1} + 4\pi j_{e2} + 4\pi j_i &= 0\,. \end{aligned} \tag{20.3}$$

The condition for homogeneous equations (20.3) to have a nontrivial solution is represented by the dispersion relation

$$D(\omega,k) \equiv (\omega^2 - \omega_i^2)(\omega - kU_{0e})^2 - \omega^2\omega_e^2 = 0\,. \tag{20.4}$$

Dispersion relation (20.4) has the same structure as dispersion relation (19.4) and goes over to the latter with the replacement $\omega_e \to \omega_b$ and $\omega_i \to \omega_p$. But these dispersion relations are very different: relation (19.4) has been investigated under the inequality $\omega_b^2 \ll \omega_p^2$, while for an electron–ion plasma, the opposite inequality, $\omega_e^2 \gg \omega_i^2$, necessarily holds.

The qualitative behavior of the dispersion curves of relation (20.4) is illustrated in Fig. 21. The curves were calculated for the ratio $\alpha = \omega_i/\omega_e = 0.02$, a value close

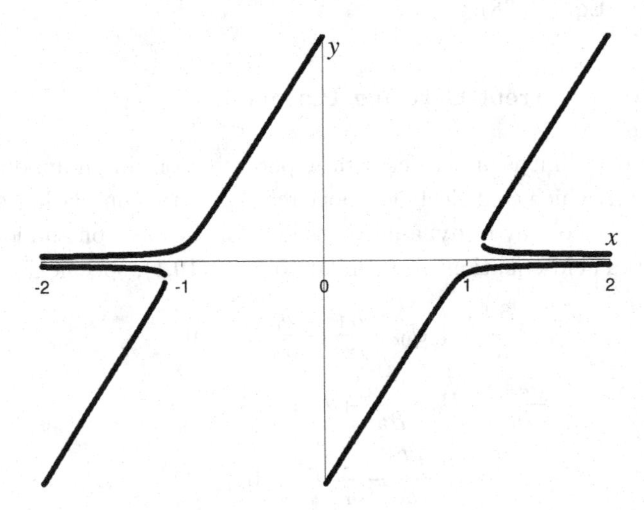

Fig. 21. Dispersion curves of dispersion relation (20.4) in the ($x = ku/\omega_e$, $y = \omega/\omega_e$) plane for $\alpha = 0.02$.

to that in a hydrogen plasma. It is worthwhile to compare Fig. 21 with Fig. 19. Since $\omega_e^2 \gg \omega_i^2$, the ion contribution to the dispersion relation can be substantial only in the range of low frequencies,

$$|\omega| \ll kU_{0e},\ \omega_e\,. \tag{20.5}$$

Let us first consider the case $kU_{0e} \neq \omega_e$ (assuming, for definiteness, that $k > 0$). In this case, dispersion relation (20.4) reads

$$(\omega^2 - \omega_i^2)k^2U_{0e}^2 - \omega^2\omega_e^2 = 0\,. \tag{20.6}$$

From this dispersion relation we find the frequency spectra:

$$\omega_{1,2} = \pm \begin{cases} i\omega_i \dfrac{kU_{0e}}{\sqrt{\omega_e^2 - k^2U_{0e}^2}}, & kU_{0e} < \omega_e\,, \\ \omega_i \dfrac{kU_{0e}}{\sqrt{k^2U_{0e}^2 - \omega_e^2}}, & kU_{0e} > \omega_e\,. \end{cases} \tag{20.7a}$$

For $kU_{0e} \neq \omega_e$, two more solutions are found by discarding the ion contribution to dispersion relation (20.4):

$$\omega_{3,4} = kU_{0e} \pm \omega_e\,. \tag{20.7b}$$

Solutions (20.7) imply that, for $kU_{0e} < \omega_e$, current-carrying plasma is unstable. This instability is called nonresonant Buneman instability.

For $kU_{0e} = \omega_e$, formulas (20.7) are inapplicable. In this case, we rewrite dispersion relation (20.4) as

$$(\omega^2 - \omega_i^2)(\omega - \omega_e)^2 - \omega^2\omega_e^2 = 0 \tag{20.8}$$

and assume that

$$\omega_i \ll |\omega| \ll \omega_e\,. \tag{20.9}$$

As a result, we arrive at the following expressions for the frequencies:

$$\begin{aligned} \omega_{1,2} &= \left(\frac{1}{2} \pm i\frac{\sqrt{3}}{2}\right)\left(\frac{\omega_i^2}{2\omega_e^2}\right)^{1/3} \omega_e\,, \\ \omega_3 &= -\left(\frac{\omega_i^2}{2\omega_e^2}\right)^{1/3} \omega_e\,. \end{aligned} \tag{20.10a}$$

The fourth frequency is given by the formula (see Fig. 21)

$$\omega_4 = kU_{0e} + \omega_e = 2\omega_e\,. \tag{20.10b}$$

The instability with the growth rate given by expressions (20.10) is called resonant Buneman instability. This instability grows at the fastest rate. It is easy to see that inequalities (20.9) reduce to the condition $\omega_e^2 \gg \omega_i^2$.

From an application viewpoint, the Buneman instability is far less important than the beam instability in plasma. That is why we do not present here a complete solution to the initial-value problem of the evolution of harmonic perturbations in a current-carrying plasma, the more so since this solution is fully analogous to solutions (19.20) and (19.27). We write out only the differential equation for the characteristic function of the state vector:

$$\left[\left(\frac{\partial^2}{\partial t^2} + \omega_i^2\right)\left(\frac{\partial}{\partial t} + U_{0e}\frac{\partial}{\partial z}\right)^2 + \omega_e^2\frac{\partial^2}{\partial t^2}\right] A(t,z) = 0\,. \tag{20.11}$$

Chapter 3

Linear Waves in Coupled Media. Slow Amplitude Method

21. Coupled Oscillator Representation and Slow Amplitude Method

A physical system in which it is possible to single out two (or more) quasi-independent subsystems can be conveniently described mathematically in terms of the interaction between them, a description that is especially effective when the interaction is weak. To illustrate, we refer again to the waveguide with an anisotropic plasma that has been considered in Sec. 13. In order to utilize Eq. (13.13) and formula (13.14), we introduce the new functions

$$A_E(t,z) = \frac{\partial A}{\partial t}\,, \quad A_P(t,z) = \frac{1}{k_\perp c}\hat{\mathrm{D}}A(t,z)\,, \tag{21.1a}$$

where $\hat{\mathrm{D}} = (\partial^2/\partial t^2 - c^2\partial^2/\partial z^2)$ is the d'Alambertian. With these functions, fourth-order equation (13.13) can be transformed to the following set of second-order equations:

$$\begin{aligned} &\left(\frac{\partial^2}{\partial t^2} - c^2\frac{\partial^2}{\partial z^2}\right)A_E(t,z) = k_\perp c\frac{\partial A_p}{\partial t}\,, \\ &\left(\frac{\partial^2}{\partial t^2} + \omega_p^2\right)A_P(t,z) - = -k_\perp c\frac{\partial A_E}{\partial t}\,, \end{aligned} \tag{21.2}$$

in which case formula (13.14) for the state vector reduces to

$$\begin{aligned} \mathbf{\Psi}(t,z) &= \{E_x(t,z), B_y(t,z), J_p(t,z), E_z(t,z)\} \\ &= \left\{\int A_E(t,z)dt,\ -\frac{1}{c}\int A_E(t,z)dz,\ -\frac{\omega_p^2}{4\pi}\frac{1}{c}\iint A_P(t,z)dzdt,\right. \\ &\quad \left. -\frac{1}{c}\int A_P(t,z)dz\right\}. \end{aligned} \tag{21.3a}$$

Let us clarify the physical meaning of the transformation provided by functions (21.1a). For $k_\perp = 0$, dispersion relation (13.12) splits into two independent relations $\omega^2 - k^2c^2 = 0$ and $\omega^2 - \omega_p^2 = 0$, which determine the frequency spectra of transverse electromagnetic waves in vacuum and of longitudinal electrostatic waves in a plasma, respectively. Moreover, for $k_\perp = 0$, these two types of waves do not interact with one another. In the case at hand, namely, $k_\perp = 0$, Eqs. (13.2), as well as Eqs. (21.2), also split into independent equations for the state vector of the electromagnetic field, $\{E_x, B_y\}$, and for the state vector of the plasma, $\{J_p, E_z\}$. The formula for $A_E(t,z)$ in (21.1) gives the characteristic function of the state vector of the electromagnetic field, and the formula for $A_P(t,z)$ in (21.1) determines the characteristic function of the state vector of the plasma. In fact, if we introduce the new functions

$$A'_E = \int A_E(t,z)dt\,, \quad A'_P = -\frac{\omega_p^2}{4\pi}\frac{1}{c}\iint A_P(t,z)dzdt\,, \tag{21.1b}$$

then we can transform formula (21.3a) to

$$\begin{aligned}\boldsymbol{\Psi}(t,z) &= \{E_x(t,z), B_y(t,z), E_z(t,z), J_z(t,z)\}\\ &= \left\{A'_E(t,z),\ -\frac{1}{c}\int\frac{\partial A'_E(t,z)}{\partial t}dz,\ A'_P(t,z),\ \frac{4\pi}{\omega_p^2}\frac{\partial A'_P(t,z)}{\partial t}\right\}.\end{aligned} \tag{21.3b}$$

The first two components of the state vector (21.3b) coincide with the components (5.10) (for $\varepsilon_0 = 1$), and the remaining two components coincide with the components (6.11).

Hence, the set of Eqs. (21.2) describes the interaction of two different physical subsystems — the electromagnetic field and the plasma. Since these subsystems are primarily subject to the wave (oscillatory) motion, their description by equations of the form (21.2) is called the coupled oscillator representation, which is especially effective when the interaction between subsystems (oscillators) is weak. Let us investigate the case of weak interaction in more detail.

For $k_\perp = 0$, the electromagnetic field and the plasma do not interact with one another. In this case, the eigenfrequencies are determined from two independent dispersion relations (for electromagnetic and plasma waves, respectively):

$$\begin{aligned}D_E(\omega,k) &\equiv -\omega^2 + k^2c^2 = 0\,,\\ D_p(\omega,k) &\equiv -\omega^2 + \omega_p^2 = 0\,.\end{aligned} \tag{21.4}$$

The interaction between subsystems is weak under the condition (see below)

$$k_\perp^2 c^2 \ll \omega_p^2\,. \tag{21.5}$$

In the zeroth approximation (when there is no interaction), the eigenfrequencies are again determined from dispersion relations (21.4). The next (first) approximation leads to frequency corrections, which are largest at the wave resonance point. The resonance point (ω_0, k_0) is determined by solving dispersion relations (21.4) simultaneously:

$$\omega_0 = \omega_p\,, \quad k_0 = \omega_p/c\,. \tag{21.6}$$

At the wave resonance point (21.6), the solution to dispersion relation (13.12) can be represented as

$$\omega = \omega_0 + \Omega\,, \quad |\Omega| \ll \omega_0\,, \tag{21.7}$$

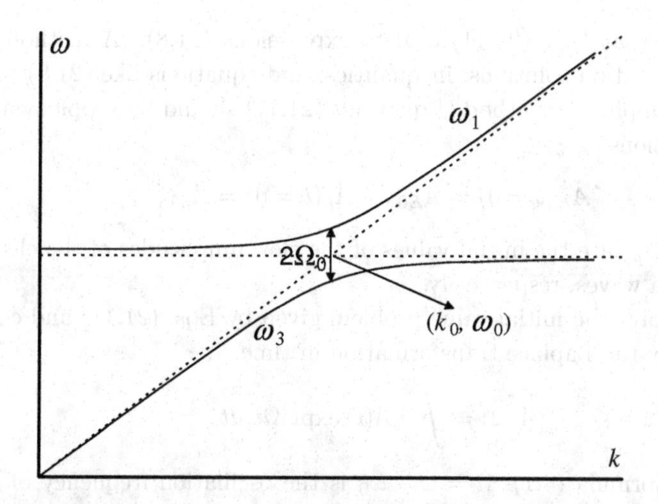

Fig. 22. Pattern of the interaction between an electromagnetic wave (ω_1) and a plasma wave (ω_3) in a waveguide. The dotted lines show the frequency spectra of the noninteracting waves.

where Ω is a correction to the resonant eigenfrequencies that comes from the interaction between waves. This correction is assumed to be small, giving a mathematical criterion that the interaction between subsystems is weak. Substituting representation (21.7) into dispersion relation (13.12) yields

$$\Omega^2 = \frac{1}{4} k_\perp^2 c^2 \equiv \Omega_0^2 \to \Omega_{1,2} = \pm\Omega_0 \,. \tag{21.8}$$

One can easily see that the condition for the corrections (21.8) to be small reduces to inequality (21.5). Hence, the interaction between subsystems gives rise to corrections to their eigenfrequencies (see Fig. 22).

In order to understand what are the physical processes occurring in the interaction between waves in a plasma waveguide, let us solve differential equations (21.2). At the wave resonance point (21.6), the solution can be sought for in the form

$$\begin{aligned} A_E(t,z) &= \tilde{A}_E(t)\exp(-i\omega_0 t + ik_0 z)\,, \\ A_p(t,z) &= \tilde{A}_p(t)\exp(-i\omega_0 t + ik_0 z)\,, \end{aligned} \tag{21.9}$$

where $\tilde{A}_E(t)$ and $\tilde{A}_p(t)$ are the slow wave amplitudes. The functions $\tilde{A}_E(t)$ and $\tilde{A}_p(t)$ vary slower than the exponentials in representation (21.9). By virtue of condition (21.5) (see also (21.7)), the slow amplitudes satisfy the inequalities

$$\left|\frac{\partial A_{E,p}}{\partial t}\right| \ll |\omega_0 A_{E,p}| \,. \tag{21.10}$$

Substituting solution (21.9) into Eqs. (21.2) and taking into account inequalities (21.10) and formulas (21.6), we obtain the following set of equations for the slow amplitudes:

$$\frac{d\tilde{A}_E}{dt} = \Omega_0 \tilde{A}_p \,, \quad \frac{d\tilde{A}_p}{dt} = -\Omega_0 \tilde{A}_E \,. \tag{21.11}$$

It is easy to see that Eqs. (21.11) lead to expressions (21.8). A method of wave theory that is based on solutions, inequalities, and equations like (21.9)–(21.11) is called the slow amplitude method. Equations (21.11) should be supplemented with the initial conditions

$$\tilde{A}_E(t=0) = A_{E0}\,, \quad \tilde{A}_p(t=0) = A_{p0}\,, \tag{21.12}$$

where A_{E0} and A_{p0} are the initial values of the slow amplitudes of the electromagnetic and plasma waves, respectively.

In order to solve the initial-value problem given by Eqs. (21.11) and conditions (21.12), we apply the Laplace transformation in time,

$$A(\Omega) = \int_0^\infty A(t)\exp(i\Omega t)dt\,. \tag{21.13}$$

Here, unlike in formula (4.1), $\Omega = \omega - \omega_0$ is the oscillation frequency of the slow amplitude and $A = \tilde{A}_{E,p}$ is any of the slow amplitudes. Applying the Laplace transformation (21.13) to Eqs. (21.11) and using property (4.2) and initial conditions (21.12) yields the following set of algebraic equations for the Laplace transformed slow amplitudes:

$$\begin{aligned} i\Omega\tilde{A}_E(\Omega) + \Omega_0\tilde{A}_p(\Omega) &= -A_{E0}\,, \\ i\Omega\tilde{A}_p(\Omega) - \Omega_0\tilde{A}_E(\Omega) &= -A_{p0}\,. \end{aligned} \tag{21.14}$$

From these equations we find

$$\tilde{A}_E(\Omega) = \frac{i\Omega A_{E0} - \Omega_0 A_{p0}}{(\Omega^2 - \Omega_0^2)}\,, \quad \tilde{A}_p(\Omega) = \frac{i\Omega A_{p0} + \Omega_0 A_{E0}}{(\Omega^2 - \Omega_0^2)}\,, \tag{21.15}$$

The inverse Laplace transformation can be written as (see formula (4.8))

$$A(t) = \frac{1}{2\pi}\int_{C(\Omega)} A(\Omega)\exp(-i\Omega t)/d\Omega\,. \tag{21.16}$$

We insert formulas (21.15) for the Laplace transformed amplitudes $\tilde{A}_E(\Omega)$ and $\tilde{A}_p(\Omega)$ into formula (21.16) in place of $A(\Omega)$ and take the resulting integral (the direction of integration along the contour $C(\Omega)$ in (21.16) should be negative, or clockwise, see Figs. 1, 2) to obtain the expressions

$$\begin{aligned} \tilde{A}_E(t) &= A_{E0}\cos\Omega_0 t + A_{p0}\sin\Omega_0 t = |A_0|\sin(\Omega_0 t + \varphi_0)\,, \\ \tilde{A}_p(t) &= A_{p0}\cos\Omega_0 t - A_{E0}\sin\Omega_0 t = |A_0|\cos(\Omega_0 t + \varphi_0)\,, \end{aligned} \tag{21.17}$$

where

$$|A_0| = \sqrt{A_{E0}^2 + A_{p0}^2}\,, \quad \varphi_0 = \arcsin(A_{E0}|A_0|^{-1})\,. \tag{21.18}$$

Solution (21.17) describes a periodic transfer of energy from the plasma wave to the electromagnetic wave, and vice versa. In this process, the total wave energy is conserved, because the quantity $\tilde{A}_E^2(t) + \tilde{A}_p^2(t) = |A_0|^2$ is constant. In oscillation theory, such processes are called wave beating. Hence, in their weak resonant interaction in a waveguide, the perturbed electromagnetic and plasma waves periodically exchange energy, while keeping their total energy unchanged.

Let us now consider the interaction between circularly polarized perturbations in a homogeneous magnetized plasma in the coupled oscillator representation. To do this, we rewrite differential equation (14.13) as

$$\left[\hat{D}\left(\frac{\partial}{\partial t} \pm i\Omega_e\right) + \omega_p^2 \frac{\partial}{\partial t}\right] A(t,z) = 0\,. \tag{21.19}$$

Introducing the new functions

$$A_C(t,z) = \hat{\mathrm{D}}A(t,z)\,, \quad A_E(t,z) = \omega_p A(t,z)\,, \tag{21.20}$$

we convert Eq. (21.19) into the following set of equations for the characteristic functions of the state vector of the electromagnetic field, $A_E(t,z)$, and of the plasma, $A_C(t,z)$:

$$\begin{aligned} \left(\frac{\partial^2}{\partial t^2} - c^2 \frac{\partial^2}{\partial z^2}\right) A_E(t,z) &= \omega_p A_C(t,z)\,, \\ \left(\frac{\partial}{\partial t} \pm i\Omega_e\right) A_C(t,z) &= -\omega_p \frac{\partial}{\partial t} A_E(t,z)\,. \end{aligned} \tag{21.21}$$

Since the functions $A_E(t,z)$ and $A_C(t,z)$ are complex, Eqs. (21.21) have a meaning different from that of Eqs. (14.15), written for real functions (the equations even have different orders in t).

For $\omega_p = 0$, the set of Eqs. (21.21) splits into two independent subsets for electromagnetic waves, $\hat{\mathrm{D}}A_E = 0$, and "single-electron" cyclotron waves, $(\partial/\partial t \pm i\Omega_e)A_C = 0$. Let us consider the weak interaction between these waves at a low frequency ω_p (see below). For $\omega_p = 0$, dispersion relation (14.7) splits into two independent relations,

$$\begin{aligned} D_E(\omega,k) &\equiv -\omega^2 + k^2c^2 = 0\,, \\ D_C(\omega,k) &\equiv -\omega \pm \Omega_e = 0\,, \end{aligned} \tag{21.22}$$

which determine the wave resonance point. Assuming that the frequency is positive (as in the above analysis of relation (14.7)), from dispersion relations (21.22) we find the resonant frequency and wavenumber:

$$\omega_0 = \Omega_e\,, \quad k_0 = \pm\Omega_e/c\,. \tag{21.23}$$

Substituting representation (21.7) into dispersion relation (14.7) (it is obvious that this same relation also follows from Eqs. (21.21)), we then obtain the corrections to the resonant frequency,

$$\Omega^2 = \frac{1}{2}\omega_p^2 \equiv \Omega_0^2 \to \Omega_{1,2} = \pm\Omega_0\,, \tag{21.24}$$

and the condition for the interaction between an electromagnetic wave and a cyclotron wave to be weak, $\Omega_0 \sim \omega_p \ll \Omega_e$.

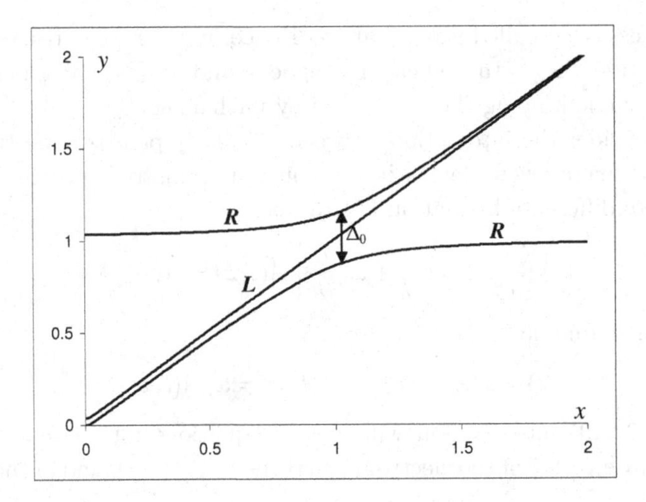

Fig. 23. Dispersion curves of transverse waves propagating in a homogeneous magnetized plasma parallel to the external magnetic field for $\omega_p/\Omega_e = 0.2$. The notation is as follows: $y = \omega/\Omega_e$, $x = kc/\Omega_e$, R denotes right polarization, L denotes left polarization, and $\Delta_0 = \sqrt{2}\omega_p/\Omega_e$.

Note that, since $\omega > 0$, expressions (21.23) have been obtained by using the second of Eqs. (21.22) with the upper sign, which corresponds to the right circular polarization (see Sec. 14). For waves with left circular polarization, resonance is impossible because left-polarized cyclotron waves do not exist (an electron in a magnetic field gyrates only in one direction, as is shown in Fig. 8). What has been just said provides a clear understanding of the nature of transverse waves propagating in a homogeneous low-density ($\omega_p \ll \Omega_e$) plasma along the external magnetic field: the left-polarized (extraordinary) wave is a conventional electromagnetic wave ($\omega \approx kc$) slightly perturbed by the low-density plasma; one of the right-polarized waves, too, is a slightly perturbed electromagnetic wave ($\omega \approx kc$); the other right-polarized wave is a slightly perturbed electron cyclotron wave ($\omega \approx \Omega_e$); and, at the resonance point (21.23), the right-polarized electromagnetic and cyclotron waves interact strongly to convert into one another (see Fig. 23).

In analogy with representation (21.9), we introduce the slow amplitudes of the electromagnetic and cyclotron waves (the requirement that the amplitude of the cyclotron wave be small is, in principle, unnecessary) to transform Eqs. (21.21) with the upper sign to the form

$$\frac{d\tilde{A}_C}{dt} = i\omega_p\Omega_e\tilde{A}_E\,, \quad \frac{d\tilde{A}_E}{dt} = \frac{1}{2}i\frac{\omega_p}{\Omega_e}\tilde{A}_C\,. \tag{21.25}$$

We will not analyze Eqs. (21.25) further because the replacement $\tilde{A}_C/(i\sqrt{2}\Omega_e) \to \tilde{A}_C$ reduces them to Eqs. (21.11) with the frequency Ω_0 given by relationship (21.24).

The coupled oscillator representation provides a description of the interaction of arbitrary wave subsystems of a certain physical system. Let the dynamics of noninteracting subsystems be described by the partial differential equations

$$\begin{aligned} D_1(\hat{\omega},\hat{k})A_1(t,z) &= 0\,, \\ D_2(\hat{\omega},\hat{k})A_2(t,z) &= 0\,, \end{aligned} \tag{21.26}$$

where $A_{1,2}(t,z)$ are the characteristic functions of the state vectors of the subsystems and $D_{1,2}(\hat{\omega},\hat{k})$ are the relevant dispersion operators. By analogy with Eqs. (21.2), the interaction between subsystems can be described by the structural mathematical relationships

$$\begin{aligned} D_1(\hat{\omega},\hat{k})A_1(t,z) &= S_1(\hat{\omega},\hat{k})A_2(t,z)\,, \\ D_2(\hat{\omega},\hat{k})A_2(t,z) &= S_2(\hat{\omega},\hat{k})A_1(t,z)\,, \end{aligned} \tag{21.27}$$

in which $S_{1,2}(\hat{\omega},\hat{k})$ are certain integrodifferential operators. Note that the order of the functions $S_{1,2}$ in the arguments $\hat{\omega}$ and $\hat{k}$ is lower than the order of the corresponding dispersion functions $D_{1,2}$.

The wave resonance point (ω_0,k_0) is determined as a solution to the set of equations

$$\begin{aligned} D_1(\omega,k) &= 0\,, \\ D_2(\omega,k) &= 0\,, \end{aligned} \tag{21.28}$$

where $D_{1,2}(\omega,k)$ are the dispersion functions of the physical subsystems under consideration. At the resonance point, the spatially harmonic solution to Eqs. (21.27) is looked for in the form (see representation (21.9))

$$\begin{aligned} A_1(t,z) &= \tilde{A}_1(t)\exp(-i\omega_0 t + ik_0 z)\,, \\ A_2(t,z) &= \tilde{A}_2(t)\exp(-i\omega_0 t + ik_0 z)\,, \end{aligned} \tag{21.29}$$

where the slow amplitudes $\tilde{A}_{1,2}(t)$ satisfy the inequalities (see (21.10))

$$\left|\frac{\partial \tilde{A}_{1,2}}{\partial t}\right| \ll |\omega_0 \tilde{A}_{1,2}|\,. \tag{21.30}$$

These inequalities imply that, when substituting solutions (21.29) into Eqs. (21.27), we can transform the operators according to the rules

$$\begin{aligned} D_{1,2}(\hat{\omega},k_0) &= D_{1,2}\left(\omega_0 + i\frac{d}{dt},\ k_0\right) \approx D_{1,2}(\omega_0,k_0) \\ &\quad + i\frac{\partial D_{1,2}}{\partial \omega}\frac{d}{dt} = i\frac{\partial D_{1,2}}{\partial \omega}\frac{d}{dt}\,, \\ S_{1,2}(\hat{\omega},k_0) &= S_{1,2}\left(\omega_0 + i\frac{d}{dt},\ k_0\right) \approx S_{1,2}(\omega_0,k_0)\,. \end{aligned} \tag{21.31}$$

Here, we have taken into account the relationships $D_{1,2}(\omega_0,k_0) = 0$ and the fact that the quantities $S_{1,2}(\omega_0,k_0)$ are nonzero. The operator d/dt is to be applied only to the slow amplitudes.

Substituting representation (21.29) into Eqs. (21.27) and using operator relationships (21.31), we obtain the following equations for the slow amplitudes:

$$\frac{d\tilde{A}_1(t)}{dt} = -iS_1(\omega_0, k_0)\left[\frac{\partial D_1(\omega_0, k_0)}{\partial\omega}\right]^{-1}\tilde{A}_2(t)\,,$$
$$\frac{d\tilde{A}_2(t)}{dt} = -iS_2(\omega_0, k_0)\left[\frac{\partial D_2(\omega_0, k_0)}{\partial\omega}\right]^{-1}\tilde{A}_1(t)\,. \tag{21.32}$$

Equations (21.32) should be supplemented with the standard initial conditions (see (21.12))

$$\tilde{A}_1(t=0) = A_{10}\,, \quad \tilde{A}_2(t=0) = A_{20}\,. \tag{21.33}$$

The initial-value problem given by Eqs. (21.32) and initial conditions (21.33) is solved by the Laplace transform method in the same way as the above problem given by Eqs. (21.11) and conditions (21.12). Let us demonstrate how Eqs. (21.32) with initial conditions (21.33) can be solved by the eigenmode method, which was used in Sec. 2 to solve the initial-value problem formulated for general equations (1.1).

We represent the solution to Eqs. (21.32) as

$$\tilde{A}_{1,2}(t) = B_{1,2}\exp(-i\Omega t)\,, \tag{21.34}$$

where $B_{1,2}$ are constants. Inserting representation (21.34) into Eqs. (21.32) yields the dispersion relation for determining the oscillation frequencies of the slow amplitudes,

$$\Omega^2 = S_1 S_2\left(\frac{\partial D_1}{\partial\omega}\frac{\partial D_2}{\partial\omega}\right)^{-1} \equiv \Omega_0^2\,, \tag{21.35}$$

and the relationship between the constants B_1 and B_2,

$$B_2 = \frac{1}{\Omega}S_2\left(\frac{\partial D_2}{\partial\omega}\right)^{-1}B_1\,, \tag{21.36}$$

Assume, for the moment, that the right-hand side of dispersion relation (21.35) is positive. In this case, there are two real eigenfrequencies, $\Omega_{1,2} = \pm\Omega_0$. With representation (21.34) and relationship (21.36), the general solution to Eqs. (21.32) can then be written as

$$\tilde{A}_1(t) = a\exp(-i\Omega_0 t) + b\exp(i\Omega_0 t)\,,$$
$$\tilde{A}_2(t) = S_2(\partial D_2/\partial\omega)^{-1}\Omega_0^{-1}(a\exp(-i\Omega_0 t) - b\exp(i\Omega_0 t)\,, \tag{21.37}$$

where a and b are constants (the coefficients in front of the linearly independent solutions $\exp(-i\Omega_0 t)$ and $\exp(i\Omega_o t)$ to Eqs. (21.32)). The formulas for calculating the coefficients a and b are found by substituting general solution (21.37) into initial conditions (21.33):

$$a = \frac{1}{2}(A_{10} + \Omega_0 S_2^{-1}(\partial D_2/\partial\omega)A_{20})\,, \quad b = \frac{1}{2}(A_{10} - \Omega_0 S_2^{-1}(\partial D_2/\partial\omega)A_{20})\,. \tag{21.38}$$

With formulas (21.38), general solution (21.37) constitutes a complete solution to the initial-value problem given by Eqs. (21.32) and initial conditions (21.33). For instance, in the particular case of electromagnetic waves in a waveguide with an anisotropic plasma, general solution (21.37) and formulas (21.38) lead to solution (21.17).

22. Beam–Plasma System in the Coupled Oscillator Representation

Let us convert dispersion relation (19.4) for a beam–plasma system into the form

$$(\omega^2 - \omega_p^2)((\omega - kU_{0b})^2 - \omega_b^2) - \omega_p^2\omega_b^2 = 0\,. \tag{22.1}$$

Here, we explicitly single out two interacting subsystems — the beam subsystem, described by the dispersion relation $(\omega - kU_{ob})^2 - \omega_b^2 = 0$, and the plasma subsystem, described by dispersion relation $\omega^2 - \omega_p^2 = 0$. With dispersion relation (22.1), we can analyze a beam–plasma system in the coupled oscillator representation. The point of resonance between the plasma wave and the beam wave, (ω_0, k_0) — in the case under consideration, namely, that of an unbounded beam and an unbounded plasma, the resonance point always exists — can be found by simultaneously solving the equations

$$\begin{aligned} D_p(\omega, k) &\equiv -\omega^2 + \omega_p^2 = 0\,, \\ D_b(\omega, k) &\equiv -(\omega - kU_{0b})^2 + \omega_b^2 = 0\,. \end{aligned} \tag{22.2}$$

The result is

$$\begin{aligned} \omega_0 &= \omega_p\,, \\ k_0 &= \omega_p/U_{0b} \mp \omega_b/U_{0b}\,. \end{aligned} \tag{22.3}$$

Here, the minus sign is for the point of resonance of the plasma wave with the fast beam wave (the frequency spectrum given by the first of expressions (18.5)), and the plus sign is for the resonance between the plasma wave and the slow beam wave (the frequency spectrum given by the second of expressions (18.5)).

We represent the solution to dispersion relation (22.1) at the resonance point (22.3) in the form (21.7) to transform the dispersion relation to

$$\Omega(\Omega^2 \pm 2\Omega\omega_b) = \frac{1}{2}\omega_p\omega_b^2\,. \tag{22.4}$$

In deriving relation (22.4), we have kept the quadratic term Ω^2 in the expansion of the beam dispersion function $D_b(\omega, k)$ in powers of Ω. The reason is that, for a beam wave, we must use not the inequality $|\Omega| \ll \omega_0$ but the inequality

$$|\Omega| \ll |\omega_0 - k_0U_{0b}|\,. \tag{22.5}$$

In fact, the quantity k_0U_{0b} is the Doppler frequency shift due to the uniform straight-line motion of an electron. This shift is unrelated to the collective oscillatory properties of the beam, which are governed by the difference $\omega - kU_{0b}$ from inequality

(22.5). The wave processes which involve an electron beam and for which inequality (22.5) is satisfied are called collective processes. The processes for which the opposite inequality is satisfied, namely,

$$|\Omega| \gg |\omega_0 - k_0 U_{0b}|\,, \tag{22.6}$$

came to be called single-particle processes. For a single-particle process due to the interaction between waves, the frequency correction substantially exceeds the frequency of the beam eigenmodes, $|\Omega| \gg \omega_b$. Consequently, in single-particle processes, the beam subsystem is not an oscillator. More precisely, the characteristic time $|\Omega|^{-1}$ of the interaction with the second subsystem is too short for the oscillatory properties of the beam to come into play. That is why single-particle processes are also called processes with strong interaction. Accordingly, collective processes are often called processes with weak interaction. Yet, the term "weak interaction" applies not only to systems with beams but also to any system described by Eq. (21.32) and dispersion relation (21.35).

It is easy to see that, under the inequality $\omega_b^2 \ll \omega_p^2$, the beam–plasma interaction is single-particle (strong), so dispersion relation (22.4) is written in the form

$$\Omega^3 = \frac{1}{2}\omega_b^2 \omega_p\,, \tag{22.7}$$

which differs radically in structure from both frequency corrections (21.8) and relation (21.35). The solution to dispersion relation (22.7) is $\Omega_{1,2,3} = \delta_{1,2,3}\Delta_0$ (see (19.11)), where the quantity Δ_0 is defined by formula (19.21) and $\delta_{1,2,3}$ are the cubic roots of unity (19.23). In the same single-particle approximation, dispersion relation (22.1) has the form

$$(\omega^2 - \omega_p^2)(\omega - kU_{0b})^2 - \omega_p^2\omega_b^2 = 0\,, \tag{22.8}$$

which is valid, however, only near the resonance point (22.3).

Let us now derive differential equations describing a beam–plasma system in the coupled oscillator representation. To do this, we turn to the basic set of Eqs. (19.1), which describes the interaction of an electron beam with a plasma. We introduce the total perturbation of the electron beam density:

$$J_b = J_{b1} + J_{b2}\,. \tag{22.9}$$

Eliminating the electric field E_z in Eqs. (19.1), we can then readily reduce the set to the following two coupled second-order differential equations:

$$\begin{aligned} &\left(\left(\frac{\partial}{\partial t} + U_{0b}\frac{\partial}{\partial z}\right)^2 + \omega_b^2\right) J_b = -\omega_b^2 J_p\,, \\ &\left(\frac{\partial^2}{\partial t^2} + \omega_p^2\right) J_p = -\omega_p^2 J_b\,. \end{aligned} \tag{22.10}$$

In the single-particle approximation, Eqs. (22.10) near the resonance point can be somewhat simplified:

$$\begin{aligned} &\left(\frac{\partial}{\partial t} + U_{0b}\frac{\partial}{\partial z}\right)^2 J_b = -\omega_b^2 J_p\,, \\ &\left(\frac{\partial^2}{\partial t^2} + \omega_p^2\right) J_p = -\omega_p^2 J_b\,. \end{aligned} \tag{22.11}$$

These same equations can also be obtained by transforming Eq. (19.28). To begin, we convert Eq. (19.28) to a form similar to dispersion relation (22.1). The simplest way to do this is to go over to the frequency and wavenumber operators (3.7) in relation (22.1). The result is

$$\left\{\left(\frac{\partial^2}{\partial t^2} + \omega_p^2\right)\left[\left(\frac{\partial}{\partial t} + U_{0b}\frac{\partial}{\partial z}\right)^2 + \omega_b^2\right] - \omega_p^2\omega_b^2\right\} A(t,z) = 0\,. \tag{22.12}$$

In the single-particle approximation, Eq. (22.12) near the resonance point reads (see (22.8))

$$\left[\left(\frac{\partial^2}{\partial t^2} + \omega_p^2\right)\left(\frac{\partial}{\partial t} + U_{0b}\frac{\partial}{\partial z}\right)^2 - \omega_b^2\omega_p^2\right] A(t,z) = 0\,. \tag{22.13}$$

In converting Eq. (22.12) into the form (22.13), we have used the following inequality, which is a generalization of condition (22.6) for the process to be single-particle in nature:

$$|\hat{L}A| \gg \omega_b |A|\,. \tag{22.14}$$

Here, $\hat{L}$ is the translation operator.

We introduce new unknown functions expressed in terms of the characteristic function of the state vector, $A'(t,z) = -\omega_b^{-2}\hat{L}^{-1}A(t,z)$ (this function, which also satisfies Eq. (22.13), is defined by formula (19.30)):

$$\begin{aligned} A_b(t,z) &= \omega_b^2 A'(t,z)\,, \\ A_p(t,z) &= -\hat{L}^2 A'(t,z)\,. \end{aligned} \tag{22.15}$$

Equations (22.13) and formulas (22.15) yield the set of equations

$$\begin{aligned} &\left(\frac{\partial}{\partial t} + U_{0b}\frac{\partial}{\partial z}\right)^2 A_b = -\omega_b^2 A_p\,, \\ &\left(\frac{\partial^2}{\partial t^2} + \omega_p^2\right) A_p = -\omega_p^2 A_b\,, \end{aligned} \tag{22.16}$$

which has the same form as the set of Eqs. (22.11). We also present the state vector (19.31) of the beam–plasma system in terms of functions (22.15):

$$\mathbf{\Psi}(t,z) = \{J_{b1}, J_{b2}, J_p, E_z\} = \left\{\hat{L}A_b, -U_{0b}\frac{\partial A_b}{\partial z}, -\omega_p^2\int A_p dt, -4\pi A_p\right\}. \tag{22.17}$$

The functions $A_b(t,z)$ and $A_p(t,z)$ can be regarded as the characteristic functions of the state vectors of the beam and plasma, respectively.

At the resonance point (22.3), the spatially harmonic solution to Eqs. (22.16) can be represented as

$$\begin{aligned} A_b(t,z) &= \tilde{A}_b(t)\exp(-i\omega_0 t + ik_0 z)\,, \\ A_p(t,z) &= \tilde{A}_p(t)\exp(-i\omega_0 t + ik_0 z)\,, \end{aligned} \tag{22.18}$$

where $\tilde{A}_{b,p}$ are the slow amplitudes. We denote by $|\Omega|^{-1}$ the characteristic time scale on which the slow amplitudes vary. Using the inequalities (see (22.6))

$$|\omega_0 - k_0 U_{0b}| \ll |\Omega| \ll \omega_0 \tag{22.19}$$

and formulas (22.3), we obtain the approximate relationships

$$\begin{aligned} \frac{\partial^2 A_p}{\partial t^2} &= -\exp(-i\omega_0 t + ik_0 z)\left(-\omega_p^2 - 2i\omega_p \frac{d\tilde{A}_p}{dt}\right), \\ \hat{L}A_b &= \exp(-i\omega_0 t + ik_0 z)\left(\mp i\omega_b + \frac{d\tilde{A}_b}{dt}\right) \approx \exp(-i\omega_0 t + ik_0 z)\frac{d\tilde{A}_b}{dt}\,. \end{aligned} \tag{22.20}$$

We then transform Eqs. (22.16) with the help of formulas (22.20) to arrive at the following equations for the slow amplitudes in the single-particle approximation (the equations describing slow amplitudes in the strong wave interaction):

$$\begin{aligned} \frac{d^2\tilde{A}_b}{dt^2} &= -\omega_b^2\tilde{A}_p\,, \\ \frac{d\tilde{A}_p}{dt} &= -\frac{1}{2}i\omega_p\tilde{A}_b\,. \end{aligned} \tag{22.21}$$

In this same single-particle approximation, the state vector (22.17) of the beam–plasma system reads

$$\begin{aligned} \boldsymbol{\Psi}(t,z) &= \{J_{b1}, J_{b2}, J_p, E_z\} \\ &= \left\{\frac{d\tilde{A}_b(t)}{dt}, -ik_0U_{0b}\tilde{A}_b(t), -i\omega_p\tilde{A}_p(t), -4\pi\tilde{A}_p(t)\right\}\exp(-i\omega_0 t + ik_0 z)\,. \end{aligned} \tag{22.22}$$

Hence, the function $\tilde{A}_b$ is proportional to the amplitude of the beam density perturbation ($J_{b2} = eU_{0b}N_b$), the derivative $d\tilde{A}_b/dt$ is proportional to the amplitude of the beam velocity perturbation ($J_{b1} = eN_{0b}U_b$), and the function $\tilde{A}_p$ determines the plasma current amplitude and the field strength.

In order to formulate the initial-value problem for Eqs. (22.21), it is sufficient to specify the values of only three quantities, $\tilde{A}_b$, $d\tilde{A}_b/dt$, and $\tilde{A}_p$, at $t = 0$. Let us consider the following initial conditions (see (19.14)):

$$\tilde{A}_b(0) = 0\,, \quad \frac{d\tilde{A}_b(0)}{dt} = J_{10}\,, \quad \tilde{A}_p(0) = 0\,. \tag{22.23}$$

Applying the Laplace transformation (21.13) to Eqs. (22.21) yields the relationships

$$\begin{aligned} &\Omega^2 \tilde{A}_b(\Omega) - \omega_b^2 \tilde{A}_p(\Omega) = -J_{10}\,, \\ &\Omega \tilde{A}_p(\Omega) - (1/2)\omega_p \tilde{A}_b(\Omega) = 0\,, \end{aligned} \tag{22.24}$$

from which we find the Laplace transformed slow amplitudes

$$\tilde{A}_b(\Omega) = -\frac{\Omega J_{10}}{\Omega^3 - \Delta_0^3}\,, \quad \tilde{A}_p(\Omega) = -\frac{(1/2)\omega_p J_{10}}{\Omega^3 - \Delta_0^3}\,. \tag{22.25}$$

Here, the quantity Δ_0 is defined by formula (19.21). We substitute expressions (22.25) into formula (21.16) for the inverse Laplace transformation to obtain a solution to the initial-value problem given by Eqs. (22.21) and initial conditions (22.23):

$$\begin{aligned} &\tilde{A}_b(t) = \frac{1}{3} i J_{10} \Delta_0^{-1} \sum_{m=1}^{3} \delta_m^* \exp(-i\delta_m \Delta_0 t)\,, \\ &\tilde{A}_p(t) = \frac{1}{3} i J_{10} \Delta_0^{-2} \frac{\omega_p}{2} \sum_{m=1}^{3} \delta_m \exp(-i\delta_m \Delta_0 t)\,, \end{aligned} \tag{22.26}$$

where $\delta_{1,2,3}$ are the cubic roots of unity (19.23). It is easy to see that, with formulas (22.22), (22.18), and (22.3), expressions (22.26) lead to solution (19.27), which has been found above by the eigenmode method.

23. Basic Equations of Microwave Electronics

The equations that generalize Eqs. (22.10) to the interaction of an electron beam with waves in an arbitrary electrodynamic system or medium can be written as

$$\begin{aligned} &\left(\left(\frac{\partial}{\partial t} + U_{0b} \frac{\partial}{\partial z} \right)^2 + \omega_b^2 \right) A_b = -\omega_b^2 S_b(\hat{\omega}, \hat{k}) A_w\,, \\ &D_w(\hat{\omega}, \hat{k}) A_w = -\omega_w^2 S_w(\hat{\omega}, \hat{k}) A_b\,. \end{aligned} \tag{23.1}$$

Here, $A_b(t,z)$ is the characteristic function of the state vector of the electron beam; $A_w(t,z)$ is the characteristic function of the state vector of the electrodynamic subsystem; $D_w(\hat{\omega}, \hat{k})$ is the dispersion operator for the electrodynamic subsystem; S_b and S_w are certain operators; and the quantity ω_w, which is introduced for convenience, has the dimension of frequency. Equations (23.1) underlie theoretical microwave electronics — a significant, rapidly developing branch of modern physics. For the beam–plasma system considered above, we have ($w = p$, $\omega_w = \omega_p$)

$$\begin{aligned} &D_p(\hat{\omega}, \hat{k}) = -\hat{\omega}^2 + \omega_p^2 = \frac{\partial^2}{\partial t^2} + \omega_p^2\,, \\ &S_b(\hat{\omega}, \hat{k}) = S_p(\hat{\omega}, \hat{k}) = 1\,. \end{aligned} \tag{23.2}$$

One of the challenges in microwave electronics is to determine the operators D_w, S_w, and S_b for particular electrodynamic systems and media. Note that, for the dispersion functions (see dispersion relations (5.4), (6.5), (7.4), (9.5), (10.6)), we have adopted here a form such that $\partial D_w/\partial\omega < 0$.

We substitute the relationships $A_b = A_{b0}\exp(-i\omega t + ikz)$ and $A_w = A_{w0}\exp(-i\omega t + ikz)$ into Eqs. (23.1) and eliminate the constants A_{b0} and A_{w0} to obtain the dispersion relation

$$D_w(\omega, k)((\omega - kU_{0b})^2 - \omega_b^2) + \omega_w^2\omega_b^2 S_w(\omega, k)S_b(\omega, k) = 0\,, \qquad (23.3)$$

which is a generalization of the beam–plasma dispersion relation (22.1) to the interaction of a beam with an arbitrary electrodynamic system or medium.

The point of resonance between the wave of the medium and the beam wave, (ω_0, k_0), can be found by simultaneously solving the equations

$$\begin{aligned} &D_w(\omega, k) = 0\,, \\ &D_b(\omega, k) \equiv 0(\omega - kU_{0b})^2 + \omega_b^2 = 0\,. \end{aligned} \qquad (23.4)$$

The presence of the resonance point implies that $\omega_0/k_0 \approx U_{0b} < c$, i.e., that the phase velocity of the wave of the medium is lower than the speed of light. Media with such waves are called slowing-down media (systems). With representation (21.7) for the solution and with the second of formulas (22.3), dispersion relation (23.3) at the resonance point can be reduced to the form (it is expedient to compare it with relation (22.4))

$$\Omega(\Omega^2 \pm 2\Omega\omega_b) = \omega_w^2\omega_b^2 S_w(\omega_0, k_0)S_b(\omega_0, k_0)\left|\frac{\partial D_w(\omega_0, k_0)}{\partial\omega}\right|^{-1} \equiv \tilde{\Omega}\omega_b^2\,, \qquad (23.5)$$

where the quantity $\tilde{\Omega}$, having the dimension of frequency, is introduced for convenience and the above condition $\partial D_w/\partial\omega < 0$ is taken into account. Physical considerations imply that $\tilde{\Omega} > 0$.

Under inequality (22.6), i.e., for $|\Omega| \gg \omega_b$, the interaction of the beam with the electromagnetic wave is single-particle in nature and corrections to the resonant frequency are described by the expressions

$$\Omega_m = \delta_m(\tilde{\Omega}\omega_b^2)^{1/3}, \quad m = 1, 2, 3\,, \qquad (23.6)$$

where δ_m are the cubic roots of unity. Inserting expressions (23.6) into inequality (22.6) yields the following criterion for the interaction process to be single-particle:

$$\omega_b \ll \tilde{\Omega}\,. \qquad (23.7)$$

Under inequality (22.5), i.e., for $|\Omega| \ll \omega_b$, the interaction of the beam with the wave of the medium is collective, in which case dispersion relation (23.5) becomes

$$\Omega^2 = \pm\frac{1}{2}\tilde{\Omega}\omega_b\,. \qquad (23.8)$$

Here, the plus sign is for the resonance of the wave of the medium with the fast beam wave and the minus sign is for the resonance with the slow beam wave. In the case of resonance with the fast beam wave, the system is stable because the roots of dispersion relation (23.8) with the plus sign are real. In this case, dispersion relation

(23.8) has the same structure as relations (21.8) and (21.35). Consequently, the interaction of the fast beam wave with the wave of a slowing-down medium results in the beating between the wave amplitudes.

In the case of resonance with the slow beam wave, the system is unstable. From dispersion relation (23.8) with the minus sign we obtain the following imaginary corrections to the resonant frequency:

$$\Omega_{1,2} = \pm i \left(\frac{1}{2} \tilde{\Omega} \omega_b \right)^{1/2} . \tag{23.9}$$

Substituting these corrections into inequality (22.5) yields the following criterion for the interaction process to be collective in nature:

$$\omega_b \gg \tilde{\Omega} . \tag{23.10}$$

Inequality (23.10) is opposite to inequality (23.7), as should obviously be the case.

For the beam–plasma system considered in Sec. 19, we have $\tilde{\Omega} = \omega_p/2$ and inequality (19.5) gives $\omega_b^2 \ll \omega_p^2$, so criterion (23.10) for the interaction process to be collective in nature cannot be satisfied. In the general case, either inequality (32.7) or inequality (23.7) can hold. In fact, dispersion relation (23.5) contain the additional parameter $S_w S_b$, called the wave coupling coefficient, which is determined by the transverse geometry of the beam and the properties of the slowing-down medium. In the general case, the wave coupling coefficient can be either large or small. For a high-density beam and a small coupling coefficient, inequality (23.10) is satisfied, in which case the interaction is weak and the interaction process is collective. For a low-density beam and large coupling coefficient, inequality (23.7) is satisfied; accordingly, the interaction is strong and we deal with a single-particle interaction process.

Introducing the slow amplitudes

$$\begin{aligned} A_b(t, z) &= \tilde{A}_b(t) \exp(-i\omega_0 t + ik_0 z) , \\ A_w(t, z) &= \tilde{A}_w(t) \exp(-i\omega_0 t + ik_0 z) , \end{aligned} \tag{23.11}$$

we convert Eqs. (23.1) at the resonant point into the form

$$\begin{aligned} \frac{d^2 \tilde{A}_b}{dt^2} \mp 2i\omega_b \frac{d\tilde{A}_b}{dt} &= -\omega_b^2 S_b(\omega_0, k_0) \tilde{A}_w , \\ \frac{d\tilde{A}_w}{dt} &= -i\omega_w^2 S_w(\omega_0, k_0) \left| \frac{\partial D_w(\omega_0, k_0)}{\partial \omega} \right|^{-1} \tilde{A}_b . \end{aligned} \tag{23.12}$$

It is easy to see that Eqs. (23.12) lead to dispersion relation (23.5).

Under the inequality

$$\left| \frac{d\tilde{A}_b}{dt} \right| \gg \omega_b |\tilde{A}_b| , \tag{23.13}$$

which is an analogue of inequality (22.6), the interaction of the beam with the wave of the slowing-down structure is single-particle. In this case, Eqs. (23.12) reduce to the equations

$$\begin{aligned} \frac{d^2\tilde{A}_b}{dt^2} &= -\omega_b^2 S_b(\omega_0, k_0)\tilde{A}_w\,, \\ \frac{d\tilde{A}_w}{dt} &= -i\omega_w^2 S_w(\omega_0, k_0)\left|\frac{\partial D_w(\omega_0, k_0)}{\partial\omega}\right|^{-1}\tilde{A}_b\,, \end{aligned} \tag{23.14}$$

which have the same structure as Eqs. (22.21). From formula (22.22), we can see that the state vector of a "beam + medium" system described by Eqs. (23.14) can be defined by the formula

$$\boldsymbol{\Psi}(t,z) = \left\{\frac{d\tilde{A}_b(t)}{dt},\ \tilde{A}_b(t),\ \tilde{A}_w(t)\right\}\exp(-i\omega_0 t + ik_0 z)\,. \tag{23.15}$$

The state vector of a system described by more general equations (23.12) can be specified by the same formula.

Under the inequality opposite to (23.13) (see inequality (22.5)), the interaction of the beam with the wave of the slowing-down medium is collective. In this case, Eqs. (23.12) reduce to the form

$$\begin{aligned} \frac{d\tilde{A}_b}{dt} &= \mp i\frac{1}{2}\omega_b S_b(\omega_0, k_0)\tilde{A}_w\,, \\ \frac{d\tilde{A}_w}{dt} &= -i\omega_w^2 S_w(\omega_0, k_0)\left|\frac{\partial D_w(\omega_0, k_0)}{\partial\omega}\right|^{-1}\tilde{A}_b\,. \end{aligned} \tag{23.16}$$

In the first equation, the upper sign is for the resonance between the wave of the medium and the fast beam wave and the lower sign is for the resonance with the slow beam wave. The state vector of a "beam + medium" system described by Eqs. (23.16) can be defined by the formula

$$\boldsymbol{\Psi}(t,z) = \{\tilde{A}_b(t),\ \tilde{A}_w(t)\}\exp(-i\omega_0 t + ik_0 z)\,. \tag{23.17}$$

For a particular beam–plasma system (see Eqs. (22.21)), equations like (23.14) have been solved in Sec. 22. That is why we consider only the solution to Eqs. (23.16). We supplement these equations with the initial conditions

$$\tilde{A}_b(0) = A_{b0}\,, \quad \tilde{A}_w(0) = A_{w0}\,. \tag{23.18}$$

Applying the Laplace transformation to Eqs. (23.16), we obtain the following set of equations for the transformed quantities:

$$\begin{aligned} \Omega\tilde{A}_b(\Omega) \mp \frac{1}{2}\omega_b S_b\tilde{A}_w(\Omega) &= iA_{b0}\,, \\ \Omega\tilde{A}_w(\Omega) - \omega_w^2 S_w|\partial D_w/\partial\omega|^{-1}\tilde{A}_b(\Omega) &= iA_{w0}\,. \end{aligned} \tag{23.19}$$

We then find the transformed quantities from Eqs. (23.19) and apply the inverse Laplace transformation to arrive at the following solution to the initial-value problem given by Eqs. (23.16) and initial conditions (23.18):

$$\begin{aligned}\tilde{A}_b(t) &= A_{b0}\left(\frac{\Omega_1}{\Omega_1-\Omega_2}\exp(-i\Omega_1 t)+\frac{\Omega_2}{\Omega_2-\Omega_1}\exp(-i\Omega_2 t)\right)\\ &\quad \pm A_{w0}(S_b/2)\left(\frac{\omega_b}{\Omega_1-\Omega_2}\exp(-i\Omega_1 t)+\frac{\omega_b}{\Omega_2-\Omega_1}\exp(-i\Omega_2 t)\right),\\ \tilde{A}_w(t) &= A_{w0}\left(\frac{\Omega_1}{\Omega_1-\Omega_2}\exp(-i\Omega_1 t)+\frac{\Omega_2}{\Omega_2-\Omega_1}\exp(-i\Omega_2 t)\right)\\ &\quad + A_{b0}S_b^{-1}\left(\frac{\tilde{\Omega}}{\Omega_1-\Omega_2}\exp(-i\Omega_1 t)+\frac{\tilde{\Omega}}{\Omega_2-\Omega_1}\exp(-i\Omega_2 t)\right).\end{aligned} \tag{23.20}$$

Here, $\Omega_{1,2}$ are the roots of dispersion relation (23.8) and the quantity $\tilde{\Omega}$ has been introduced in the comments on dispersion relation (23.5). Recall that, in (23.8), (23.16), and (23.20), the upper sign refers to the resonance between the wave of the medium and the fast beam wave. For the resonance with the slow beam wave, one must use the lower sign.

We choose the upper sign in dispersion relation (23.8) and solution (23.20) to transform the latter to

$$\begin{aligned}\tilde{A}_b(t) &= \frac{1}{2}(A_{b0}+(\omega_b S_b\Omega_0^{-1}/2)A_{w0})\exp(-i\Omega_0 t)\\ &\quad + \frac{1}{2}(A_{b0}-(\omega_b S_b\Omega_0^{-1}/2)A_{w0})\exp(i\Omega_0 t),\\ \tilde{A}_w(t) &= \frac{1}{2}(A_{w0}+(\tilde{\Omega}S_b^{-1}\Omega_0^{-1})A_{b0})\exp(-i\Omega_0 t)\\ &\quad + \frac{1}{2}(A_{w0}-(\tilde{\Omega}S_b^{-1}\Omega_0^{-1})A_{b0})\exp(i\Omega_0 t),\end{aligned} \tag{23.21}$$

where $\Omega_0 = \sqrt{(1/2)\tilde{\Omega}\omega_b}$. Solution (23.21) is completely equivalent in structure to solution (21.37) with coefficients (21.38). Consequently, the resonant interaction of the wave of the slowing-down medium with the fast beam wave is accompanied by the beating between the wave amplitudes (see above). That this process is stable is clear from Sec. 21.

Now, we choose the lower sign in dispersion relation (23.8) and solution (23.20). In this case, the solution becomes

$$\begin{aligned}\tilde{A}_b(t) &= \frac{1}{2}(A_{b0}+i(\omega_b S_b\Omega_0^{-1}/2)A_{w0})\exp(\Omega_0 t)\\ &\quad + \frac{1}{2}(A_{b0}-i(\omega_b S_b\Omega_0^{-1}/2)A_{w0})\exp(-\Omega_0 t),\\ \tilde{A}_w(t) &= \frac{1}{2}(A_{w0}-i(\tilde{\Omega}S_b^{-1}\Omega_0^{-1})A_{b0})\exp(\Omega_0 t)\\ &\quad + \frac{1}{2}(A_{w0}+i(\tilde{\Omega}S_b^{-1}\Omega_0^{-1})A_{b0})\exp(-\Omega_0 t).\end{aligned} \tag{23.22}$$

That solution (23.22) contains exponentially increasing terms clearly shows that the resonant interaction between the wave of the slowing-down structure and the slow wave of the electron beam is an unstable process.

24. Resonant Buneman Instability in a Current-Carrying Plasma in the Coupled Oscillator Representation

The description of the resonant instability of a current-carrying plasma in the coupled oscillator representation differs from that of the instabilities considered in the last two sections. In the resonant Buneman instability, the role of the slowing-down medium with which an electron beam interacts is played by plasma ions. By virtue of the left inequality in (20.9), the characteristic time scale on which the instability develops is too short for the oscillatory properties of the ions to come into play, so it is impossible to introduce the slow amplitude of the ion wave. Consequently, in the case of Buneman instability, the second of Eqs. (23.12) should be second-order in t. As for the electrons, the right inequality in (20.9) implies that their interaction with the ion subsystem is collective in nature.

Let us rewrite dispersion relation (20.4) in a form similar to the beam–plasma dispersion relation (22.1):

$$(\omega^2 - \omega_i^2)((\omega - kU_{0e})^2 - \omega_e^2) - \omega_i^2\omega_e^2 = 0\,. \tag{24.1}$$

Relation (24.1) has the form of general dispersion relation (23.3) with $D_w = D_i = -\omega^2 + \omega_i^2$, $\omega_w = \omega_i$, and $S_w = S_b = 1$ (the subscript "b" is replaced by "e"). In dispersion relation (24.1), we have explicitly singled out two interacting subsystems: the electron subsystem, described by the dispersion relation $D_e = (\omega - kU_{0e})^2 - \omega_e^2 = 0$, and the ion subsystem, described by the dispersion relation $D_i = 0$. With the left inequality in (20.9), dispersion relation (24.1) can be written as the relation

$$\omega^2((\omega - kU_{0e})^2 - \omega_e^2) - \omega_i^2\omega_e^2 = 0\,, \tag{24.2}$$

the solution to which is given by expressions (20.10), by virtue of the right inequality in (20.9).

The differential equations describing the resonant Buneman instability in the coupled oscillator representation can be derived by switching to the operators in dispersion relation (24.1). But a simpler way is to immediately write general equations (23.1) for a current-carrying plasma:

$$\begin{aligned} \left(\left(\frac{\partial}{\partial t} + U_{0e}\frac{\partial}{\partial z}\right)^2 + \omega_e^2\right) A_e &= -\omega_e^2 A_i\,, \\ \left(\frac{\partial^2}{\partial t^2} + \omega_i^2\right) A_i &= -\omega_i^2 A_e\,. \end{aligned} \tag{24.3}$$

Here, A_e and A_i are the characteristic functions of the state vectors of the electrons and ions, respectively. We represent the solutions to Eqs. (24.3) as

$$\begin{aligned} A_e(t,z) &= \tilde{A}_e(t)\exp(-i\omega_0 t + ik_0 z)\,, \\ A_i(t,z) &= \tilde{A}_i(t)\exp(-i\omega_0 t + ik_0 z)\,, \end{aligned} \tag{24.4}$$

where the resonant frequency and wavenumber, $\omega_0 = \omega_i$ and $k_0 = (\omega_e + \omega_i)/U_{0e}$, are determined from the dispersion relations $D_e = D_i = 0$. From inequalities (20.9) we have

$$\omega_i|\tilde{A}_{e,i}| \ll \left|\frac{d\tilde{A}_{e,i}}{dt}\right| \ll \omega_e|\tilde{A}_{e,i}|\,. \tag{24.5}$$

Substituting representation (24.4) into Eqs. (24.3) and taking into account inequalities (24.5) and the inequality $\omega_i^2 \ll \omega_e^2$, we obtain the following equations for the amplitudes $\tilde{A}_{e,i}(t)$:

$$\begin{aligned} \frac{d\tilde{A}_e}{dt} &= \frac{1}{2}i\omega_e\tilde{A}_i\,, \\ \frac{d^2\tilde{A}_i}{dt^2} &= -\omega_i^2\tilde{A}_e\,. \end{aligned} \tag{24.6}$$

The general solution to Eqs. (24.6) is a superposition of three harmonic waves with frequencies (20.10a). The set of Eqs. (24.6) coincides in structure with the sets of Eqs. (23.14) and (22.21), differing from them only in the sign of the right-hand side of the first-order equation. Physically, this difference is expressed by the relationships $U_{0e} \gg \mathrm{Re}\,\omega_1/k_0 = U_{0e}(1/2)(\omega_i^2/2\omega_e^2)^{1/3} > 0$, as may be seen from formulas (20.10a). Consequently, in the resonant Buneman instability, the electric field of the growing wave resonantly ($\mathrm{Re}\,\omega_1/k_0 \approx U_{0i} = 0$) accelerates plasma ions in the direction of motion of the electrons. On the other hand, for the beam instability, from solution (19.11) we have $\mathrm{Re}\,\omega_1/k_0 = U_{0b}(1-(1/2)(\omega_b^2/2\omega_p^2)^{1/3})$. That is, the electric field of the wave resonantly ($\mathrm{Re}\,\omega_1/k_0 \approx U_{0b}$) decelerates beam electrons. Of course, in the course of the Buneman instability, the electrons are decelerated too, but not under the Cherenkov resonance conditions, since $U_{0e} \gg \mathrm{Re}\,\omega_1/k_0$.

25. Dispersion Function and Wave Absorption in Dissipative Systems

Here, we consider the absorption of waves in a dissipative system described by the dispersion relation $D(\omega,k) = 0$. The dispersion function of a dissipative system is complex even for real ω and k (see, e.g., dispersion relation (6.16)), so we have

$$D(\omega,k) = D'(\omega,k) + iD''(\omega,k)\,, \tag{25.1}$$

where D' and D'' are, respectively, the real and imaginary parts of the dispersion function. When the dissipation is weak, complex frequencies (1.16) of the eigenmodes of the system are determined as follows. To zero order in the absorption, the imaginary part of the frequency, $\omega'_m(k)$, is found from the equation

$$D'(\omega', k) = 0\,. \tag{25.2}$$

To the next order, we have

$$D(\omega' + i\omega'', k) = D'(\omega', k) + i\omega'' \frac{\partial D'}{\partial \omega}(\omega', k) + iD''(\omega', k) = 0\,. \tag{25.3}$$

From Eq. (25.2) we find the imaginary part of the frequency, $\omega' = \omega'_m(k)$, and from Eq. (23.3) we obtain the damping rate

$$\omega''_m(k) = -D''(\omega'_m, k)\left(\frac{\partial D'_m}{\partial \omega}\right)^{-1}, \quad \frac{\partial D'_m}{\partial \omega} = \frac{\partial D'}{\partial \omega}(\omega, k)\bigg|_{\omega=\omega'_m(k)}\,. \tag{25.4}$$

For an equilibrium dissipative medium, we should have $\omega''_m(k) < 0$ for all the branches.

Assume that a certain external source excites harmonic oscillations with an arbitrary frequency ω_0 and arbitrary wavenumber k. Substituting the external force $F(t, z) = f_0 \exp(-i\omega_0 t + ikz)$ and the characteristic function of the state vector, $A(t, z) = A_0(t) \exp(-i\omega_0 t + ikz)$, into Eq. (3.12) yields the equation

$$D\left(\omega_0 + i\frac{d}{dt},\ k\right) A_0(t) = f_0\,. \tag{25.5}$$

When the constant f_0 is sufficiently small, the amplitude $A_0(t)$ is slow and Eq. (25.5) can be cast into the form

$$i\frac{\partial D_0}{\partial \omega}\frac{dA_0}{dt} + D_0 A_0 = f_0\,, \tag{25.6}$$

where

$$D_0 = D(\omega_0, k)\,, \quad \frac{\partial D_0}{\partial \omega} = \frac{\partial D}{\partial \omega}(\omega, k)\bigg|_{\omega=\omega_0}\,. \tag{25.7}$$

Moreover, within the accuracy adopted for calculating damping rate (25.4), it is necessary to set $\partial D_0/\partial\omega = \partial D'_0/\partial\omega$.

We multiply Eq. (25.6) by A_0^* and take the sum of the resulting equation with its complex conjugate to obtain the relationship

$$i\frac{\partial D'_0}{\partial \omega}\frac{dW}{dt} + (D_0 - D_0^*)W = f_0 A_0^* - f_0^* A_0\,, \tag{25.8}$$

where the quantity $W = |A_0|^2$ describes the wave energy in the medium. For $f_0 = 0$, relationship (25.8) becomes

$$\frac{dW}{dt} + 2\delta\omega W = 0\,, \quad \delta\omega = -\frac{1}{2}i\left(\frac{\partial D'_0}{\partial \omega}\right)^{-1}(D_0 - D_0^*) = D''_0\left(\frac{\partial D'_0}{\partial \omega}\right)^{-1}. \tag{25.9}$$

For $f_0 = 0$, oscillations of the medium are free and thereby occur at the eigenfrequencies, so the frequency ω_0 is one of the frequencies $\omega'_m(k)$. Relationships (25.9) and (25.4) then yield $\delta\omega = -\omega''_m(k) > 0$, and we have $W = W_0 \exp(-2\delta\omega t)$, which is certainly the case for a wave damped at the rate (25.4).

Let us convert relationship (25.8) into the form

$$\frac{dW}{dt} + 2\delta\omega W = -i\left(\frac{\partial D_0'}{\partial \omega}\right)^{-1} (f_0 A_0^* - f_0^* A_0) \equiv Q\,, \tag{25.10}$$

where Q is the fraction of the power of the external source that is deposited in the medium. In a steady state, the energy of the source is completely absorbed. We express the amplitude A_0 from the equation $D_0 A_0 = f_0$ — a time-independent version of Eq. (25.6) — to find the fraction of the source energy that is deposited in the medium per unit time:

$$Q = \left(\frac{\partial D_0'}{\partial \omega}\right)^{-1} \frac{D_0''}{(D_0')^2 + (D_0'')^2} |f_0|^2 = \frac{\delta\omega}{(D_0')^2 + (D_0'')^2} |f_0|^2\,, \tag{25.11}$$

where the frequency correction $\delta\omega$ is defined in relationships (25.9). If the frequency of the external source, ω_0, coincides with one of the eigenfrequencies $\omega_m'(k)$, then formula (25.11) becomes

$$Q = \left(\frac{\partial D_m'}{\partial \omega}\right)^{-2} \frac{1}{|\omega_m''(k)|} |f_0|^2\,. \tag{25.12}$$

In order to illustrate how formula (25.11) can be applied in practice, let us consider the absorption of the energy of an external source that excites longitudinal electrostatic perturbations in a cold electron plasma. In this case, the dispersion function is given by relationship (6.16). Using this relationship, we transform formula (25.11) to

$$Q = \frac{1}{2} \frac{v_{en}}{(\omega_0^2 - \omega_p^2)^2 + v_{en}^2 \omega_0^2} |f_0|^2\,. \tag{25.13}$$

When the source frequency ω_0 is sufficiently close to ω_p, formula (25.13) can be written as

$$Q = \frac{1}{4\omega_p^2} \frac{v_{en}/2}{(\omega_0 - \omega_p)^2 + (v_{en}/2)^2} |f_0|^2\,. \tag{25.14}$$

For $|\omega_0 - \omega_p| \ll v_{en}$, formula (25.14) becomes

$$Q = \frac{1}{2\omega_p^2 v_{en}} |f_0|^2\,. \tag{25.15}$$

This latter formula can also be obtained from formula (25.12). When the absorption is negligibly weak, formula (24.14) reduces to the expression

$$Q = \frac{\pi}{4\omega_p^2} \delta(\omega_0 - \omega_p) |f_0|^2\,, \tag{25.16}$$

which describes the energy expended by an external source to resonantly excite eigenmodes in a plasma.

Hence, the energy of the external source is deposited in the medium not only by dissipation but also by exciting eigenmodes.

As another example, let us consider the absorption of the energy of an external source that excites high-frequency transverse electromagnetic waves in an isotropic collisional electron plasma under conditions corresponding to inequality (7.16). In this case, the dispersion function is given by relationship (7.20). Using this relationship, we transform formula (25.11) to

$$Q = \frac{1}{2\omega_0^2} \frac{v_{en}\omega_p^2}{(\omega_0^2 - \omega_s^2)^2 + v_{en}^2\omega_p^2(\omega_p^2/\omega_0^2)} |f_0|^2 , \tag{25.17}$$

where $\omega_S = \sqrt{\omega_p^2 + k^2c^2}$ is the eigenfrequency of an electromagnetic wave in a collisionless plasma. When the frequency of the external source, ω_0, is sufficiently close to ω_S, formula (25.17) can be written in a form analogous to (25.14):

$$Q = \frac{1}{4\omega_S^2} \frac{v'/2}{(\omega_0 - \omega_S)^2 + (v'/2)^2} |f_0|^2 , \quad v' = v_{en} \frac{\omega_p^2}{\omega_S^2} . \tag{25.18}$$

And finally, let us calculate the energy lost by the source as it excites ion acoustic waves in a nonisothermal plasma. When the ion collision frequency v_{in} is low, the dispersion function of the ion acoustic waves (see dispersion relation (12.12)) can be represented as

$$D(\omega, k) \equiv -\omega^2 + \omega_S^2 - iv_{in}\frac{\omega_S^2}{\omega} , \quad \omega_S^2 = \omega_i^2 \frac{k^2 r_{De}^2}{1 + k^2 r_{De}^2} , \tag{25.19}$$

where ω_S is the ion acoustic frequency in a collisionless plasma (see (12.6)). Substituting dispersion function (25.19) into formula (25.11) yields

$$Q = \frac{1}{2\omega_0^2} \frac{v_{in}\omega_S^2}{(\omega_0^2 - \omega_S^2)^2 + v_{in}^2\omega_S^2(\omega_S^2/\omega_0^2)} |f_0|^2 . \tag{25.20}$$

For $\omega_0 \sim \omega_S$, we obtain the formula

$$Q = \frac{1}{4\omega_S^2} \frac{v_{in}/2}{(\omega_0 - \omega_S)^2 + (v_{in}/2)^2} |f_0|^2 . \tag{25.21}$$

Although formulas (25.13), (25.17), and (25.20), as well as their simpler versions (25.14), (25.18), and (25.21), are similar, there are differences among them due to the dispersion of the plasma medium. In formula (25.13), the eigenfrequency $\omega_S = \omega_p$ is a constant independent of the wavenumber k. In the remaining two formulas, the eigenfrequencies ω_S depend, and even differently, on k. If the external source is nonharmonic and contains components with different k and ω_0 values, then the total energy input from the source into the medium is described by the integral (nonharmonic perturbations in dispersive media will be considered in detail in the next sections, see also Appendix 1)

$$\iint Q(\omega_0, k) d\omega_0 dk , \tag{25.22}$$

in which the symbol $Q(\omega_0, k)$ stands for functions (25.13), (25.17), and (25.20) and for functions similar to them (here, the function $f_0 = f_0(\omega_0, k)$ is determined by the external source). It is obvious that the value of the integral (25.22) depends strongly on the frequencies $\omega_S(k)$ of the eigenmodes of the medium.

26. Some Effects in the Interaction between Waves in Coupled Systems

Let us consider some particular cases of the interaction between waves in coupled subsystems. Without pretending to give a rigorous and complete description, we restrict ourselves to a few specific examples.

1. In the coupled oscillator representation, the interaction between oblique Alfvén and magnetosonic waves is described by dispersion relation (16.16). The wave coupling is characterized by the quantity $k_\perp^2 U_0^2$, and the set of Eqs. (21.28) is written as

$$\begin{aligned} D_A(\omega, k) &\equiv (\omega^2 - (k^2 + k_\perp^2)V_A^2) = 0\,, \\ D_M(\omega, k) &\equiv (\omega^2 - k^2 U_0^2) = 0\,. \end{aligned} \tag{26.1}$$

In turn, Eqs. (26.1) yield the following expressions for the resonant wavenumber and frequency:

$$k_0 = k_\perp \sqrt{V_A^2 (U_0^2 - V_A^2)^{-1}}\,, \quad \omega_0 = k_\perp U_0 \sqrt{V_A^2 (U_0^2 - V_A^2)^{-1}}\,. \tag{26.2}$$

We can see that the frequency ω_0 is proportional to the wave coupling parameter, so the frequency correction due to the wave interaction is always on the order of the frequency itself. In other words, in magnetic hydrodynamics, the interaction between waves is always strong. The only exception is the velocity resonance

$$U_0 = V_A\,. \tag{26.3}$$

Under the resonance condition (26.3), the hydrodynamic and magnetic pressures in the fluid are the same and dispersion relation (16.16) has the solution

$$\omega = kU_0 \pm (1/2)k_\perp U_0\,, \tag{26.4}$$

where the longitudinal wavenumber varies within a semi-infinite range, $k_\perp < |k| < +\infty$. That is, under the velocity resonance condition (26.3), the waves are in resonance everywhere except in the longest wavelength range, in which solution (26.4) is inapplicable. In Fig. 24, the heavy lines show dispersion curves calculated for $U_0 = V_A$ by solving dispersion relation (16.21) with $\alpha = 1$ (in terms of dimensionless variables (16.20), solution (26.4) is approximated by $y = x \pm 1/2$). The light line in the figure shows the dispersion law (16.19) for a conventional Alfvén wave, specifically, $\omega = kV_A = kU_0$ (in terms of variables (16.20), we have $y = x$). Hence, we are dealing with a characteristic example of resonant wave interaction within a finite range of wavenumbers rather that at the only wavenumber $k = k_0$. This is why the dispersion curves of the waves deviate from one another not at the only point (as is shown in Figs. 22, 23) but everywhere (as is shown in Fig. 24). The dispersion relation for waves that are subject to such a "distributed" interaction has the structure $(\omega - kU_1)(\omega - kU_2) = \Omega_0^2$, with $U_1 = U_2$. An important point is that, although their dispersion laws are the same (or are close to one another),

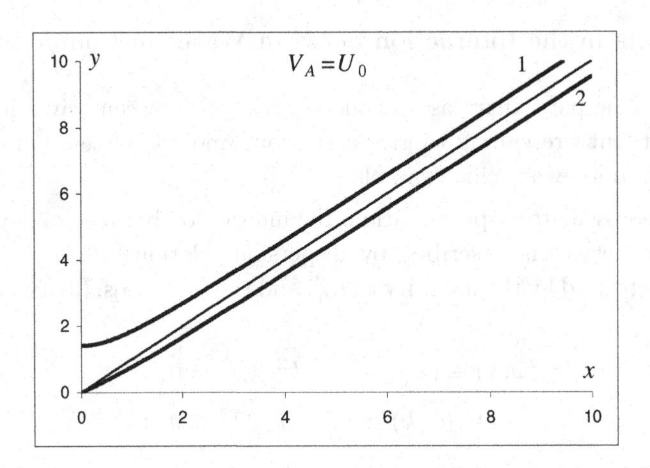

Fig. 24. Dispersion curves for the resonant interaction of oblique MHD waves in a conducting fluid.

the interacting waves are of different physical nature, like magnetosonic and Alfvén waves in the example just examined.

2. Even if the resonance conditions are not satisfied, weak interaction between subsystems can give rise to important physical effects. Let the interaction of subsystems be described by Eqs. (21.27). Assume that a certain external source resonantly excites the eigenmode of the first subsystem, namely,

$$A_1(t,z) = A_{10} \exp(-i\omega_1 t + ikz) . \tag{26.5}$$

Here, A_{10} is a constant amplitude; k is the wavenumber set by the source; and the eigenfrequency of the first subsystem, ω_1, is determined from the dispersion relation $D_1(\omega_1, k) = 0$. If, in this case, we have $D_2(\omega_1, k) \neq 0$, then, in the second subsystem, the source excites a noneigenmode wave,

$$A_2(t,z) = A_{20} \exp(-i\omega_1 t + ikz) . \tag{26.6}$$

For weakly coupled subsystems, the amplitude of this noneigenmode wave can be calculated from the formula

$$A_{20} = \frac{S_2(\omega_1, k)}{D_2(\omega_1, k)} A_{10} . \tag{26.7}$$

Let us consider the example from Sec. 13 — the excitation of a noneigenmode longitudinal Langmuir wave in an anisotropic plasma in a strong external magnetic field. In terms of the frequency ω and longitudinal wavenumber k, the set of Eqs. (21.2) reads

$$\begin{aligned} (\omega^2 - k^2c^2)A_E &= i\omega k_\perp c A_p \,, \\ (\omega^2 - \omega_p^2)A_p &= -i\omega k_\perp c A_E \,. \end{aligned} \tag{26.8}$$

If the external source excites an electromagnetic wave with the frequency $\omega = kc \neq \omega_p$ that propagates in the plasma exactly along the external magnetic field (in particular, such a wave can be injected into the plasma through its boundary), then we have

$$A_E = A_{E0}\exp(-ik(ct - z))\,, \quad A_P = 0\,. \tag{26.9}$$

But when the angle between the wave propagation direction and the magnetic field is nonzero, we deal with the excitation of a noneigenmode longitudinal Langmuir wave in the plasma,

$$A_P = i\frac{kk_\perp c^2}{\omega_p^2 - k^2c^2}A_{E0}\exp(-ik(ct - z))\,. \tag{26.10}$$

Formula (26.10) implies that $k_\perp^2 \ll k^2$.

3. The excitation of noneigenmode waves can lead to an additional damping of eigenmodes. Thus, in the above case of an electromagnetic wave in an anisotropic plasma in a strong external magnetic field, we have the dispersion relation

$$D(\omega, k) \equiv \left(1 - \frac{\omega_p^2}{\omega(\omega + i v_{en})}\right)(\omega^2 - k^2c^2) - k_\perp^2 c^2 = 0\,. \tag{26.11}$$

Here, v_{en} is the frequency of collisions involving plasma electrons (this frequency has been introduced in Sec. 6; see dispersion relation (6.16) and the comments on it). For $k_\perp^2 \ll k^2$, dispersion relation (26.11) yields the following frequency spectrum of a high-frequency electromagnetic wave:

$$\omega = kc - \frac{1}{2}i v_{en}\frac{k_\perp^2}{k^2}\frac{\omega_p^2}{k^2c^2}\,. \tag{26.12}$$

This spectrum has been obtained under the assumption

$$k_\perp c\,, \quad v_{en}\,, \quad \omega_p \ll kc\,. \tag{26.13}$$

In frequency spectrum (26.12), the damping rate of a transverse electromagnetic wave is governed by its nonresonant interaction with a noneigenmode longitudinal plasma wave that is dissipated by electron collisions.

A similar dissipative effect has been considered in Sec. 17, devoted to acoustic waves in crystals. The damping rate (17.17) of longitudinal acoustic waves is determined by the nonresonant excitation of an acoustic wave by a noneigenmode plasma wave that is damped by collisions of electrons with the crystal lattice. It should be noted that dispersion relations (17.14) and (26.11) have similar structures. As for the solutions to these dispersion relations, they have been obtained, respectively, under inequalities (17.15) and (26.13), which are opposite to each other.

4. To conclude this section, note that, in the presence of a constant external electric field, which sets the charge carriers into drift motion, the plasma damping of an acoustic wave in a crystal can change sign and the wave can begin to grow via the plasma mechanism. Thus, if the external electric field is directed along the OZ

axis and causes electrons to drift parallel to this axis with the velocity U_0, then dispersion relation (17.14) becomes

$$D(\omega, k) \equiv \left(1 - \frac{\omega_p^2}{(\omega - kU_0)(\omega - kU_0 + i\nu_{eff})}\right)(\omega^2 - k^2 C_\parallel^2) - \frac{4\pi\beta_3^2}{\rho} k^2 = 0 \,. \tag{26.14}$$

So, instead of dispersion relation (17.16), we arrive at following relation for the acoustic wave:

$$D(\omega, k) \equiv \omega^2 + i\frac{4\pi\beta_3^2 \nu_{eff} k^2}{\rho\omega_p^2}(\omega - kU_0) - k^2 C_\parallel^2 = 0 \,. \tag{26.15}$$

We can see that, under the condition $kU_0 > \omega \approx kC_\parallel$, the imaginary term in (26.15) changes sign and the imaginary part of the frequency is given not by formula (17.17) but by the formula

$$\omega'' = -\frac{2\pi\beta_3^2 \nu_{eff} k^2}{\rho\omega_p^2}\left(1 - \frac{U_0}{C_\parallel}\right). \tag{26.16}$$

An acoustic wave begins to grow ($\omega'' > 0$) for $U_0 > C_\parallel$, providing evidence that the instability has a Cherenkov nature. The instability in turn stems from the nonresonant interaction of the acoustic wave with noneigenmode plasma oscillations of charge carriers moving in a crystal.

27. Waves and their Interaction in Periodic Structures

Up to this point, we have considered systems that are spatially homogeneous in the wave propagation direction. For systems whose parameters vary gradually in space, wave theory is constructed in the geometrical-optics approximation — an issue that is now beyond the scope of our analysis. Waves in systems whose parameters vary in a jumplike manner, i.e., those in which there are boundaries between the media, require a separate study. There is a particular case, however, that has to be analyzed separately: media and systems with periodically varying parameters. Although the theory of waves in periodic systems has much in common with the wave theory for spatially homogeneous systems, it has its own specific features and makes it possible to reveal new physical effects. Let us now present some aspects of this theory, using as an example a simple system that has been investigated in Sec. 5.

To do this, we consider linearly polarized electromagnetic waves propagating along the OZ axis in an unbounded medium whose permittivity is periodically modulated in the same direction. We begin with basic differential equations (5.1), in which the permittivity ε_0 is now assumed to be a periodic function of z, say

$$\varepsilon_0(z) = \varepsilon + h \cos \chi z \,. \tag{27.1}$$

Here, ε, h, and χ are positive constants (such that $2\pi/\chi$ is the spatial modulation period of the medium). We also assume that the medium is modulated only slightly, which corresponds to the inequality

$$\mu = h/\varepsilon \ll 1 \,. \tag{27.2}$$

Equations (5.1) with permittivity (27.1) can be solved exactly by the methods of the theory of differential equations with periodic coefficients. In particular, we can obtain the dispersion relation $D(\omega, k) = 0$ in the form of an infinite determinant (the Hill determinant) from which to find the eigenfrequencies of electromagnetic waves in a periodic dielectric. But we restrict ourselves to finding an approximate solution in the form of an expansion in the small parameter (27.2), a much simpler task.

We seek the solution in the form

$$\boldsymbol{\Psi}(t,z) = \{E_x(t,z), B_y(t,z)\} = \{e_x(\omega,k,z), b_y(\omega,k,z)\}\exp(-i\omega t + ikz)\,, \quad (27.3)$$

which differs from solution (5.2) in that the components e_x and b_y of the complex state vector depend also on the spatial coordinate z. The theory of differential equations with periodic coefficients makes use of the following representations:

$$\begin{aligned} e_x(\omega,k,z) &= \sum_{n=-\infty}^{\infty} e_n(\omega,k)\exp(in\chi z)\,, \\ b_y(\omega,k,z) &= \sum_{n=-\infty}^{\infty} b_n(\omega,k)\exp(in\chi z)\,. \end{aligned} \quad (27.4)$$

In what follows, the arguments of the functions $e_n(\omega,k)$ and $b_n(\omega,k)$ will be omitted for brevity. In the theory of waves in periodic systems, functions like (27.7), having the spatial period $2\pi/\chi$, are called Bloch functions.

To zero order in the small parameter μ, only one term in the sums in representations (27.4) is nonzero — e.g., the term with $n = 0$. In this case, solution (27.3) goes over to solution (5.2). To the next order in μ, it is necessary to account for the terms with $n = \pm 1$, and so on. Restricting ourselves to the first approximation, we write

$$\begin{aligned} e_x(\omega,k,z) &= e_{-1}\exp(-i\chi z) + e_0 + e_1\exp(i\chi z)\,, \\ b_x(\omega,k,z) &= b_{-1}\exp(-i\chi z) + b_0 + b_1\exp(i\chi z)\,. \end{aligned} \quad (27.5)$$

Taking into account representations (27.5), we substitute formulas (27.1) and (27.3) into Eqs. (5.1) and equate to zero the coefficients of the functions $\exp(in\chi z)$ with $n = 0, \pm 1$ to obtain the set of equations

$$\begin{aligned} \omega e_{-1} - (k-\chi)(c/\varepsilon)b_{-1} + \omega(\mu/2)e_0 &= 0\,, \\ \omega b_{-1} - (k-\chi)ce_{-1} &= 0\,, \\ \omega e_0 - k(c/\varepsilon)b_0 + \omega(\mu/2)e_{-1} + \omega(\mu/2)e_1 &= 0\,, \\ \omega b_0 - kce_0 &= 0\,, \\ \omega e_1 - (k+\chi)(c/\varepsilon)b_1 + \omega(\mu/2)e_0 &= 0\,, \\ \omega b_1 - (k+\chi)ce_1 &= 0\,. \end{aligned} \quad (27.6)$$

For $\mu = 0$, the set of Eqs. (27.6) splits into three independent subsets for the pairs of coefficients (e_{-1}, b_{-1}), (e_0, b_0), and (e_1, b_1). In particular, the subset for (e_0, b_0)

coincides with the set of Eqs. (5.3). But the remaining two subsets have the same sense as the subset for (e_0, b_0). In fact, for a spatially homogeneous system ($\mu = 0$), the replacement $k \to k \pm \chi$ has no physical meaning and serves only to redefine the wavenumber. But this is not so for $\mu \neq 0$.

Eliminating the quantities $b_{-1,0,1}$ in Eqs. (27.6) yields the set of equations

$$\begin{aligned} &[\omega^2 - (k-\chi)^2 c_0^2]e_{-1} + \omega^2(\mu/2)e_0 = 0\,, \\ &[\omega^2 - k^2 c_0^2]e_0 + \omega^2(\mu/2)e_{-1} + \omega^2(\mu/2)e_1 = 0\,, \\ &[\omega^2 - (k+\chi)^2 c_0^2]e_1 + \omega^2(\mu/2)e_0 = 0\,, \end{aligned} \tag{27.7}$$

where $c_0^2 = c^2/\varepsilon$. And by eliminating the amplitudes $e_{-1,0,1}$ in Eqs. (27.7), we obtain the following dispersion relation for determining the eigenfrequencies:

$$D(\omega, k) \equiv D_{-1}D_0D_1 - \omega^4\left(\frac{\mu}{2}\right)^2 (D_{-1} + D_1) = 0\,. \tag{27.8}$$

Here,

$$D_n = \omega^2 - (k + n\chi)^2 c_0^2\,.$$

For $\mu \to 0$ (when the permittivity is unmodulated), dispersion relation (27.9) splits into the independent relations $D_n = 0$, which yield the spectra

$$\omega = \pm(k + n\chi)c_0\,, \tag{27.9}$$

with n being any integer.[1] Spectra (27.9) are identical since the replacement $k + n\chi \to k$ transforms them into each other. But this is not so for $\mu \neq 0$ because of the interaction between waves having frequencies (27.9) with different n values. Let us first consider a nonresonant interaction such that, e.g., $D_0(\omega, k) \approx 0$ and $D_{\mp 1}(\omega, k) \neq 0$. Taking into account inequality (27.2), from dispersion relation (27.8) we find

$$\omega^2 = k^2 c_0^2 \left[1 + \frac{1}{2}\left(\frac{\mu}{2}\right)^2 \left(1 - \frac{\chi^2}{4k^2}\right)^{-1}\right]. \tag{27.10}$$

Extracting the square root gives

$$\omega_{1,2} = \pm k c_0 \left[1 + \frac{1}{4}\left(\frac{\mu}{2}\right)^2 \left(1 - \frac{\chi^2}{4k^2}\right)^{-1}\right]. \tag{27.11}$$

We then express the amplitudes e_{-1} and e_1 from the first and third of Eqs. (27.7) and also determine the functions $b_{-1,0,1}$ from Eqs. (27.6) and substitute them into relationships (27.5). As a result, we obtain the following formulas for the components $e_x(\omega, k, z)$ and $b_y(\omega, k, z)$ of the complex state vector:

$$\begin{aligned} e_x(\omega, k, z) &= e_0\left[1 - \frac{\mu}{2}\frac{k^2}{\chi(2k-\chi)}\exp(-i\chi z) + \frac{\mu}{2}\frac{k^2}{\chi(2k+\chi)}\exp(i\chi z)\right], \\ b_y(\omega, k, z) &= e_0\frac{kc}{\omega}\left[1 - \frac{\mu}{2}\frac{k(k-\chi)}{\chi(2k-\chi)}\exp(-i\chi z) + \frac{\mu}{2}\frac{k(k+\chi)}{\chi(2k+\chi)}\exp(i\chi z)\right]. \end{aligned} \tag{27.12}$$

[1] Dispersion relation (27.8) contains only the terms with $n = 0, \pm 1$, as a consequence of approximation (27.2). It can be shown that an exact dispersion relation in the form of the Hill determinant contains all integer values of n.

And finally, we arrive at the following expression for the state vector (27.3) of harmonic electromagnetic waves in a periodically modulated dielectric:

$$\psi(t,z) = \begin{Bmatrix} E_x(t,z) \\ B_y(t,z) \end{Bmatrix}$$
$$= \sum_{m=1}^{2} A_m \begin{Bmatrix} \left[1 - \dfrac{\mu}{2}\dfrac{k^2}{\chi(2k-\chi)}\exp(-i\chi z) + \dfrac{\mu}{2}\dfrac{k^2}{\chi(2k+\chi)}\exp(i\chi z)\right] \\ \dfrac{kc}{\omega}\left[1 - \dfrac{\mu}{2}\dfrac{k(k-\chi)}{\chi(2k-\chi)}\exp(-i\chi z) + \dfrac{\mu}{2}\dfrac{k(k+\chi)}{\chi(2k+\chi)}\exp(i\chi z)\right] \end{Bmatrix}$$
$$\times \exp(-i\omega_m t + ikz)\,. \tag{27.13}$$

Here, the frequencies ω_m are given by formulas (27.11) and A_m are complex amplitudes. In expression (27.13), we have set $e_0 = 1$. It is easy to see that, for $\mu \to 0$, expression (27.13) goes over to expression (5.6a).

Solution (27.13) has a fairly clear physical meaning. The first term in the square brackets — the unity — describes a conventional electromagnetic wave. Such a wave polarizes a periodic dielectric to give rise to modes with amplitudes varying as $\sim \exp(\mp i\chi z)$ — see the second and third terms in the square brackets. Going to higher orders in the parameter μ leads to higher modes. In the case at hand, all the spatial modes are noneigenmode (nonresonant) waves that are intentionally induced by the zeroth mode, which is exactly the eigenmode (resonant) electromagnetic wave of the dielectric. The excitation of spatial modes is accounted for by corrections to eigenfrequencies (27.11).

Formulas (27.10)–(27.13) are inapplicable for $k = \pm\chi/2$. But this is just the resonant case in which the pairs of the functions D_0 and D_{-1} or D_0 and D_1 can simultaneously become zero. Let us consider the wave resonance under the conditions

$$D_0(\omega,k) \equiv \omega^2 - k^2c_0^2 = 0\,,$$
$$D_{-1}(\omega,k) \equiv \omega^2 - (k-\chi)^2c_0^2 = 0\,. \tag{27.14}$$

An analogue of dispersion relations (27.14) is, in particular, represented by dispersion relations (21.4). Under inequality (27.2), the interaction of waves described by dispersion relations (27.14) is weak. The wave resonance point is found by simultaneously solving relations (27.14), specifically,

$$\omega_0 = \chi c_0/2\,, \quad k_0 = \chi/2\,. \tag{27.15}$$

In this case, dispersion relation (27.8) can be written as

$$D(\omega,k) \equiv D_{-1}D_0 - \omega^4(\mu/2)^2 = 0\,. \tag{27.16}$$

Substituting the solution in the form (21.7) into dispersion relation (27.16) yields

$$\Omega^2 = \frac{1}{4}\omega_0^2(\mu/2)^2 \equiv \Omega_0^2 \to \Omega_{1,2} = \pm\Omega_0\,. \tag{27.17}$$

Solution (27.17) has exactly the same structure as solution (21.8). Consequently, under the resonance condition (27.5), the interacting waves are subject to beating; i.e., they transfer energy among themselves according to the formulas (see (21.17))

$$A_0 = \tilde{A}_0(t)\exp(-i\omega_0 t + ik_0 z) = \tilde{A}_0(t)\exp\left(i\frac{\chi}{2}(z - c_0 t)\right),$$

$$A_{-1} = \tilde{A}_{-1}(t)\exp(-i\omega_0 t + i(k_0 - \chi)z) = \tilde{A}_{-1}(t)\exp\left(-i\frac{\chi}{2}(z + c_0 t)\right), \quad (27.18)$$

$$\tilde{A}_0(t) = B_0\sin(\Omega_0 t + \varphi_0)\,, \quad \tilde{A}_{-1}(t) = B_0\cos(\Omega_0 t + \varphi_0)\,.$$

Here, B_0 and φ_0 are constants, which are to be found from the initial conditions.

Waves with amplitudes (27.18) propagate towards one another. In this connection, we should make the following important remark. Solution (27.18) has been obtained in formulating the initial-value problem. But the solution obtained in formulating the boundary-value problem of the interaction between waves is of a different nature: instead of periodic beating between waves, one deals with a kind of damping: the forward wave is completely converted into the backward wave. This phenomenon is called Bragg reflection.

The dispersion curves of dispersion relation (27.8) are illustrated in Fig. 25, which shows the results calculated for a system with the parameters $\chi = 1$ cm^{-1}, $\mu = 0.4$, and $c_0 = 2.5 \cdot 10^{10}$ cm/s. The dispersion curves are indicated by the numbers of the spatial modes. The gap between the curves is due to the resonant interaction between the waves. The gap width is $2\Omega_0$, where the frequency Ω_0 is given by formula (27.17).

A quick look at Fig. 25 may give the wrong impression that waves with anomalous dispersion can exist in a periodically modulated dielectric. In fact, it can be seen that, on some portions of the dispersion curves, the phase velocity ω/k and group velocity $d\omega/k$ have opposite signs. But in actuality, there are no waves with

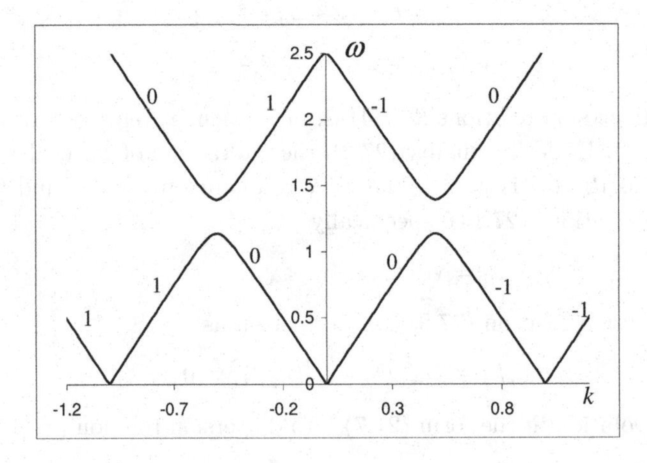

Fig. 25. Dispersion curves of electromagnetic waves in a periodically modulated dielectric.

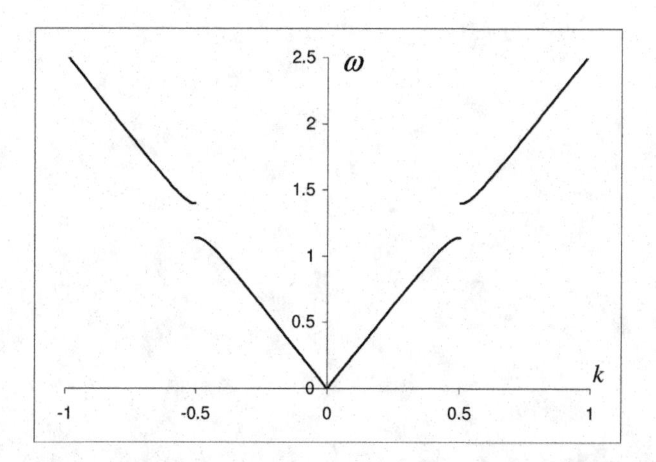

Fig. 26. Spectrum of electromagnetic waves in a periodic dielectric, plotted as an energy spectrum of a quantum particle in a field with a periodic potential.

anomalous dispersion in conventional periodic structures. The group velocity is always calculated from the formula $V_{gr} = d\omega/dk$ (for more detail on group velocities, see Sec. 29), while the phase velocity should be calculated not from formula (1.6) but from the formula

$$V_ph = \omega/(k + n\chi)\,. \tag{27.19}$$

For instance, for the left branch of the spatial mode (the branch with $n = -1$), we have $V_{gr} < 0$ and $V_{ph} < 0$, and vice versa for the right branch of the mode ($n = 1$). Often (especially in quantum mechanics, when the energy of an electron in a field with a periodic potential is considered), the dispersion curves of waves in periodic structures are plotted as shown in Fig. 26. In this way, no problems arise with the nature of wave dispersion. Figure 26 is obtained from Fig. 25 by deleting all the "redundant" (doubling) portions of the dispersion curves.

In Fig. 26, the dispersion curves are discontinuous at $k = \pm\chi/2$ — the wavenumber that corresponds to the Bragg reflection gap at the lowest frequency in the resonant interaction between the spatial modes with $n = 0, -1$ (the discontinuity at $k = +\chi/2$) and the modes with $n = 0, +1$ (the discontinuity at $k = -\chi/2$). There are also reflection gaps at higher frequencies in the interaction between modes with $n = 0, \mp 2$ (the discontinuities being at $k = \pm 2(\chi/2)$), $n = 0, \mp 3$ (the discontinuities being at $k = \pm 3(\chi/2)$), and so on. In order to take such reflection gaps into account, it is necessary to write dispersion relation (27.8) to within terms on the order of μ^{2m}, where m is the number of the highest of the gaps. In quantum mechanics, Bragg reflection gaps are called forbidden energy gaps of an electron in a field with a periodic potential.

Chapter 4

Nonharmonic Waves in Dispersive Media

28. General Solution to the Initial-Value Problem

In the previous sections of the monograph, we have considered waves excited in a medium by an initial, spatially harmonic perturbation of the state vector or by the same spatially harmonic perturbation from an external source. Let us now investigate wave perturbations with an arbitrary spatial structure. To do this, we begin by presenting the most important harmonic solutions to the differential equation for the characteristic function of the state vector. The harmonic solution to homogeneous equation (3.11) has the form

$$A(t,z) = A(t,k)\exp(ikz)\,. \tag{28.1}$$

Here, the amplitude $A(t,k)$ is calculated from one of the following equivalent formulas:

$$A(t,k) = \frac{1}{2\pi}\int_{C(\omega)} \frac{P_{n-1}(\omega,k)}{D(\omega,k)}\exp(-i\omega t)d\omega\,, \tag{28.2a}$$

$$A(t,k) = \sum_{m=1}^{n}[A_m(k)\exp(-i\omega_m(k)t)]\,. \tag{28.2b}$$

The functions $P_{n-1}(\omega,k)$ and $A_m(k)$ are determined from the initial conditions (more details on the use of the initial conditions are given in Sec. 4 and at the end of the present section). Assume for the moment that the functions $P_{n-1}(\omega,k)$ and $A_m(k)$ are known. Recall that formula (28.2b) is derived by calculating the integral in relationship (28.2a).

The harmonic solution to inhomogeneous equation (3.12) with $F(t,z) = F(t,k)\exp(ikz)$ again has the form (28.1), but with the amplitude $A(t,k)$ given by the Duhamel integral

$$A(t,k) = \int_0^t G(\tau,k)F(t-\tau,k)d\tau = \int_0^t G(t-\tau,k)F(\tau,k)d\tau\,, \tag{28.3}$$

where

$$G(t,k) = \frac{1}{2\pi}\int_{C(\omega)} \frac{1}{D(\omega,k)} \exp(-i\omega t)d\omega \tag{28.4}$$

is the unit instantaneous point source function.

Differential equation (3.11) is linear. Consequently, by the superposition principle, its most general solution is a linear combination of spatially harmonic solutions (28.1) with different values of the wavenumber k. That is, in the most general case, we have

$$A(t,z) = \frac{1}{2\pi}\int_{-\infty}^{+\infty} A(t,k)\exp(ikz)dk\,. \tag{28.5}$$

We multiply both sides of relationship (28.5) by $\exp(-ikz)$ and integrate over the z coordinate from minus to plus infinity. In doing so, we replace the variable of integration k in (28.5) with k' and use the familiar integral representation for the delta function:

$$\delta(k-k') = \frac{1}{2\pi}\int_{-\infty}^{+\infty} \exp[i(k-k')z]dz\,. \tag{28.6}$$

The integration yields

$$A(t,k) = \int_{-\infty}^{\infty} A(t,z)\exp(-ikz)dz\,. \tag{28.7}$$

We can thus see that relationships (28.5) and (28.7) are the inverse and direct spatial Fourier transforms of the characteristic function of the state vector. Function (28.7) is called the Fourier transform of the function $A(t,z)$ (28.5) in the spatial variable z. In turn, the integral in relationship (28.5) is called the inverse Fourier transform.

Substituting formulas (28.2) into relationship (28.5), we obtain the following equivalent expressions for the general solution to linear homogeneous equation (3.11) in the initial-value problem:

$$A(t,z) = \frac{1}{(2\pi)^2}\int_{-\infty}^{+\infty} dk \int_{C(\omega)} \frac{P_{n-1}(\omega,k)}{D(\omega,k)} \exp(-i\omega t + ikz)d\omega\,, \tag{28.8}$$

$$A(t,z) = \frac{1}{(2\pi)}\int_{-\infty}^{+\infty} \sum_{m=1}^{n} [A_m(k)\exp(-i\omega_m(k)t + ikz)]dk\,. \tag{28.9}$$

Let us now turn to the general solution to inhomogeneous equation (3.12). Substituting expression (28.3) into relationship (28.5), we get

$$A(t,z) = \frac{1}{2\pi}\int_{-\infty}^{+\infty} \int_0 G(\tau,k)F(t-\tau,k)\exp(ikz)d\tau dk\,. \tag{28.10}$$

In order to transform the integral in (28.10), we use the convolution theorem (which is analogous to property 4 of the Laplace transformation). This theorem tells us that, if the functions $\varphi_{1,2}(k)$ are the Fourier transforms of the functions $\varphi_{1,2}(z)$, then the product $\varphi_1(k)\varphi_2(k)$ is the Fourier transform of the convolution, specifically,

$$\int_{-\infty}^{+\infty} \varphi_1(z-x)\varphi_2(x)dx = \int_{-\infty}^{+\infty} \varphi_1(x)\varphi_2(z-x)dx$$

$$= \frac{1}{2\pi}\int_{-\infty}^{+\infty} \varphi_1(k)\varphi_2(k)\exp(ikz)dk\,. \qquad (28.11)$$

Applying formula (28.11) to relationship (28.5) yields

$$A(t,z) = \int_0^t d\tau \int_{-\infty}^{+\infty} dx G(\tau,x)F(t-\tau,z-x)\,, \qquad (28.12)$$

where $F(t,z)$ is the right-hand side of inhomogeneous equation (3.12) and

$$G(t,z) = \frac{1}{2\pi}\int_{-\infty}^{\infty} G(t,k)\exp(ikz)dk$$

$$= \frac{1}{(2\pi)^2}\int_{-\infty}^{+\infty} dk \int_{C(\omega)} \frac{1}{D(\omega,k)}\exp(-i\omega t + ikz)d\omega\,. \qquad (28.13)$$

From relationship (28.5) we can see that function (28.13) is the inverse Fourier transform of the unit point source function $G(t,k)$ (28.4).

Let us now show that function (28.13) is a solution to the following inhomogeneous differential equation with a special right-hand side, supplemented with zero initial conditions:

$$D(\hat{\omega},\hat{k})G(t,z) = \delta(t)\delta(z)\,. \qquad (28.14)$$

We multiply both sides of Eq. (28.14) by $\exp(i\omega t - ikz)$ and integrate the result over t from zero to infinity and over z from minus to plus infinity. Taking into account the properties of the delta function, the definition of the Laplace and Fourier transforms, and the homogeneity of the initial conditions for Eq. (28.14), we find

$$D(\omega,k)G(\omega,k) = 1\,, \qquad (28.15)$$

where the solution $G(\omega,k)$ to Eq. (28.14) is Laplace transformed in t and Fourier transformed in z. Applying the inverse transformations (4.8) and (28.5) to the function $G(\omega,k) = D^{-1}(\omega,k)$ leads to an expression that coincides with (28.13). Hence, the function $G(t,z)$ (28.13) is that of a unit point source, instantaneous in time and zero-dimensional in space, because the right-hand side of Eq. (28.14) is the product of the delta functions, $\delta(t)\delta(z)$. This function is usually called Green's function. The explicit form of the Green's function is determined only by the roots of the dispersion relation $D(\omega,k) = 0$. Consequently, Green's function characterizes the physical system itself, rather than the method of producing perturbations in the system.

Let us also refine the formulation of the initial-value problem of determining the characteristic function of the state vector of an initial perturbation that depends arbitrarily on the spatial coordinate z. We begin with expression (28.9), which gives the characteristic function as a solution to homogeneous equation (3.11). In this expression, it is necessary to determine the quantities $A_m(k)$, $m = 1, 2, \ldots, n$. Doing so requires that we know the initial conditions.

Assume that, at the initial instant $t = 0$, the characteristic function and its derivatives of orders $s = 1, 2, \ldots, n-1$ are expressed by the formulas

$$A(t,z)|_{t=0} \equiv A_0^{(0)}(z) = \frac{1}{2\pi}\int_{-\infty}^{+\infty} A_0^{(0)}(k)\exp(ikz)dk\,,$$

$$\left.\frac{d^{(s)}A(t,z)}{dt^{(s)}}\right|_{t=0} \equiv A_0^{(s)}(z) = \frac{1}{2\pi}\int_{-\infty}^{+\infty} A_0^{(s)}(k)\exp(ikz)dk,\ \ s = 1, 2, \ldots, n-1\,. \tag{28.16}$$

Here, $A_0^{(s)}(z)$ are given functions of the z coordinate; $A_0^{(s)}(k)$ are their Fourier transforms ($s = 0, 1, 2, \ldots, n-1$); and the number of eigenmode branches, n, is equal to the order of differential equations (1.1) in t. We substitute solution (28.9) into relationships (28.16), multiply both their sides by $\exp(-ikz)$, integrate them over z from minus to plus infinity, and take into account formula (28.6). As a result, we obtain the following set of equations for the functions $A_m(k)$:

$$\sum_{m=1}^{n}(\omega_m(k))^s A_m(k) = (i)^s A_0^{(s)}(k), \quad s = 0, 1, 2, \ldots, n-1\,. \tag{28.17}$$

The principal determinant of the set of Eqs. (28.17), called the Vandermonde determinant, is calculated from the following formula of linear algebra:

$$D_V = \begin{Vmatrix} 1 & 1 & \cdots & 1 \\ \omega_1 & \omega_2 & \cdots & \omega_n \\ \omega_1^2 & \omega_2^2 & \cdots & \omega_n^2 \\ \cdots & \cdots & \cdots & \cdots \\ \omega_1^{n-1} & \omega_2^{n-1} & \cdots & \omega_n^{n-1} \end{Vmatrix}$$

$$= (\omega_2 - \omega_1)(\omega_3 - \omega_1)\cdots(\omega_n - \omega_1)(\omega_3 - \omega_2)\cdots(\omega_n - \omega_2)\cdots(\omega_n - \omega_{n-1})\,. \tag{28.18}$$

If there are no multiple roots among the roots $\omega_m(k)$ of the dispersion relation, then the determinant is nonzero, $D_V \neq 0$. In this case, Eqs. (28.17) have a unique solution, which can be written as

$$A_m(k) = D_V^{-1}\sum_{S=1}^{n}(-1)^{s+m}(i)^{s-1}\tilde{D}_{sm}A_0^{(s-1)}(k), \quad m = 1, 2, \ldots, n\,. \tag{28.19}$$

Here, $\tilde{D}_{sm}$ is the algebraic supplement of the element at the intersection of the sth row and the mth column of the Vandermonde determinant (28.18). It can be shown that formulas (28.19) and (4.21) are identical.

Note that, in solving particular problems, the initial conditions are usually formulated not for the characteristic function and its derivatives, which are somewhat abstract quantities, but for the state vector components, which have a clear physical meaning. That is, in place of conditions (28.16), it is necessary to use the initial conditions

$$\begin{pmatrix} 1 \\ L_2(\hat{\omega}, \hat{k}) \\ \vdots \\ L_n(\hat{\omega}, \hat{k}) \end{pmatrix} A(t, z)|_{t=0} = \begin{pmatrix} b_1(z) \\ b_2(z) \\ \vdots \\ b_n(z) \end{pmatrix}, \tag{28.20}$$

which generalize conditions (4.14) to a nonharmonic initial perturbation. We carry out the same manipulations with conditions (28.20) as those done with conditions (28.16) in deriving Eqs. (28.17) and rewrite the former for the spatial Fourier transforms of $A(0, k)$ and of the quantities

$$b_s(k) = \int_{-\infty}^{+\infty} b_s(z) \exp(-ikz)dz, \quad s = 1, 2, \dots, n. \tag{28.21}$$

The result is

$$\begin{pmatrix} 1 \\ L_2(\hat{\omega}, \hat{k}) \\ \vdots \\ L_n(\hat{\omega}, \hat{k}) \end{pmatrix} A(t, k)|_{t=0} = \begin{pmatrix} b_1(k) \\ b_2(k) \\ \vdots \\ b_n(k) \end{pmatrix}, \tag{28.22}$$

where the function $A(0, k)$ is given by formula (28.2b) with $t = 0$. The initial conditions (28.22) for the Fourier transforms of the state vector components are analogues of conditions (28.16) for the Fourier transforms of the characteristic function and its derivatives. It is necessary that initial conditions (28.16) and (28.22) be mathematically equivalent. That this ought to be the case stems from the fact that the operators $L_s(\hat{\omega}, k)$, $s = 1, 2, \dots, n$, contain operations of differentiation with respect to time in all orders from 0 to $n - 1$. This latter assertion is generally difficult to prove, but its validity is confirmed by the differential equations for the characteristic function and by the corresponding expressions for the state vectors of the particular physical systems that have been considered in Chapter 2. Expressing the functions $A_0^{(s)}(k)$, $s = 0, 1, 2, \dots, n - 1$, from initial conditions (28.22), we again arrive at initial conditions (28.16). Hence, in any formulation of the initial conditions, the integrands in solutions (28.8) and (28.9) are determined uniquely.

29. Quasi-Harmonic Approximation. Group Velocity

Here, we investigate solution (28.9). We assume that the functions $A_m(k)$ have been found from the initial conditions and that the functions $|A_m(k)|$ satisfy all the requirements for the existence of the integral in this solution (e.g., the functions approach zero as $k \to \pm\infty$). The integrand in (28.9) is a sum of equivalent terms. It is therefore clear that, in order to investigate the properties of the integral, it is sufficient to consider only one of the terms, which corresponds to a certain fixed branch of the eigenfrequencies. That is why, for the moment, we represent the solution to the initial-value problem as

$$A(t,z) = \frac{1}{2\pi}\int_{-\infty}^{\infty} A(k)\exp(-i\omega(k)t + ikz)dk\,, \tag{29.1}$$

where we have omitted the number m of the branch for simplicity in writing the formulas to follow. The task of investigating the integral in solution (29.1) runs into serious difficulties, the main of which is that the exponential contains the function $\omega(k)$, generally complex and nonlinear in k. In addition, the integrand in solution (28.9) oscillates increasingly rapidly as t increases.

Setting $t = 0$ in solution (29.1) we get

$$A(0,z) = \frac{1}{2\pi}\int_{-\infty}^{+\infty} A(k)\exp(ikz)dk \equiv A_0(z)\,, \tag{29.2}$$

where we have introduced the notation $A_0(z)$ for the initial perturbation of the characteristic function. Applying the Fourier transformation to (29.2), we find the Fourier transformed initial perturbation:

$$A(k) = \int_{-\infty}^{+\infty} A_0(z)\exp(-ikz)dz \equiv A_0(k)\,. \tag{29.3}$$

The function $A_0(k)$ is called the spectral density of the perturbation. One notable property of the functions $A_0(z)$ and $A_0(k)$ is expressed by the relationship

$$\Delta z \cdot \Delta k \sim 2\pi\,. \tag{29.4}$$

Here, Δz is the interval on the coordinate axis z, $z \in (-\infty, +\infty)$, where the perturbation is localized and Δk is the interval on the wavenumber axis k, $k \in (-\infty, +\infty)$, where the spectral density of the perturbation is localized. By the localization region of a function $f(z)$ is meant a continuous region — an interval on the number axis (the z axis) — within which the absolute value of the function, $|f(z)|$, exceeds a certain minimum value (say, the half-height). For any $t \geq 0$, functions (28.5) and (28.7), too, possess this same property (29.4).

Property (29.4) is a mere mathematical consequence of relationships (29.2) and (29.3). In quantum physics, this property is known as the Heisenberg uncertainty principle. Its physical meaning is as follows: the more harmonic the perturbation, the wider the spatial region where it is localized, and vice versa. Let us consider as an example a so-called delta-shaped initial perturbation, i.e., a point perturbation localized within an infinitely narrow interval on the spatial coordinate axis z: $A_0(z) = C_0\delta(z - z_0)$, where z_0 and C_0 are constants. From relationship (29.3), we have $A_0(k) = C_0\exp(-ikz_0)$ and therefore $|A_0(k)| = C_0 = \text{const}$, so the width of the spectral density of the perturbation along the wavenumber axis k is infinitely large. Conversely, let the width of the spectral density along the wavenumber axis k be infinitely small, $A_0(k) = 2\pi C_0\delta(k - k_0)$, where k_0 is a constant. In this case,

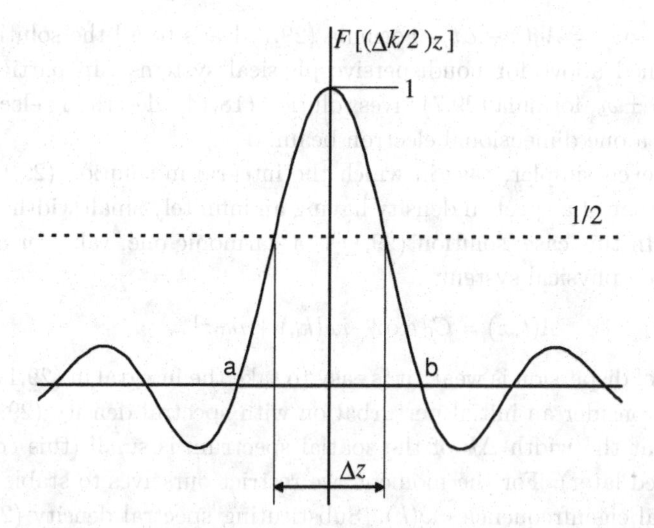

Fig. 27. Envelope of perturbation (29.6). The perturbation is localized within the interval Δz.

according to relationship (29.2), we deal with a plane wave $A_0(z) = C_0 \exp(ik_0 z)$ having an infinite spatial width.

Consider another example. Let the spectral density be constant within a finite range Δk of wavenumbers k and be identically zero outside this range,

$$A_0(k) = C_0 \frac{2\pi}{\Delta k} \begin{cases} 1, & k \in [k_0 - \Delta k/2,\ k_0 + \Delta k/2], \\ 0, & k \notin [k_0 - \Delta k/2,\ k_0 + \Delta k/2]. \end{cases} \tag{29.5}$$

Substituting (29.5) into formula (29.2) and performing simple integration, we obtain the following expression for the perturbation $A_0(z)$:

$$A_0(z) = F[(\Delta k/2)z] \cdot C_0 \exp(ik_0 z), \quad F(\xi) = \frac{\sin(\xi)}{\xi}. \tag{29.6}$$

Perturbation (29.6) is localized in the vicinity of the main maximum of the envelope $F[(\Delta k/2)z]$ at $z = 0$ (see Fig. 27). At the half-height of the main maximum, the width of the localization region is close to half the distance between the roots of the function $F[(\Delta k/2)z]$ at $z = \pm 2\pi/\Delta k$ (see Fig. 27, points a, b). The consequence is again relationship (29.4).

For zero dispersion, the integral in solution (29.1) can be taken exactly. Inserting the eigenfrequency $\omega(k)$ given by formula (1.14) into solution (29.1) yields

$$A(t, z) = f(z - \alpha t) \exp(-i\beta t), \tag{29.7}$$

where

$$f(z - \alpha t) = \frac{1}{2\pi} \int A(k) \exp(-ik(z - \alpha t))dk. \tag{29.8}$$

There is no need to calculate the integral in (29.8). In fact, setting $t = 0$ in formula (29.7) and comparing the result with (29.2), we find $f(z) = A(0, z) =$

$A_0(z)$, i.e., $f(z-\alpha t) = A_0(z-\alpha t)$. Formula (29.7) leads to all the solutions that have been obtained above for nondispersive physical systems. In particular, for $\alpha = U_{0b}$ and $\beta = \pm\omega_b$, formula (29.7) gives solution (18.14), describing electrostatic perturbations in a one-dimensional electron beam.

The second, even simpler, case in which the integral in solution (29.1) can be taken exactly is that of a spectral density having an infinitely small width, $A_0(k) = 2\pi C_0 \delta(k-k_0)$. In this case, solution (29.1) is a harmonic one, valid for any wave dispersion law in a physical system:

$$A(t,z) = C_0 \exp[-i\omega(k_0) + ik_0 z] . \qquad (29.9)$$

When the effect of dispersion is weak, it is easy to take the integral in (29.1) approximately. Let us consider an initial perturbation with spectral density (29.5) under the condition that the width Δk of the spatial spectrum is small (this conditions will be formulated later). For the moment, we restrict ourselves to stable physical systems with real eigenfrequencies $\omega(k)$. Substituting spectral density (29.5) into solution (29.1), we find

$$A(t,z) = C_0 \frac{1}{\Delta k} \int_{k_0-\Delta k/2}^{k_0+\Delta k/2} \exp[-i\omega(k)t + ikz] dk . \qquad (29.10)$$

We expand the frequency $\omega(k)$ in a Taylor series around the central point k_0 of spectral density (29.5):

$$\omega(k) = \omega(k_0) + (k-k_0)\left(\frac{d\omega}{dk}\right)_{k=k_0} + \frac{1}{2}(k-k_0)^2 \left(\frac{d^2\omega}{dk^2}\right)_{k=k_0} + \cdots . \qquad (29.11)$$

Since spectral density (29.5) has a small width, we can take the integral in expression (29.10) by keeping only the first two terms in expansion (29.11). As a result, we obtain

$$A(t,z) = F[(\Delta k/2)(z - V_g t)] \cdot C_0 \exp[-i\omega(k_0)t + ik_0 z] , \quad F(\xi) = \frac{\sin(\xi)}{\xi} , \qquad (29.12)$$

where

$$V_g = \left(\frac{d\omega}{dk}\right)_{k=k_0} \qquad (29.13)$$

is the quantity that has came to be called the group velocity.

Solution (29.12) describes a harmonic wave (29.9) with the amplitude modulated envelope of the function $F(\xi) = F\lfloor(\Delta k/2)(z - V_g t)\rfloor$. Under the inequality

$$|\Delta k/k_0| \ll 1 , \qquad (29.14)$$

the characteristic spatial scale on which the envelope varies (on the order of Δk^{-1}) is large in comparison with the length of the harmonic wave (29.9) (on the order of k_0^{-1}). That is the envelope $F(\xi)$ can be considered as the slow amplitude of a nonharmonic perturbation. For $\Delta k \to 0$, the function $F(\xi)$ approaches unity and solution (29.12) goes over to the harmonic wave (29.9).

For $t = 0$, formula (29.12) reduces to (29.6). Therefore, perturbation (29.12) can be treated as the endpoint of the time evolution of an initial perturbation (29.6). In this context, it should be kept in mind that formula (29.6) is the exact one, while solution (29.12) has been obtained by retaining only the first two terms in expansion (29.11). Consequently, formula (29.12) has a limited range of applicability (which will be determined below).

Solution (29.12) partially answers the question of what is the propagation velocity of wave perturbations in dispersive systems and media. Earlier, we have demonstrated (with sufficient examples) that, in the general case, the phase velocity of a harmonic wave does not characterize the propagation velocity of perturbations in a medium. Moreover, the very question of the propagation velocity of a purely harmonic perturbation is incorrect. In fact, consider the perturbation energy as a physical quantity transported by a perturbation. The energy of perturbations in a medium is determined by a certain linear combination of the squares of the state vector components, a combination that can be expressed as the quantity (see also Sec. 25)

$$W(t,z) = |A(t,z)|^2 , \tag{29.15}$$

where $A(t,z)$ is the characteristic function of the state vector. For a harmonic wave (29.9), the quantity (29.15) is (identically) constant, so we deal with a spatially uniform energy state against the background of which the energy is not transported in actuality. This indicates that, for a plane layer $z_0 \leq z \leq z_0 + \Delta z$ in the medium, the energy flux in a harmonic wave through the layer boundaries $z = z_0$ and $z = z_0 + \Delta z$ is the same for any z_0 and Δz (the layer as a kind of measurement unit). For a purely harmonic wave, only the propagation velocity of the constant-phase points (1.5), i.e., the phase velocity (1.6), is meaningful. As for the physically measurable quantities (such as amplitudes, energies, momenta, etc.), they are not transported by a harmonic wave.

For a nonharmonic solution (29.12), the quantity (29.15) is given by the formula

$$W(t,z) = \left| C_0 \frac{\sin[(\Delta k/2)(z - V_g t)]}{[(\Delta k/2)(z - V_g t)]} \right|^2 , \tag{29.16}$$

which describes a spatial formation propagating along the z axis with the group velocity V_g. The envelope of solution (29.12) propagates with the same velocity. From formula (29.16) we can see that the points $W(t,z) = \text{const}$, in particular, the maximum $W(t,z) = C_0^2$, also propagate with the velocity V_g. It is the group velocity V_g that is considered as the propagation velocity of nonharmonic wave perturbations in dispersive systems. Note that the shape of the envelope (29.12) of a perturbation propagating with the group velocity remains unchanged.

Solution (29.12) and the notion of the group velocity V_g can also be derived by considering a perturbation with a more general spectral density than that given by formula (29.5). Let $A_0(k)$ be a finite function, i.e., a function that is identically zero outside a finite interval of values of its argument. Specifically,

$$A_0(k) = C_0 \frac{2\pi}{\Delta k}\varphi(k)\begin{cases} 1, & k \in [k_0 - \Delta k/2,\ k_0 + \Delta k/2], \\ 0, & k \notin [k_0 - \Delta k/2,\ k_0 + \Delta k/2]. \end{cases} \tag{29.17}$$

Here, $\varphi(k)$ is a continuous function on the closed interval $[k_0 - \Delta k/2,\ k_0 + \Delta k/2]$. Without any loss of generality, we can obviously set $|\varphi(k)|_{\max} = 1$ on this interval, and it is not necessary that the central point $k = k_0$ be the maximum point of the function $\varphi(k)$. If the width of spectrum (29.17) is small (the corresponding condition will be presented below), then the integral in (29.1) can be calculated with the same accuracy as solution (29.12). The result is

$$A(t, z) = \Phi\lfloor\Delta k(z - V_g t)\rfloor \cdot C_0 \exp[-i\omega(k_0)t + ik_0 z]\,,$$

$$\Phi(\lambda) = \int_{-1/2}^{+1/2} \varphi(k_0 + \Delta kq)\exp(i\lambda q)dq\,, \quad \lambda = \Delta k(z - V_g t)\,. \tag{29.18}$$

Here, the envelope $\Phi\lfloor\Delta k(z - V_g t)\rfloor$, too, propagates in space with the phase velocity (29.13) without distortion. In contrast to solution (29.12), the envelope in the solution (29.18) is a fairly arbitrary (see below) function of the argument $z - V_g t$. For $\varphi = \text{const}$, we have $\Phi(\lambda) = F(\lambda/2)$, where the function F is defined in (29.12). In this case, solution (29.18) goes over to solution (29.12).

Perturbations (29.12) and (29.18) are called wave packets or wave pulses. The approach used to obtain solutions (29.12) and (29.18) is called the quasi-harmonic approximation. Hence, in the quasi-harmonic approximation, waves exist as wave packets propagating with group velocities without distortion of the shape of their envelopes. The applicability condition of the quasi-harmonic approximation will be formulated below.

Let us calculate the asymptotic behavior of the function $\Phi(\lambda)$ in (29.18) for large λ values. To do this, we turn to the Riemann–Lebesgue lemma, which states that if $\tilde{\varphi}(q)$ is a piecewise constant function on a closed interval $[a, b]$, then

$$\int_a^b \tilde{\varphi}(q)\exp(i\lambda q)dq \to 0 \quad \text{for } |\lambda| \to \infty\,. \tag{29.19}$$

Integrating by parts once, we convert the expression for $\Phi(\lambda)$ into the form

$$\Phi(\lambda) = \frac{i}{\lambda}(S + \varepsilon)\,,$$

$$S = \varphi(k_0 - \Delta k/2)\exp(-i\lambda/2) - \varphi(k_0 + \Delta k/2)\exp(i\lambda/2)\,, \tag{29.20}$$

$$\varepsilon = \int_{-1/2}^{+1/2} \varphi'(k_0 + \Delta kq)\exp(i\lambda q)dq\,.$$

Since $|\varphi(k)|_{\max}$, the quantity $|S|$ in (29.20) does not exceed two. Since φ' is the derivative of a continuous function φ and, as such, satisfies the conditions of the Riemann–Lebesgue lemma (namely, the function φ' is piecewise constant), we have $\varepsilon \to 0$ for $|\lambda| \to \infty$. Consequently, for large λ values, we find

$$\Phi(\lambda) \sim \lambda^{-1}\,. \tag{29.21}$$

This result enables us to determine how the envelopes of wave packets (29.12) and (29.18) behave at large distances from their central point, which moves according to the law $z = V_g t$. In fact, since $\lambda = \Delta k(z - V_g t)$, from asymptotic expression (29.21) we get

$$\Phi[\Delta k(z - V_g t)] \sim \frac{1}{\Delta k(z - V_g t)}, \quad \text{for } |z - V_g t| \gg \frac{1}{\Delta k}. \tag{29.22}$$

Of course, solution (29.12), being a particular version of the general solution (29.18), has an envelope exhibiting the asymptotic behavior (29.22).

Hence, we have shown that, if the spectral density is a piecewise continuous, finite function of k (just like function (29.17)), then the envelope of the wave packet has the asymptotic behavior (29.22). Moreover, the asymptotic decrease according to the law z^{-1} (29.22) is most gradual for all piecewise continuous spectral densities. Recall that the function that is identically zero outside a finite interval of values of its argument is called a finite function and the function that is continuous at all the points of an interval except at a finite number of removable discontinuity points or simple discontinuity points is called a piecewise continuous function on the interval. If the spectral density $A_0(k)$ is not a piecewise continuous function but has at least one nonremovable discontinuity, then the envelope of the wave packet decreases more gradually than according to the law (29.22). On the other hand, if a piecewise continuous spectral density is not a finite function, then the envelope of the packet decreases more sharply than according to the law (29.22). For instance, for

$$A_0(k) = 2\pi C_0 \frac{1}{\pi} \frac{\Delta k}{(\Delta k)^2 + (k - k_0)^2}, \tag{29.23}$$

we deal with an exponential decrease in the envelope:

$$A(t, z) = \exp\lfloor -\Delta k |z - V_g t| \rfloor \cdot C_0 \exp[-i\omega(k_0)t + ik_0 z]. \tag{29.24}$$

A finite function is naturally the most sharply decreasing one among all the functions. In fact, instead of being decreasing, a finite function is identically zero outside a finite interval of values of its argument. Let us find out whether the evolution of a perturbation that is finite in the z direction can be described in the quasi-harmonic approximation. Note first of all that the direct and inverse Fourier transformations (28.5) and (28.7), as well as (29.2) and (29.3), are mutually complementary, i.e., symmetric in z and k. In particular, if a piecewise continuous, finite spectral density $A_0(k)$ generates an initial perturbation $A_0(z)$ that decreases as $|z|^{-1}$ at large z values, then the contrary is also true: a piecewise continuous, finite initial perturbation $A_0(z)$ has a spectral density $A_0(k)$ that decreases as $|k|^{-1}$ at large k values. But for such a spectral density, there is no compact localization interval Δk to which the integration in solution (29.1) is to be limited. The third term in expansion (29.11) in the argument of the exponential in the integrand makes a contribution on the order of $(\Delta k)^2(d^2\omega/dk^2)t$. Since the integration interval Δk is large (in fact, this is the entire number axis), the contribution $(\Delta k)^2(d^2\omega/dk^2)t$ is not small even at

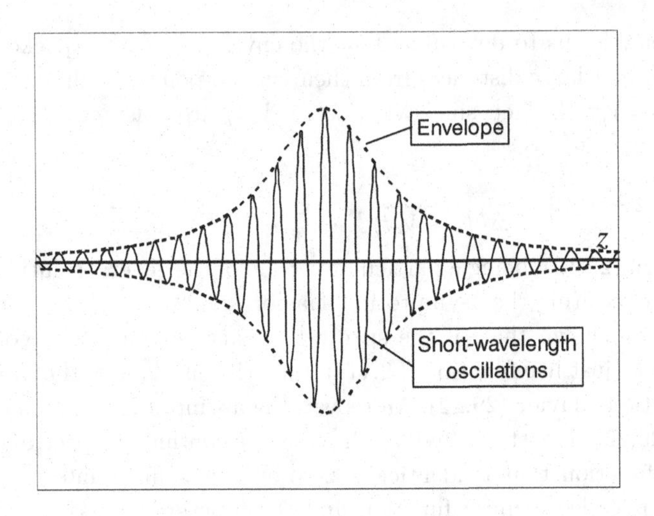

Fig. 28. Representative quasi-harmonic perturbation — wave packet.

arbitrarily small t values. Hence, the quasi-harmonic approximation is incapable of describing the evolution of a perturbation in the form of a piecewise continuous, finite function of the z coordinate. A representative wave perturbation that can be described in the quasi-harmonic approximation is shown in Fig. 28. Such a perturbation, called a wave packet (pulse), propagates with the group velocity V_g without change of form. Figure 29 exemplifies a perturbation that cannot be described in the quasi-harmonic approximation. Such a perturbation, being a spatially localized object with sharp boundaries (points A and B in Fig. 29), is called a signal. The velocity at which the entire signal (and, in particular, its boundary points) propagates in space is by no means the group velocity. The propagation of signals should be considered by retaining more than two terms in expansion (29.11). In this way, it is possible to overcome many artificial problems in the theory of waves in dispersive media, such as the problem of superluminal transport of the perturbation energy.

In the theory of quasi-harmonic waves, an approach equivalent to that just presented is based not on transforming the Fourier integral in solution (29.1) but on using Eq. (3.11) and the slow amplitude method. Knowing the dependence $\omega(k)$, we can calculate the frequency range $\Delta\omega$ corresponding to the spatial spectral interval Δk:

$$\Delta\omega \approx V_g \Delta k\,, \quad or \quad \frac{\Delta\omega}{\omega_0} \approx \frac{k_0 V_g}{\omega_0}\frac{\Delta k}{k_0}\,. \tag{29.25}$$

Here, $\omega_0 = \omega(k_0)$. If the parameter $k_0 V_g/\omega_0$ is small, then inequality (29.14) and relationship (29.25) yield

$$|\Delta\omega/\omega_0| \ll 1\,. \tag{29.26}$$

With inequalities (29.14) and (29.26), the solution to Eq. (3.11) can be represented as

$$A(t,z) = \Phi(t,z)\exp(-i\omega_0 t + ik_0 z)\,, \tag{29.27}$$

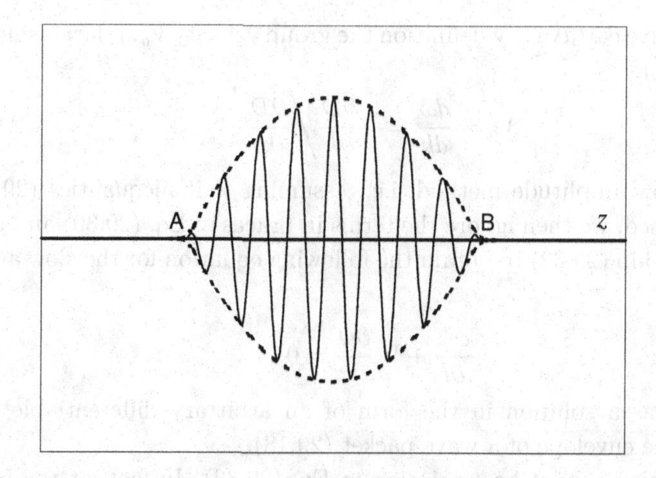

Fig. 29. Non-quasi-harmonic perturbation.

where $\Phi(t, z)$ is the slow amplitude. Moreover, since $\omega_0 = \omega(k_0)$, we have $D(\omega_0, k_0) = 0$.

According to definition (3.7) of the frequency and wavenumber operators, we can write

$$\begin{aligned}\hat{\omega}A(t,z) &= \exp(-i\omega_0 + ik_0 z)\left(\omega_0 + i\frac{\partial}{\partial t}\right)\Phi(t,z)\,,\\ \hat{k}A(t,z) &= \exp(-i\omega_0 + ik_0 z)\left(k_0 - i\frac{\partial}{\partial z}\right)\Phi(t,z)\,.\end{aligned} \tag{29.28}$$

We substitute relationships (29.28) into differential equation (3.11) and cancel the exponential factor to arrive at the following differential equation for the amplitude $\Phi(t, z)$:

$$D\left(\omega_0 + i\frac{\partial}{\partial t},\ k_0 - i\frac{\partial}{\partial z}\right)\Phi(t,z) = 0\,. \tag{29.29}$$

Like basic equation (3.11), Eq. (29.29) is exact. Expanding the dispersion function in Eq. (29.29) in a Taylor series in powers of the operators involved, we obtain the equation

$$\frac{\partial D}{\partial \omega}\frac{\partial \Phi}{\partial t} - \frac{\partial D}{\partial k}\frac{\partial \Phi}{\partial z} + i\left\{\frac{1}{2}\frac{\partial^2 D}{\partial \omega^2}\frac{\partial^2 \Phi}{\partial t^2} + \frac{1}{2}\frac{\partial^2 D}{\partial k^2}\frac{\partial^2 \Phi}{\partial z^2} - \frac{\partial^2 D}{\partial k \partial \omega}\frac{\partial^2 \Phi}{\partial t \partial z} + \mathfrak{R}\right\} = 0\,. \tag{29.30}$$

Here, the symbol $\mathfrak{R}$ denotes terms containing the third- and higher order derivatives of the function $\Phi(t, z)$ and the derivatives of the dispersion function are taken at the point (ω_0, k_0). In deriving Eq. (29.30), we have used the equality $D(\omega_0, k_0) = 0$.

In order to transform Eq. (29.30) further, we differentiate the identity $D(\omega(k), k) = 0$ with respect to the wavenumber:

$$\frac{dD}{dk} = \frac{\partial D}{\partial \omega}\frac{d\omega}{dk} + \frac{\partial D}{\partial k} = 0\,, \tag{29.31}$$

Since the derivative $d\omega/dk$ is by definition the group velocity V_g, relationship (29.31) can be rewritten as

$$V_g = \frac{d\omega}{dk} = -\frac{\partial D}{\partial k}\bigg/\frac{\partial D}{\partial \omega}\,. \tag{29.32}$$

Applying the slow amplitude method, i.e., assuming that inequalities (29.14) and (29.26) are satisfied, we then ignore the terms in braces in Eq. (29.30) and take into account relationship (29.32) to obtain the following equation for the slow amplitude $\Phi(t, z)$:

$$\frac{\partial \Phi}{\partial t} + V_g \frac{\partial \Phi}{\partial z} = 0\,. \tag{29.33}$$

This equation has a solution in the form of an arbitrary differentiable function $\Phi(z - V_g t)$ — the envelope of a wave packet (29.18).

A finite function cannot be a solution to Eq. (29.33). In fact, at the boundary values of its argument, at which the finite function Φ becomes zero (points A and B in Fig. 29), Eq. (29.33) itself is inapplicable. In the vicinities of these boundary points, it is necessary to take into account the terms in braces in Eq. (29.30). Moreover, it is these terms that predominate in the vicinities of the points in question. Hence, perturbations of the form of a signal cannot be described in terms of Eq. (29.33), as well as in the language of the integral in solution (29.18).

From general equation (29.30) we can see that Eq. (29.33) for the envelope of the wave packet is inapplicable at the points where

$$\frac{\partial D}{\partial \omega} = 0\,. \tag{29.34}$$

At these points, the group velocity V_g becomes infinite, as follows from relationship (29.32). It is evident that, at the points where equality (29.34) is satisfied, expansion (29.11) is also inapplicable, as well as the relevant solutions (29.12) and (29.18). In order to determine the type of points (ω_0, k_0) where equality (29.34) holds, we expand the dispersion function in a Taylor series:

$$D(\omega, k) = D(\omega_0, k_0) + \frac{\partial D}{\partial \omega}(\omega - \omega_0) + \frac{\partial D}{\partial k}(k - k_0) + \frac{1}{2}\frac{\partial^2 D}{\partial \omega^2}(\omega - \omega_0)^2 + \cdots\,. \tag{29.35}$$

Taking into account the equalities $D(\omega_0, k_0) = 0$ and $D(\omega, k) = 0$ and assuming that equality (29.34) is met, we obtain from series expansion (29.35) the following representation for the eigenfrequency:

$$\omega(k) = \omega_0 \pm \left(2\frac{\partial D}{\partial k}\bigg/\frac{\partial^2 D}{\partial \omega^2}\right)^{1/2} \sqrt{k - k_0} + \cdots\,. \tag{29.36}$$

We thus see that the sought-for points are the branch points of the function $\omega(k)$. It is in the vicinity of the branch points of the eigenfrequency $\omega(k)$ that the quasi-harmonic approximation is inapplicable. Near its branch point, the function $\omega(k)$ behaves approximately as shown in Fig. 30. Hence, the branch point (point O in the figure) is the stability boundary of the system, a result that also follows from

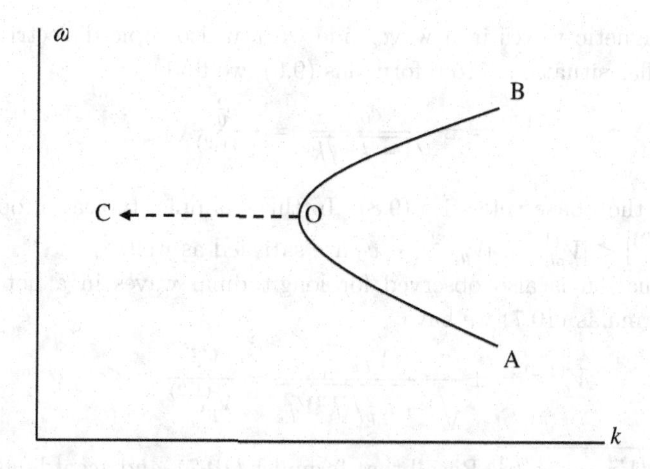

Fig. 30. Function $\omega(k)$ (curve AOB) near the branch point (ω_0, k_0) (point O): OC is the instability region.

formula (29.36). Further analysis of the branch points of the functions $\omega(k)$ will be given later.

Let us calculate the group velocities of waves in some of the physical systems considered in Chapter 2. For transverse electromagnetic waves in an isotropic dielectric, expressions (5.5) for the frequency spectra of the eigenmodes yield

$$V_g^{(1,2)} = \pm c_0 \,. \tag{29.37}$$

One can see that the group velocities are constant and coincide with the phase velocities, a simplest situation typical of acoustic waves in a nondispersive system.

For longitudinal electrostatic waves in a cold plasma, frequency spectra (6.6) give

$$V_g^{(1,2)} = 0 \,, \tag{29.38}$$

while the corresponding phase velocities can take arbitrary values, depending on the wavenumber. Thus, we again deal with the simplest example of nondispersive waves, a situation typical of longitudinal waves in a plasma and of other short-wavelength optical waves in plasmalike media.

For transverse electromagnetic waves in an isotropic plasma, from frequency spectra (7.5) we have

$$V_g^{(1,2)} = \pm \frac{c}{\sqrt{1 + \omega_p^2/k^2c^2}} = \frac{c^2}{V_{ph}^{(1,2)}} \,, \tag{29.39}$$

where $V_{ph}^{(1,2)}$ are the phase velocities (7.7). The group velocities (29.39) are always less in absolute value than the corresponding phase velocities, and, moreover, they are always less than the speed of light c. This situation is typical of optical waves in dispersive media.

For electromagnetic waves in a waveguide with an isotropic dielectric, we are faced with a similar situation. From formulas (9.6), we find

$$V_g^{(1,2)} = \pm \frac{c_0}{\sqrt{1 + k_\perp^2/k^2}} = \frac{c_0^2}{V_{ph}^{(1,2)}}\,, \tag{29.40}$$

where $V_{ph}^{(1,2)}$ are the phase velocities (9.8). In this example, typical of optics, the inequalities $|V_g^{(1,2)}| < |V_{ph}^{(1,2)}|$, $|V_g^{(1,2)}| < c_0$ are satisfied as well.

A similar situation is also observed for longitudinal waves in a hot electron plasma. From formulas (10.7) we have

$$V_g^{(1,2)} = \pm \frac{V_{Te}}{\sqrt{1 + \omega_p^2/(k^2 V_{Te}^2}} = \frac{V_{Te}^2}{V_{ph}^{(1,2)}}\,, \tag{29.41}$$

where $V_{ph}^{(1,2)} = \sqrt{V_{Te}^2 + \omega_p^2/k^2}$. Recall that formulas (10.7), and accordingly formulas (29.41), are valid only for $k^2 V_{Te}^2 < \omega_p^2$.

For ion acoustic waves in a nonisothermal plasma, formulas (12.6) lead to the following group velocities:

$$V_g^{(1,2)} = \pm \frac{V_S}{(1 + k^2 r_{D_e}^2)^{3/2}} = \frac{V_{ph}^{(1,2)}}{(1 + k^2 r_{D_e}^2)}\,. \tag{29.42}$$

For $k \to 0$, the group and phase velocities of ion acoustic waves coincide, while in the short-wavelength range, the group velocities of these waves are lower than their phase velocities. This situation is typical of acoustic (in particular, ion acoustic) waves.

As for waves in a waveguide with an anisotropic plasma, their frequencies are given by expressions (13.6). In this case, the formulas for the group and phase velocities are rather involved, so we restrict ourselves to a graphical analysis. Figure 31 shows the group (solid curves) and phase (dashed curves) velocities calculated as functions of the wavenumber k by using formulas (13.6) for a particular plasma waveguide. Curves 1g and 1ph characterize an electromagnetic wave with the frequency $\omega_1 = \Omega_1(k)$, which is an optical wave. As for curves 3g and 3ph, they characterize a plasma wave with the frequency $\omega_3 = \Omega_2(k)$, which is an acoustic wave. The numbering of the waves in the figure is the same as in Fig. 7. We can see that the group velocities of waves of any type are lower than the speed of light, $c \approx 3 \cdot 10^{10}$ cm/s, and also that the phase velocity of an optical wave is higher than c, while the phase velocity of an acoustic wave is lower than c.

For the frequencies of electromagnetic waves propagating in a magnetized plasma along the external magnetic field, we have not presented analytic formulas because, in the general case, the solution to dispersion relation (14.7) is very lengthy. The type to which such waves belong has been indicated in Sec. 14. Here, we present only the velocities of a helicon whose frequency is given by formula (14.11):

$$V_g = 2kc^2 \frac{\Omega_e}{\omega_p^2} = 2V_{ph}\,. \tag{29.43}$$

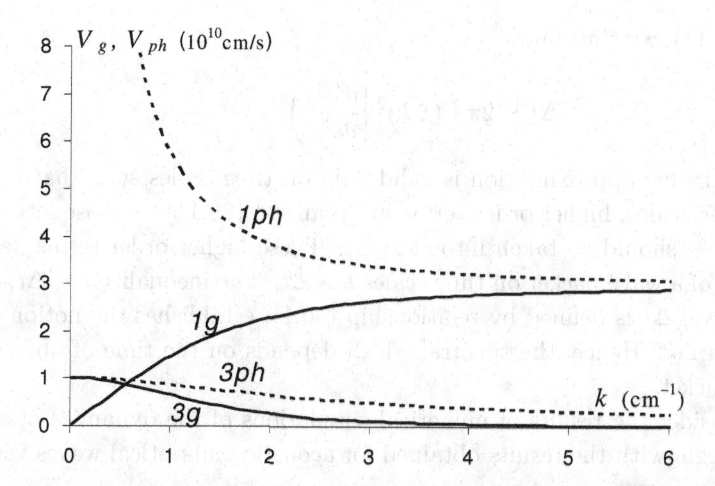

Fig. 31. Group and phase velocities of waves in a waveguide with an anisotropic plasma for $N_{0p} = 10^{11}$ cm^{-3} and $L = 2$ cm (see Fig. 7): *1g* is for the group velocity of an electromagnetic wave with the frequency ω_1, *1ph* is for the phase velocity of an electromagnetic wave with the frequency ω_1, *3g* is for the group velocity of a plasma wave with the frequency ω_3, and *3ph* is for the phase velocity of a plasma wave with the frequency ω_3.

In contrast to optical waves, for which $V_g V_{ph}$ = const (see formulas (29.39)–(29.41)), and to acoustic waves, for which V_g/V_{ph} = const < 1 (see formula (29.42)), the group and phase velocities of a helicon satisfy another relationship: V_g/V_{ph} = const > 1. Consequently, it is expedient to consider helicons as a specialized type of wave.

For longitudinal electrostatic waves in a one-dimensional electron beam, from expressions (18.5) we have

$$V_g^{(1,2)} = U_{0b} \,. \tag{29.44}$$

That is, the group velocities of the beam waves are constant (the waves are nondispersive) and coincide with the unperturbed beam velocity, while the phase velocities (see (18.7)) can take arbitrary values, depending on the wavenumber.

30. Pulse Spreading in Equilibrium Dispersive Media

In deriving solutions (29.12) and (29.18), we have used expansion (29.11) in which the third- and higher order terms have been omitted. In the initial integral in solution (29.1), the terms that have been discarded (we denote them by $\tilde{\omega}$) enter the exponential in the form of the product $i\tilde{\omega}t$. Consequently, there exists a certain time interval Δt in which the quasi-harmonic approximation is applicable. This time interval is determined from the relationship $|\tilde{\omega}|\Delta t \sim \pi$, which implies that, in the argument of the exponential, the term $i\tilde{\omega}t$ certainly cannot be ignored. With expansion (29.11), we thus find

$$\Delta t \sim 2\pi \left((\Delta k)^2 \left| \frac{d^2\omega}{dk^2} \right| \right)^{-1} . \tag{30.1}$$

The quasi-harmonic approximation is valid only on time scales such that $t < \Delta t$. For longer time scales, higher order terms in expansion (29.11) — those introduced by dispersion — should be taken into account. These higher order terms describe the spreading of a wave packet on time scales $t > \Delta t$. The inequality $t < \Delta t$, where the time interval Δt is defined by relationship (30.1), establishes the notion of the spectral width Δk. Hence, the spectral width depends on the time of observation of the wave packet.

Let us consider the results of numerical simulations of the dynamics of a wave packet. We begin with the results obtained for acoustic and optical waves with the eigenfrequency spectra

$$\omega(k) = \frac{kV_0}{\sqrt{1 + k^2V_0^2/\Omega_0^2}} \quad \text{— acoustic type}\,, \tag{30.2a}$$

$$\omega(k) = \sqrt{1 + k^2V_0^2/\Omega_0^2} \quad \text{— optical type}\,, \tag{30.2b}$$

where Ω_0 is a certain frequency and V_0 is a certain velocity. Formula (30.2a) provides a fairly good description of the dispersion of a plasma wave in a waveguide with an anisotropic plasma (see the frequency $\omega_3(k)$ in spectra (13.6)). Moreover, in the potential approximation ($\omega^2 \ll k^2c^2$), formula (30.2a) exactly describes the dispersion of a plasma wave with the frequency $\omega_3(k)$, in which case we have $\Omega_0 = \omega_p$ and $V_0 = \omega_p/k_\perp$. Also, this same formula exactly describes the dispersion of ion acoustic waves with frequency spectrum (12.6), in which case we have $\Omega_0 = \omega_i$ and $V_0 = V_S$. Formula (30.2b) describes well the dispersion of an electromagnetic wave in a waveguide with an anisotropic plasma (see the frequency $\omega_1(k)$ in spectra (13.6)). In addition, this formula is exact for transverse electromagnetic waves in an isotropic plasma (see the frequency $\omega_1(k)$ in spectra (7.5) with $\Omega_0 = \omega_p$ and $V_0 = c$) and for electromagnetic waves in a waveguide with an isotropic dielectric (see the frequency $\omega_1(k)$ in spectra (9.6) with $\Omega_0 = k_\perp c_0$ and $V_0 = c_0$).

Since waves whose dispersion is described by formulas like (30.2) are objects of various branches of physics, it is worthwhile to summarize once again the differential equations describing physical systems with such waves. For waves with dispersion law (30.2a), the equation has the form

$$\left(\frac{\partial^2}{\partial t^2} - V_0^2 \frac{\partial^2}{\partial z^2} - V_0^2 \Omega_0^{-2} \frac{\partial^2}{\partial t^2} \frac{\partial^2}{\partial z^2} \right) A(t, z) = 0\,. \tag{30.3}$$

With the corresponding values of Ω_0 and V_0, this equation describes longitudinal waves in elastic rods, shallow water waves, potential waves in a waveguide with an anisotropic plasma, ion acoustic waves in a nonisothermal plasma, and many other types of waves. In different areas of physics, this equation is known as the linear

Boussinesq equation or the Love equation. Among the equations of this type, we can also mention Eq. (13.13) for nonpotential waves in a waveguide with an anisotropic plasma. For waves with dispersion law (30.2b), we have the differential equation

$$\left(\frac{\partial^2}{\partial t^2} - V_0^2 \frac{\partial^2}{\partial z^2} + \Omega_0^2\right) A(t, z) = 0 . \tag{30.4}$$

This is the familiar Klein–Gordon equation, which we have already dealt with in considering transverse waves in an isotropic plasma (see Eq. (7.10)) and waves in a waveguide with an isotropic dielectric (see Eq. (9.11)). Equations like (30.4) describe the spatial propagation of various electromagnetic and elastic perturbations and also the free motion of a spin-zero particle in relativistic quantum mechanics.

Let the spectral density of an initial perturbation be given by the formula

$$A_0(k) = C_0 \frac{\pi}{\Delta k} \begin{cases} 1, & k \in [-\Delta k, \Delta k] , \\ 0, & k \notin [-\Delta k, \Delta k] . \end{cases} \tag{30.5}$$

We introduce the dimensionless variables and quantities

$$\tau = \Omega_0 t , \quad \xi = \frac{\Omega_0}{V_0} z , \quad a(\tau, \xi) = C_0^{-1} A(t, z) ,$$
$$x = \frac{k V_0}{\Omega_0} , \quad \Delta = (\Delta k) \frac{V_0}{\Omega_0} . \tag{30.6}$$

Substituting formulas (30.2) and (30.5) into representation (29.1) and taking into account relationships (30.6), we obtain the following dimensionless solutions to the initial-value problem:

$$a(\tau, \xi) = \frac{1}{\Delta} \int_0^{\Delta} \cos\left(\frac{x}{\sqrt{1+x^2}} \tau - x\xi\right) dx \text{ — acoustic type} , \tag{30.7a}$$

$$a(\tau, \xi) = \frac{1}{\Delta} \int_0^{\Delta} \cos(\sqrt{1+x^2} \tau - x\xi) dx \text{ — optical type} . \tag{30.7b}$$

Formulas (30.7) describe perturbations propagating in the positive direction of the z axis. Perturbations that propagate in the negative direction are not considered here because they are described by formulas similar to (30.7), but with the replacement of the minus sign in front of the term $x\xi$ with the plus sign.

The quasi-harmonic approximation is valid under the inequality $\Delta^2 \ll 1$, when it is sufficient to retain only the terms of zero and first orders in the dimensionless wavenumber x in the integrand in solutions (30.7) over the entire interval of integration. In this case, solutions (30.7) reduce to

$$a(\tau, \xi) = \frac{\sin[\Delta(\xi - \tau)]}{\Delta(\xi - \tau)} \text{ — acoustic type} , \tag{30.8a}$$

$$a(\tau, \xi) = \frac{1}{\Delta\xi} [\sin \Delta\xi \cos \tau + (1 - \cos \Delta\xi) \sin \tau] \approx \frac{\sin \Delta\xi}{\Delta\xi} \cos \tau \text{ — optical type} . \tag{30.8b}$$

Formula (30.8a) describes a wave packet with a high group velocity, while formula (30.8b) describes a wave packet with a zero group velocity. This result agrees with formulas (30.2), from which the dimensionless group velocities for $\Delta \ll 1$ can be estimated by

$$\frac{V_g}{V_0} \approx \begin{cases} 1 & \text{— acoustic type}\,, \\ x \ll 1 & \text{— optical type}\,. \end{cases} \tag{30.9}$$

According to formula (30.8a), the perturbation amplitude at the central point $\xi = \tau$ of an acoustic wave packet oscillates at a zero frequency, $a(\xi = \tau) = 1$, while the perturbation amplitude at the central point $\xi = 0$ of an optical wave packet oscillates at a unit circular frequency, $a(\tau, 0) = \cos\tau$. This result is consistent with formulas (30.2), from which the dimensionless frequencies for $\Delta \ll 1$ can be estimated by

$$\frac{\omega}{\Omega_0} \approx \begin{cases} x \ll 1 & \text{— acoustic type}\,, \\ 1 & \text{— optical type}\,. \end{cases} \tag{30.10}$$

Note that if we expand the integrand in formula (30.7b) in powers of x and keep the zero- and first-order terms in the expansion, then we arrive at the function $\cos(\tau - x\xi)$. Integrating this function leads to formula (30.8b), a consequence of the fact that, for $k = 0$, an optical wave with frequency (30.2b) has a zero group velocity.

Figure 32 shows the spatial structure of an acoustic perturbation with $\Delta = 0.5$ for several values of the dimensionless time τ (the vertical axis in the figure is the dimensionless coordinate ξ). Since $\Delta^2 = 0.25$, the quasi-harmonic approximation is satisfactory, and the figure confirms this conclusion completely. The main, central part of the wave packet propagates essentially without distortion with a dimensionless velocity close to unity (see the first of estimates (30.9)). The maximum amplitude of the packet is nearly constant (see the first of estimates (30.10)). A slight distortion of the pulse — a break in the symmetry of its leading and trailing edges — is observed only on a time scale of $\tau \sim 20$ and at later times. It is on this time scale that the quasi-harmonic approximation is applicable in the case at hand, in accordance with relationship (30.1). In general, the pulse evolution shown in Fig. 32 agrees quite well with formula (30.8a), except for some minor details.

Figure 33 shows the spatial structure of an optical perturbation with the same value $\Delta = 0.5$ for several values of the dimensionless time τ. In contrast to the previous case, agreement with the results of the quasi-harmonic approximation (see formula (30.8b)) is somewhat worse. The trailing edge of the pulse is distorted most strongly. But the main, central part of the pulse and its leading edge are again distorted slightly and propagate with a low velocity (see the second of estimates (30.9)). In addition, the amplitude of the pulse at its central point varies in an oscillatory fashion (an analysis of the results obtained with a small time step shows that the oscillation frequency agrees with the second of estimates (30.10)). It should be noted that the propagation of an optical wave pulse, illustrated in Fig. 33, is

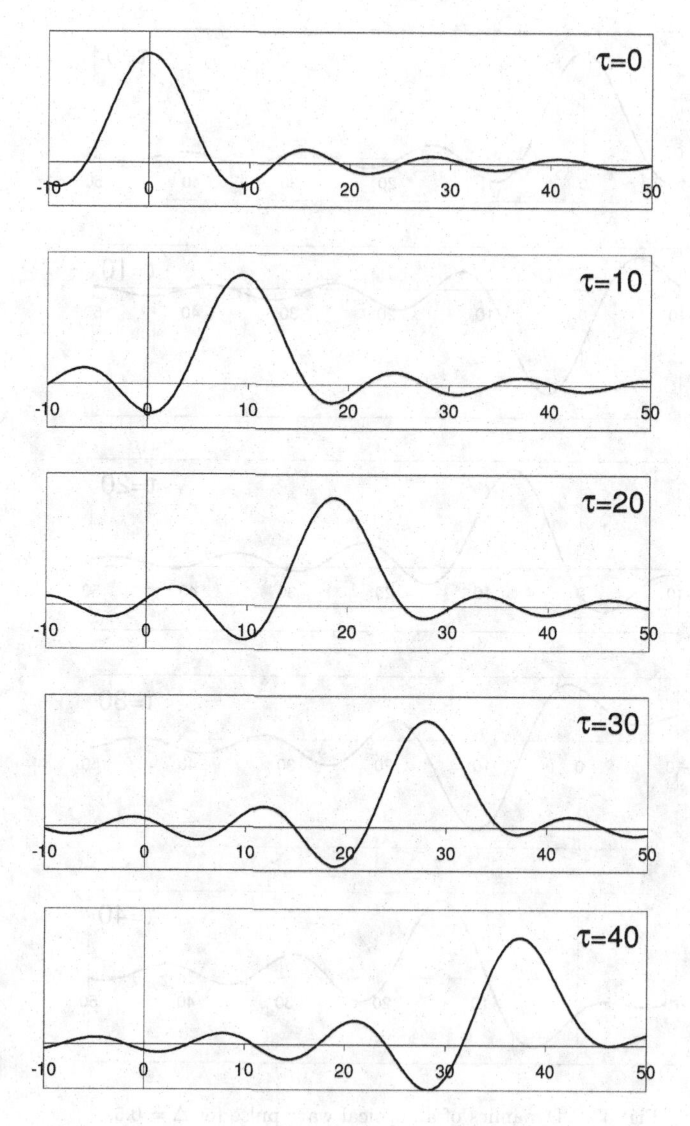

Fig. 32. Dynamics of an acoustic wave pulse for $\Delta = 0.5$.

described by the quadratic term proportional to $\sim x^2\tau$ in the expansion of the integrand in formula (30.7b) in powers of x (the linear term proportional to $\sim x\tau$ is not present in the expansion). That is, the terms describing the propagation of the pulse and the distortion of its shape are of the same order — a situation that goes beyond the applicability limits of the quasi-harmonic approximation.

Let us now consider a perturbation with a wide spectral density. To do this, we set $\Delta = 10$. For such a large Δ value, it is illegitimate to expand the integrands in formulas (30.7) in powers of x. Moreover, for $\Delta = 10$, relationship (29.4) implies

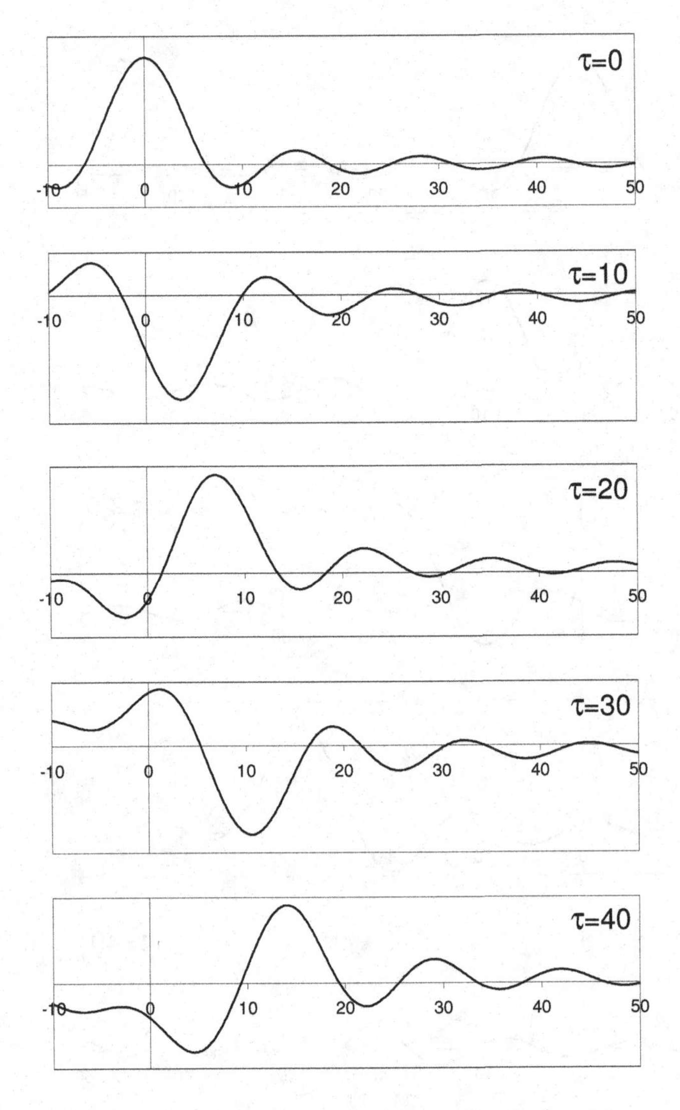

Fig. 33. Dynamics of an optical wave pulse for $\Delta = 0.5$.

that the dimensionless size of the region where the initial perturbation is localized in space is on the order of a few tenths (an almost delta-shaped initial perturbation).

Figure 34 shows the spatial structure of an acoustic perturbation with $\Delta = 10$ for several values of the dimensionless time τ. We can see that, on the whole, the main part of pulse remains immobile, with its centre in the vicinity of the point $\xi \approx 0$. The amplitude of the pulse at its central point varies in an oscillatory fashion. To the right of the main part of the pulse, a sinusoidal perturbation propagates, whose length increases and whose amplitude decreases. In the region $\xi < 0$, the structure of the perturbation remains essentially unchanged with time.

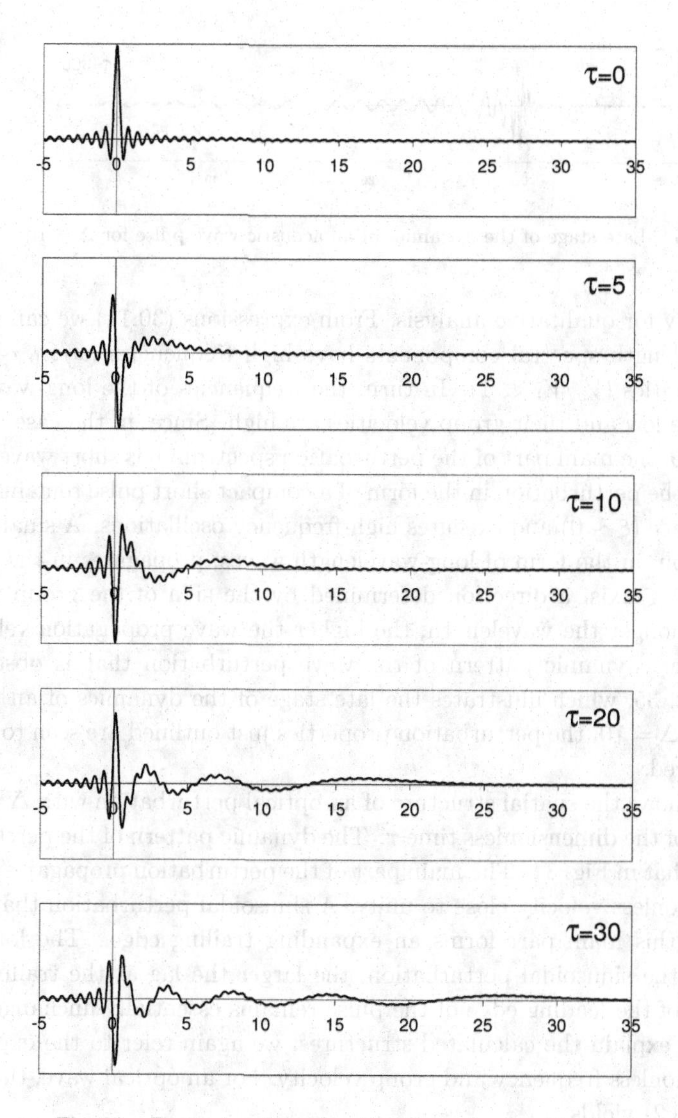

Fig. 34. Dynamics of an acoustic wave pulse for $\Delta = 10$.

In order to explain the calculated structures, we present expressions for the dimensionless frequency and group velocity of an acoustic wave:

$$\frac{\omega}{\Omega_0} = \frac{x}{\sqrt{1+x^2}}, \quad \frac{V_g}{V_0} = \frac{1}{(1+x^2)^{3/2}}. \tag{30.11}$$

The spectrum of the perturbation can be divided into two parts — long- and short-wavelength. The long-wavelength part consists of the components with wavenumbers lying in the range $0 < x < \tilde{x}$, and the short-wavelength part contains all the remaining spectral components, those with wavenumbers lying in the range $\tilde{x} < x < \Delta$. Here, $\tilde{x} \sim 1$ is a conditional wavenumber, which needs not to be

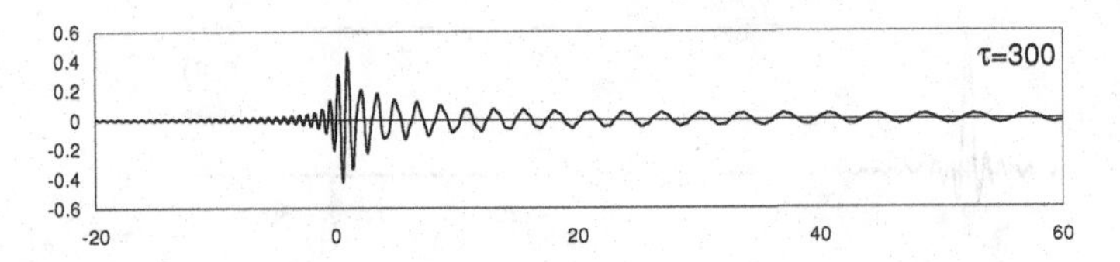

Fig. 35. Late stage of the dynamics of an acoustic wave pulse for $\Delta = 10$.

specified exactly for qualitative analysis. From expressions (30.11) we can see that the short-wavelength spectral components have high frequencies ($\omega/\Omega_0 \sim 1$) and low group velocities ($V_g/V_0 \ll 1$). In turn, the frequencies of the long-wavelength components are low and their group velocities are high. Since, in the case at hand, we have $\Delta = 10$, the main part of the perturbation spectrum has short wavelengths. Consequently, the perturbation in the form of a compact short pulse remains mostly at its initial place ($\xi \sim 0$) and executes high-frequency oscillations. A small part of the perturbation, in the form of long-wavelength waves, propagates in the positive direction of the ξ axis, a direction determined by the sign of the group velocity. Moreover, the longer the wavelength, the higher the wave propagation velocity in space. It is this dynamic pattern of the wave perturbation that is observed in Fig. 34. In Fig. 35, which illustrates the late stage of the dynamics of an acoustic wave pulse for $\Delta = 10$, the perturbation properties just outlined are seen to be even more pronounced.

Figure 36 shows the spatial structure of an optical perturbation with $\Delta = 10$ for several values of the dimensionless time τ. The dynamic pattern of the perturbation is opposite to that in Fig. 34. The main part of the perturbation propagates in space with a dimensionless velocity close to unity. A sinusoidal perturbation that is seen to lag behind this main part forms an expanding trailing edge. The longer the wavelength of the sinusoidal perturbation, the larger the lag at the trailing edge. The structure of the leading edge of the pulse remains essentially unchanged.

In order to explain the calculated structures, we again refer to the expressions for the dimensionless frequency and group velocity. For an optical wave, the second of formulas (30.2) yields

$$\frac{\omega}{\Omega_0} = \sqrt{1+x^2}\,, \quad \frac{V_g}{V_0} = \frac{x}{\sqrt{1+x^2}}\,. \tag{30.12}$$

From expressions (30.12) we can see that the short-wavelength ($\tilde{x} < x < \Delta$) spectral components have very high frequencies ($\omega/\Omega_0 \gg 1$) and very high group velocities ($V_g/V_0 \sim 1$). As for the long-wavelength ($0 < x < \tilde{x}$) spectral components, they have lower frequencies ($\omega/\Omega_0 \sim 1$) and low group velocities ($V_g/V_0 \ll 1$). Consequently, the main part of the perturbation, in the form of a compact short pulse, propagates rapidly in space with the maximum group velocity $V_g = V_0$ (so that $\xi = \tau$) and executes high-frequency oscillations. The shorter wavelength part lags behind, forming a sinusoidal trailing edge with an increasing wavelength and decreasing amplitude. Since the group velocities of optical waves do not exceed V_0,

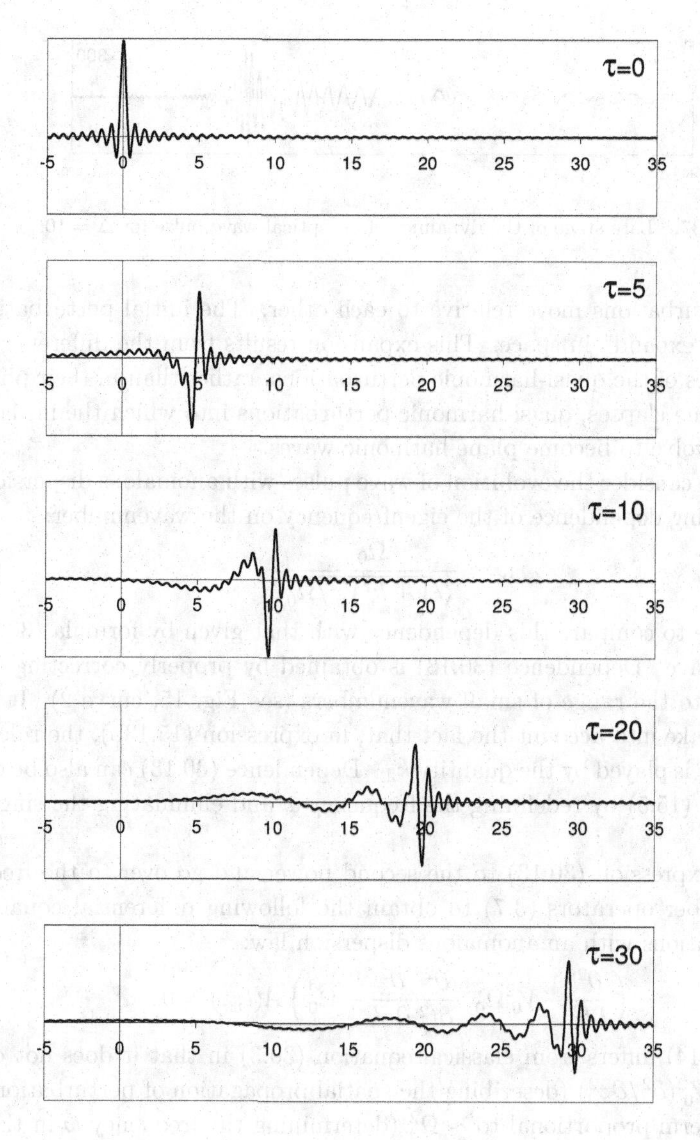

Fig. 36. Dynamics of an optical wave pulse for $\Delta = 10$.

there are no marked changes ahead of the front of the main part of the perturbation. In Fig. 37, which illustrates the late stage of the dynamics of an optical wave pulse for $\Delta = 10$, the perturbation properties just outlined are seen to be even more pronounced.

Hence, as a wave pulse propagates in a dispersive medium, it spreads out in space. The spreading mechanism can be explained as follows. Any spatially localized perturbation can be described as a superposition of less localized, quasi-harmonic perturbations. Since their phase velocities are different, the quasi-

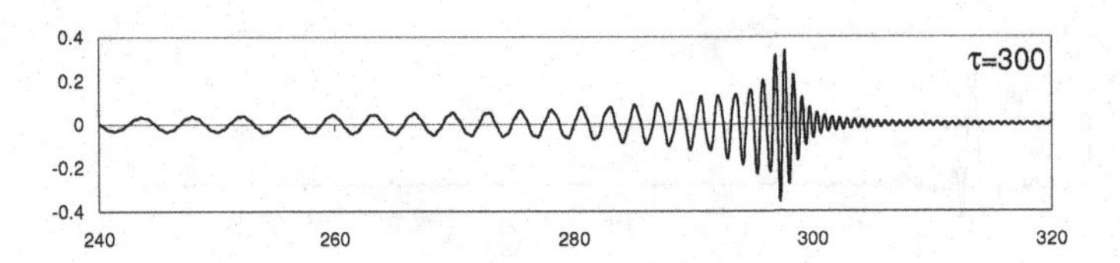

Fig. 37. Late stage of the dynamics of an optical wave pulse for $\Delta = 10$.

harmonic perturbations move relative to each other. The initial perturbation behaves as if it "expand" in space. This expansion results from the difference in the group velocities of the quasi-harmonic perturbations, rather than in their phase velocities. As time elapses, quasi-harmonic perturbations into which the initial pulse decomposes evolve to become plane harmonic waves.

Let us now consider the evolution of wave pulses with anomalous dispersion. We use the following dependence of the eigenfrequency on the wavenumber:

$$\omega = \frac{\Omega_0}{\sqrt{1 + k^2 V_0^2 / \Omega_0^2}} . \tag{30.13}$$

It makes sense to compare this dependence with that given by formula (30.2a) for an acoustic wave. Dependence (30.13) is obtained by properly correcting expression (15.13b) to the range of small wavenumbers (see Fig. 15, curve 2). It is only necessary to take into account the fact that, in expression (15.13b), the role of the wavenumber k is played by the quantity $k_\perp$. Dependence (30.13) can also be derived from formulas (15.5) by redefining the frequency ω and eliminating the singularity at $k \to 0$.

We raise expression (30.13) to the second power and go over to the frequency and wavenumber operators (3.7) to obtain the following differential equation for wave perturbations with an anomalous dispersion law:

$$\left(\frac{\partial^2}{\partial t^2} - V_0^2 \Omega_0^{-2} \frac{\partial^2}{\partial t^2} \frac{\partial^2}{\partial z^2} + \Omega_0^2 \right) A(t, z) = 0 . \tag{30.14}$$

Equation (30.14) differs from classical equation (30.3) in that it does not contain the operator $V_0^2 (\partial^2 / \partial z^2)$ (describing the spatial propagation of perturbations) and includes the term proportional to $\sim \Omega_0^2$ (determining the frequency ω in the long-wavelength range).

In terms of dimensionless variables and quantities (30.6), we write the solution to the initial-value problem for Eq. (30.14) as

$$a(\tau, \xi) = \frac{1}{\Delta} \int_0^\Delta \cos \left(\frac{\tau}{\sqrt{1 + x^2}} - x\xi \right) dx , \tag{30.15}$$

where the spectral density of the initial perturbation is assumed to be given by formula (30.5). The frequency and group velocity of the waves under consideration — those with anomalous dispersion — have the form

$$\frac{\omega}{\Omega_0} = \frac{1}{\sqrt{1 + x^2}} , \quad \frac{V_g}{V_0} = - \frac{x}{(1 + x^2)^{3/2}} . \tag{30.16}$$

In the quasi-harmonic approximation, which is valid under the inequality $\Delta^2 \ll 1$, we can retain only the zero- and first-order terms in the dimensionless wavenumber x in the integrand in solution (30.15). In this case, solution (30.15) reduces to expression (30.8b). The reason is clear: for $k = 0$, a wave with frequency (30.13) has a zero group velocity, as well as an optical wave with frequency (30.2b). Consequently, for small Δ values, the characteristic features of the propagation of a wave pulse with frequency (30.15) can only be described by taking into account the terms proportional to $\sim x^2\tau$, i.e., by going beyond the scope of the quasi-harmonic approximation.

Figure 38 shows the spatial structure of a wave pulse with anomalous dispersion and with $\Delta = 0.5$ for several values of the dimensionless time τ. The dynamic

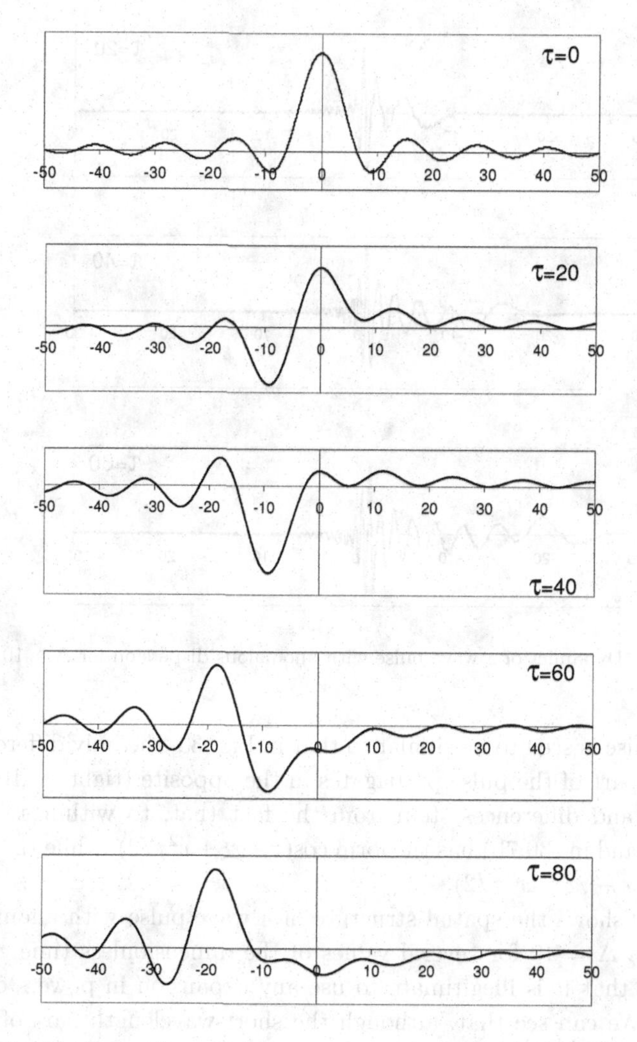

Fig. 38. Dynamics of a wave pulse with anomalous dispersion for $\Delta = 0.5$.

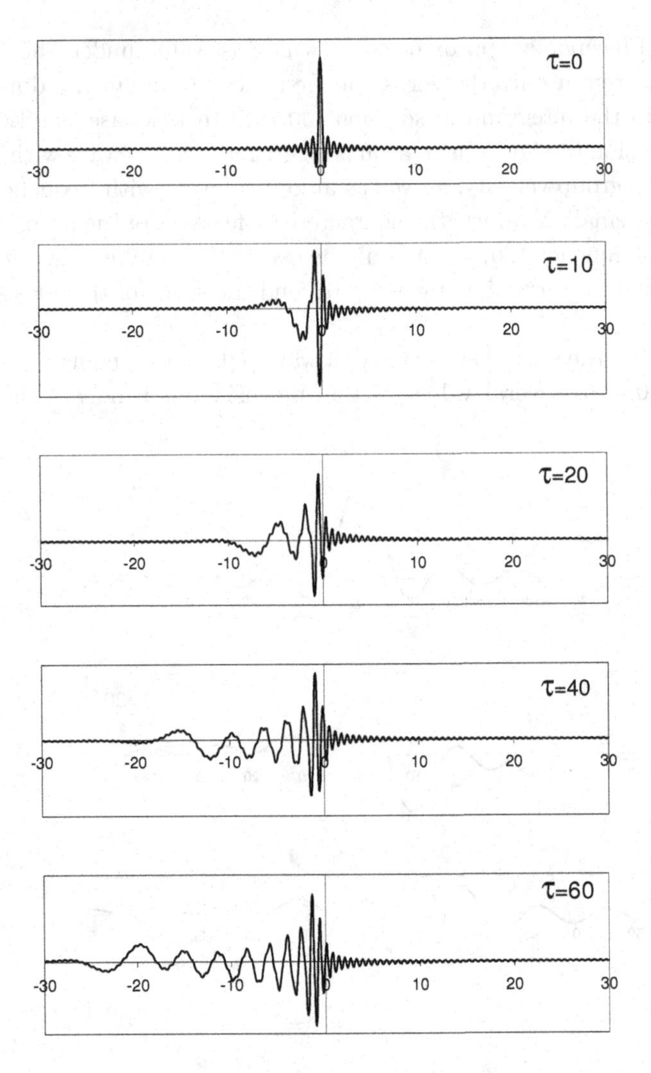

Fig. 39. Dynamics of a wave pulse with anomalous dispersion for $\Delta = 10$.

pattern of the pulse is seen to be similar to that in Fig. 33, the only difference being that the central part of the pulse propagates in the opposite (rightward) direction. The similarities and differences stem from the fact that, to within second-order terms, the integrand in (30.7b) has the form $\cos(\tau - x\xi + x^2\tau/2)$, while the integrand in (30.15) is $\cos(\tau - x\xi - x^2\tau/2)$.

Next, Fig. 39 shows the spatial structure of a wave pulse with anomalous dispersion and with $\Delta = 10$ for several values of the dimensionless time τ. This Δ value is so large that it is illegitimate to use any expansion in powers of x in solution (30.15). We can see that, although the short-wavelength part of the pulse is displaced slightly leftward, it can nevertheless be considered as remaining essen-

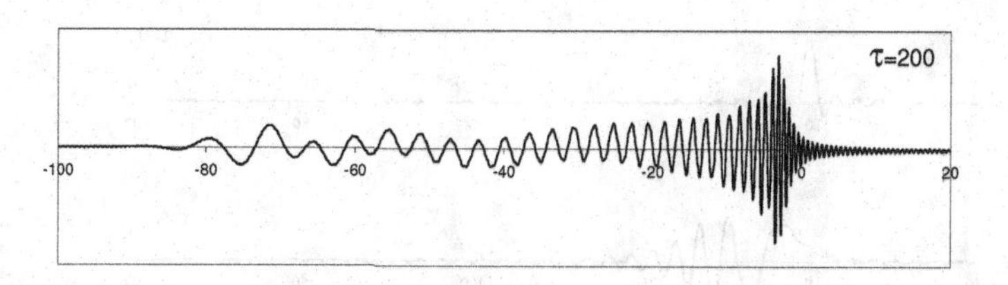

Fig. 40. Late stage of the dynamics of a wave pulse with anomalous dispersion for $\Delta = 10$.

tially in the same, initial place ($\xi = 0$). In contrast, the longer wavelength part is markedly displaced in the negative direction of the ξ axis. And finally, the longest wavelength part of the pulse is absent. In Fig. 40, which illustrates the late stage of the dynamics of a wave pulse with anomalous dispersion, the pulse properties just outlined are seen to be especially pronounced. These results are easy to understand by referring to the second of formulas (30.16). The group velocity is low in both the long-wavelength ($x \ll 1$) and short-wavelength ($x > 1$) ranges. Consequently, the long- and short-wavelength parts of the pulse remain essentially in the same places (the long-wavelength perturbations near the point $\xi = 0$ are not seen against the background of short-wavelength perturbations). The absolute value of the group velocity is maximum at $x = 1/\sqrt{2}$, a value that corresponds to a dimensionless wavelength of about $2\pi/x \approx 9$. In this case, the group velocity is $V_g/V_0 \approx -0,4$. It is the perturbations with this group velocity that move the largest distance in the negative direction of the ξ axis.

Let us also consider the dynamics of a helicon pulse. To do this, we adopt the dispersion law

$$\omega = k^2 \frac{V_0^2}{\Omega_0}, \tag{30.17}$$

in which case the spatiotemporal dynamics of the pulse is described by the formula

$$a(\tau, \xi) = \frac{1}{\Delta} \int_0^{\Delta} \cos(x^2 \tau - x\xi) dx, \tag{30.18}$$

and the dimensionless frequency and group velocity are given by the relationships

$$\frac{\omega}{\Omega_0} = x^2, \quad \frac{V_g}{V_0} = 2x. \tag{30.19}$$

It is obviously meaningless (ineffective) to expand the integrand in formula (30.18) in powers of x. The reason is that x^2 is a sharply increasing function and integrals of rapidly oscillating functions of the form $\cos(x^2\tau)$ are difficult to calculate. That is why, in formula (30.18), we set $\Delta = 1$. Figure 41 illustrates the results of calculating the pulse dynamics from formula (30.18). In complete agreement with the second of formulas (30.19), shorter wavelength perturbations propagate with higher velocities: the shorter the wavelength, the higher the propagation velocity. From the first of the formulas, we can see that the shorter the wavelength of the perturbations, the higher their frequency. Consequently, the higher the frequency

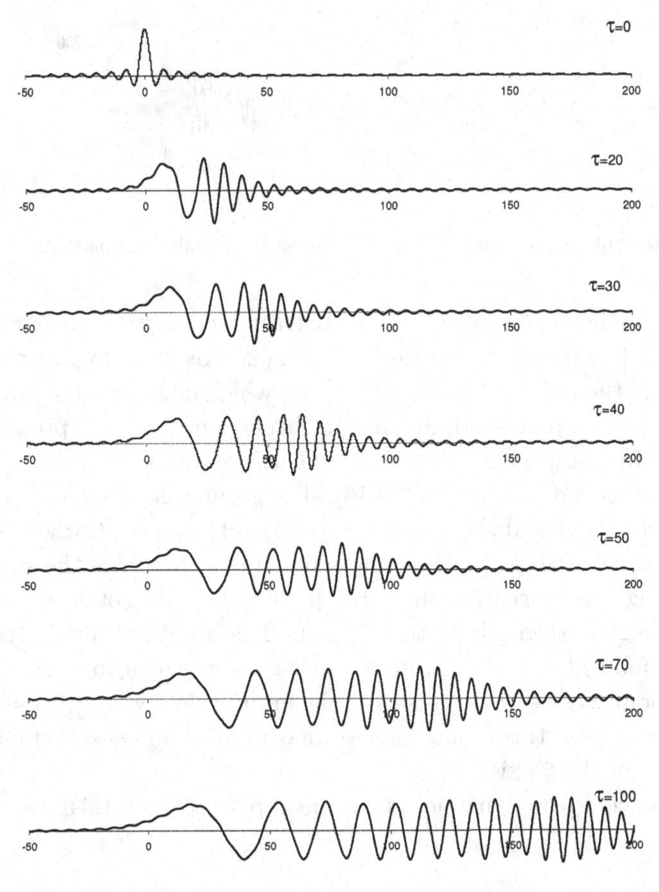

Fig. 41. Dynamics of a helicon pulse.

of the perturbations, the higher their propagation velocity. It is for this reason that radio-frequency helicons in the Earth's ionosphere came to be called whistling atmospherics.

31. Stationary-Phase Method

The asymptotic behavior of a wave packet (pulse) at $t \to \infty$ can be investigated analytically by using the familiar stationary-phase method. This method is intended to estimate integrals of the form

$$I(\lambda) = \int_a^b P(q) \exp[i\lambda\Psi(q)]dq \tag{31.1}$$

for $\lambda \to \infty$. Here, λ is a real parameter; $P(q)$ is a continuous function, $\Psi(q)$ is a twice continuously differentiable real function; and the limits of integration, a and b, can be finite or infinite. For large λ values, the integrand in formula (31.1) is a rapidly oscillating function and the oscillations cancel each other over most of the integration interval. It can therefore be expected that, as the parameter λ increases, integral (31.1) approaches zero sufficiently rapidly. Because of symmetry breaking, the oscillations of the integrand are not cancelled out at the points a and b (if they are not at infinity). Also, they are not cancelled out in the vicinities of the zeros of the derivative $\Psi'(q)$. In the vicinities of these points (called stationary points), at which $\Psi'(q) = 0$, the phase $\Psi(q)$ varies slightly. Consequently, the oscillations of the integrand are synchronous and, being added together, make a large contribution to integral (31.1). The method for estimating the contribution from the stationary points to the integral came to be known as the stationary-phase method.

We begin by considering the case in which there are no stationary points, so we have $\Psi'(q) \neq 0$ over the entire integration interval. Introducing the new variable

$$x = \Psi(q), \tag{31.2}$$

we reduce integral (31.1) to

$$I(\lambda) = \int_{\Psi(a)}^{\Psi(b)} (P(x)/\Psi'(x)) \exp(i\lambda x) dx\,. \tag{31.3}$$

Since $\Psi'(x) \notin 0$, the integrand in (31.3) is continuous over the integration interval. Integrating by parts once and applying the Riemann–Lebesgue lemma (in the same way as in deducing the asymptotic behavior of the function $\Psi(\lambda)$ (29.18)), we find $I(\lambda) \sim \lambda^{-1}$, $\lambda \to \infty$ (provided of course that there are no stationary points). It will be clear later that, in the presence of stationary points, quantities of this order of magnitude can be ignored quite legitimately.

Let us determine the contribution of the stationary points to integral (31.1). Assume that there is only one stationary point, $q = q_0 \in (a, b)$. At this point, we have $\Psi'(q_0) = 0$. In the vicinity of the stationary point, the integrand in (31.1) for large λ values has the form

$$P(q_0) \exp\left[i\lambda\left(\Psi(q_0) + \frac{1}{2}(q - q_0)^2\Psi''(q_0)\right)\right], \tag{31.4}$$

provided that the quantities $\Psi(q_0)$ and $\Psi''(q_0)$ are nonzero. Substituting expression (31.4) into integral (31.1), we can see that the problem of finding the asymptotic expansion of the integral $I(\lambda)$ reduces to that of calculating the integral

$$I_0(\lambda) = \int_a^b \exp[i\lambda\Psi''(q_0)(q - q_0)^2/2] dq\,. \tag{31.5}$$

For large λ values, a substantial contribution to integral (31.5) comes only from the vicinity of the point $q = q_0$. Consequently, in formula (31.5), we can replace the limits of integration by $-\infty$ and $+\infty$ and use the familiar relationship

$$\int_{-\infty}^{+\infty} \exp(\pm ix^2) dx = \sqrt{\pi} \exp\left(\pm i\frac{\pi}{4}\right). \tag{31.6}$$

As a result, we obtain

$$I_0(\lambda) = \left|\frac{2\pi}{\lambda\Psi''(q_0)}\right|^{1/2} \exp\left(\pm i\frac{\pi}{4}\right). \tag{31.7}$$

Here, the plus sign is for $\lambda\Psi'' > 0$ and the minus sign is for the opposite case. Inserting expressions (31.4) and (31.7) into integral (31.1), we finally arrive at the following asymptotic representation:

$$I(\lambda) = \left|\frac{2\pi}{\lambda\Psi''(q_0)}\right|^{1/2} P(q_0)\exp\left(i\lambda\Psi(q_0) \pm i\frac{\pi}{4}\right). \tag{31.8}$$

Hence, the contribution from the stationary point to integral (31.1) is on the order of $\lambda^{-1/2}$ and is therefore dominant. If there are several stationary points, their contributions should simply be summed.

Let us apply the stationary-phase method to investigate the dynamics of a wave packet on long time scales. To do this, it is necessary to estimate the integral in solution (29.1) from the contribution of the stationary points. We first convert the integral into the form

$$A(t,z) = \frac{1}{2\pi}\int_{-\infty}^{+\infty} A_0(k)\exp[it\Psi(k)]dk\,, \quad \Psi(k) = k\frac{z}{t} - \omega(k)\,. \tag{31.9}$$

In the case at hand, the role of the large parameter λ is played by the time t and the stationary points are determined from the relationship

$$\frac{d\Psi}{dk} = \frac{z}{t} - V_g(k) = 0\,, \tag{31.10}$$

where $V_g(k) = d\omega/dk$ is the group velocity.

Relationship (31.10) is the equation for finding the stationary point $k_0 = k_0(\xi)$, with $\xi = z/t$. For certain values of $\xi = z/t$, this equation may not have a solution. If, for some ξ values, the solution does not exist, the stationary point is absent. In this case, within the accuracy of the stationary-phase method, integral (31.9) should be considered to be zero. This implies that the perturbation does not manage to approach the points at which Eq. (31.10) has no solutions.

Let us consider solutions to Eq. (31.10) for waves whose dispersion is described by formulas (30.2). For acoustic waves, we have

$$k_0(\xi) = \frac{\Omega_0}{V_0}\left[\left(\frac{V_0}{\xi}\right)^{2/3} - 1\right]^{1/2}, \tag{31.11}$$

and, for optical waves, the stationary point is

$$k_0(\xi) = \frac{\Omega_0}{V_0}\left[\left(\frac{V_0}{\xi}\right)^{2} - 1\right]^{-1/2}. \tag{31.12}$$

Formulas (31.11) and (31.12) are meaningful only for $|\xi| < V_0$ or $-V_0 t < z < V_0 t$. The reason is that the perturbation simply does not have enough time to enter the other regions because the group velocities of the waves under consideration are lower in absolute value than V_0.

Formulas (31.11) and (31.12) are qualitatively different, reflecting physical differences in the dynamics of acoustic and optical wave pulses. Thus, in the region $0 < \xi < V_0$, function (31.11) decreases monotonically from $+\infty$ to zero; in contrast, in the same ξ region, function (31.12) is monotonically increasing from zero to $+\infty$. The result is that, as an acoustic wave pulse propagates and spreads out, the long-wavelength part of the perturbation ($k_0 \sim 0$) becomes localized near the pulse's leading edge ($\xi \sim V_0$) and the short-wavelength part of the perturbation ($k_0 \to \infty$) becomes localized near the pulse's trailing edge ($\xi \sim 0$). For an optical wave pulse, the situation is opposite. These are the features that we have mentioned earlier in discussing the results presented in Figs. 34–37.

In order to transform integral (31.9) to its final version, we need only to slightly rewrite formula (31.8) by replacing λ with t, q with k, and $P(q)$ with $A_0(k)/2\pi$ and by substituting the function $\Psi(k)$ defined in (31.9). As a result, for $t \to \infty$, we obtain the following asymptotic formula:

$$A(t,z) = \frac{A_0(k_0)}{\sqrt{2\pi t|G(k_0)|}} \exp\left[-i\omega(k_0)t + ik_0 z \pm i\frac{\pi}{4}\right], \quad G(k_0) = \left(\frac{d^2\omega}{dk^2}\right)_{k=k_0}. \tag{31.13}$$

Here, the plus sign is for $G(k_0) < 0$ and the minus sign is for the opposite case. In formula (31.13), the quantity k_0 is the root of Eq. (31.10), i.e., k_0 is a function of $\xi = z/t$. Hence, solution (31.13) describes a quasi-harmonic perturbation with slowly varying amplitude, wavenumber, and frequency. The perturbation amplitude decreases with time as $t^{-1/2}$. From Eq. (31.10) we can see that the rate at which the wavenumber of a quasi-harmonic perturbation described by solution (31.13) changes in space has the form

$$\frac{dk_0}{dz} = \frac{1}{t|G(k_0)|}. \tag{31.14}$$

As t increases, the rate of change (31.14) becomes smaller, indicating that function (31.13) can be considered as harmonic on larger spatial scales. The initial perturbation behaves as if it were "decomposed" into sinusoidal components that become increasingly wider in space but progressively smaller in amplitude.

32. Some Problems for Wave Equations with a Source

Here, we consider some problems for inhomogeneous equation (3.12) that are of great practical interest. We begin with the problem of calculating the energy lost by a point electric charge q in its uniform motion along a straight line in an isotropic transparent dispersive medium. The velocity of the charge motion, $\mathbf{U} = (0, 0, u)$, is assumed to be nonrelativistic and directed parallel to the OZ coordinate axis. A charge moving in a medium can emit electromagnetic waves. The charge loses energy due to the work done on it by the fields of the emitted waves. Consequently, in order to solve the problem as formulated, it is necessary to find the electromagnetic field created by the charge in the medium — a task that reduces to that of obtaining and solving an equation of the form (3.12).

The charge density of a point charge q moving along the OZ axis can be written as

$$\rho_0(t, \mathbf{r}) = q\delta(x)\delta(y)\delta(z - ut)\,. \tag{32.1}$$

Here, (x, y, z) are Cartesian coordinates and $\mathbf{r} = (x, y, z)$ is the position vector. Charge density (32.1) is a source of electromagnetic radiation: it should be substituted into the set of equations for the electromagnetic field and dynamic equations for the material medium. Under the above assumption that the charge motion is nonrelativistic, the field created by the charge can be treated with good accuracy as being potential. Accordingly, we can set

$$\mathbf{E} = -\boldsymbol{\nabla}\Phi\,, \quad \mathbf{B} = 0\,, \tag{32.2}$$

where $\mathbf{E}$ and $\mathbf{B}$ are the electromagnetic field strength and induction and $\Phi(t, \mathbf{r})$ is the scalar potential.

The properties of an isotropic medium can be characterized by the permittivity operator $\hat{\varepsilon} = \varepsilon(\hat{\omega})$, where $\hat{\omega}$ is the frequency operator in (3.7). For the moment, the explicit form of ε is not specified. We substitute the electric field induction $\mathbf{D} = \hat{\varepsilon}\mathbf{E}$ and charge density ρ_0 (32.1) into the field equation

$$\boldsymbol{\nabla} \cdot \mathbf{D} = 4\pi\rho_0\,. \tag{32.3}$$

Expressing the electric field $\mathbf{E}$ in terms of the potential Φ, we then obtain the inhomogeneous differential equation

$$\boldsymbol{\nabla} \cdot [\varepsilon(\hat{\omega})\boldsymbol{\nabla}\Phi] = 4\pi q\delta(x)\delta(y)\delta(z - ut)\,, \tag{32.4}$$

which is a particular version of Eq. (3.12) for the problem to be solved in the present section.

In contrast to Eq. (3.12), Eq. (32.4) is three-dimensional in spatial coordinates. In addition, in the case in question, the formulation of the initial conditions differs somewhat from that in the conventional initial-value problem. Assuming that the charge began to move in the infinite past, we set $\Phi(t, \mathbf{r}) \to 0$ for $t \to -\infty$. In this case, Eq. (32.4) can conveniently be solved by applying the Fourier transformation in time and in spatial coordinates. Specifically,

$$\begin{aligned} \Phi(t, \mathbf{r}) &= \int \Phi(\omega, \mathbf{k}) \exp(-i\omega t + i\mathbf{k}\mathbf{r}\; d\omega d\mathbf{k}\,, \\ \Phi(\omega, \mathbf{r}) &= \frac{1}{(2\pi)^4} \int \Phi(t, \mathbf{r}) \exp(i\omega t - i\mathbf{k}\mathbf{r}\; dt d\mathbf{k}\,. \end{aligned} \tag{32.5}$$

Here, $\mathbf{k} = (k_x, k_y, k_z)$ is the wave vector, k_z is the wave vector component along the OZ axis, and $d\mathbf{r} = dxdydz$ and $d\mathbf{k} = dk_x dk_y dk_z$.

We multiply both sides of Eq. (32.4) by $\exp(i\omega t - i\mathbf{k}\mathbf{r})$ and integrate over t and $\mathbf{r}$ to find the following expression for the Fourier transformed scalar potential:

$$\Phi(\omega, \mathbf{k}) = \frac{4\pi q}{(2\pi)^3}\delta(\omega - k_z u)\frac{1}{k^2\varepsilon(\omega)}. \tag{32.6}$$

Using the relationship between $\mathbf{E}(t, \mathbf{r})$ and $\Phi(t, \mathbf{r})$ and the first of formulas (32.5), we also get an expression for the Fourier transformed electric field strength:

$$\mathbf{E}(\omega, \mathbf{k}) = -i\mathbf{k}\Phi(\omega, \mathbf{k}) = -i\frac{4\pi q}{(2\pi)^3}\delta(\omega - k_z u)\frac{\mathbf{k}}{k^2\varepsilon(\omega)}. \tag{32.7}$$

By applying the inverse Fourier transformation to expression (32.7), one can deduce the strength of the electric field, $\mathbf{E}(t, \mathbf{r})$, created in a medium by a uniformly moving charge. But the problem as formulated does not require calculating the field, the more so since the function $\varepsilon(\omega)$ has not yet been specified.

Let us now immediately turn to calculating the energy the moving charge loses by radiation. The energy loss per unit length of the charge trajectory is determined by the work done by the electric field $\mathbf{E}(t, \mathbf{r})$ on the charge:

$$W = q\mathbf{u}\mathbf{E}|_{z=ut,\, x=y=0} = quE_z(t, z = ut, x = y = 0).$$

The work W includes the contribution from the projection of the radiation electric field strength at the point charge onto the direction of the velocity $\mathbf{u}$. With expression (32.7), the formula for the work W becomes

$$W = -i\frac{q^2 u}{2\pi^2}\int d\mathbf{k}\frac{k_z}{k^2\varepsilon(k_z u)} = -i\frac{q^2 u}{2\pi^2}\int\frac{k_z dk_x dk_y dk_z}{(k_x^2 + k_y^2 + k_z^2)\varepsilon(k_z u)}. \tag{32.8}$$

We can see that the energy loss is determined by the poles of the integrand in formula (32.8), i.e., by the points where $\varepsilon(k_z u) = 0$. On the other hand, the frequencies of longitudinal waves in the medium are determined from the dispersion relation $\varepsilon(\omega) = 0$ (see Sec. 6) and the relationship $\omega = k_z u$ is the condition for Cherenkov emission from a charged particle at the frequency ω. Consequently, from formula (32.8) we can see that the energy lost by a charge in its uniform motion along a straight line in an isotropic transparent medium with the permittivity $\varepsilon(\omega)$ is governed by the emission of longitudinal waves in the medium. Note that a charge moving uniformly along a straight line in an isotropic medium can also emit transverse waves — an issue that can be studied theoretically in a way quite similar to that just presented (for more details on this phenomenon, which is called Cherenkov radiation, see Chapter 7).

For a cold collisionless plasma as a medium (see Sec. 6), we have

$$\varepsilon(\omega) = 1 - \frac{\omega_p^2}{\omega^2}, \tag{32.9}$$

in which case the integral in formula (32.8) can be taken exactly,

$$W = \frac{q^2\omega_p^2}{2u}\int_0^\infty\frac{dk^2}{k^2 + \omega_p^2/u^2} = \frac{q^2\omega_p^2}{2u^2}[\ln(k^2 + \omega_p^2/u^2)]_0^\infty. \tag{32.10}$$

Having obtained the desired result — formula (32.10) — we find that there is a divergence. The reason is that expression (32.9) is only applicable to a limited range of wavenumbers, $k^2 < \omega_p^2/V_{Te}^2$ (see Sec. 10 above). Substituting the maximum value of the wavenumber, $k^2 = k_{\max}^2 = \omega_p^2/V_{Te}^2$, into formula (32.10), we finally obtain

$$W = \frac{q^2\omega_p^2}{2u} \ln\left(1 + \frac{u^2}{V_{Te}^2}\right) \approx \frac{q^2\omega_p^2}{u} \ln\frac{u}{V_{Te}} . \tag{32.11}$$

In Chapter 7, we will derive formula (32.11) in a different way and also consider other problems of the excitation (emission) of electromagnetic waves by a moving electric charge.

Now, we are going to consider the problem of the excitation of acoustic waves by an oscillating electric charge in a piezocrystal. To do this, we utilize Eq. (17.10) in which the electric field $\mathbf{E}(t, \mathbf{r})$ is assumed to be produced by an external source. In this case, the field $\mathbf{E}(t, \mathbf{r})$ can be calculated from Eq. (32.3) in which ρ_0 is the charge density of the external source, $\mathbf{D} = \varepsilon_0\mathbf{E}$, and ε_0 is the crystal permittivity. Equations (17.10) and (32.3) constitute a complete set of equations for solving the problem of the excitation of acoustic waves by a moving electric charge.

The problem at hand can be simplified in the same way as in Sec. 17, namely, by considering a piezocrystal with hexagonal lattice symmetry and with the principal axis oriented along the wave propagation direction, i.e., along the OZ axis. In this case, the set of Eqs. (17.10) and (32.3) reduces to a single one-dimensional equation for the longitudinal component $U_\parallel(t, z)$ of the displacement vector (see the third of Eqs. (17.11)):

$$\rho\frac{\partial^2 U_\parallel}{\partial t^2} - \lambda_\parallel \frac{\partial^2 U_\parallel}{\partial z^2} = 4\pi\beta_3\varepsilon_0^{-1}\rho_0(t, z) . \tag{32.12}$$

We assume that the external source is an infinitely thin, charged plane perpendicular to the OZ axis. Let the surface charge of the plane oscillate in time. We have

$$\rho_0(t, z) = \sigma_0\delta(z)\exp(-i\omega_0 t) , \tag{32.13}$$

where σ_0 is the surface charge density and ω_0 is the oscillation frequency of the surface charge.

We substitute relationship (32.13) into Eq. (32.12) and apply the one-dimensional Fourier transformation in accordance with formulas (32.5). Simple manipulations then lead us to the following expression for the longitudinal component of the displacement vector:

$$\begin{aligned} U_\parallel(t, z) &= -\frac{2\beta_3\sigma_0}{\rho\varepsilon_0}\exp(-i\omega_0 t)\int_{-\infty}^{+\infty} \frac{\exp(ikz)}{\omega_0^2 - k^2C_\parallel^2} dk \\ &= -i\frac{\pi\beta_3\sigma_0}{\omega_0\rho\varepsilon_0 C_\parallel}\exp(-i\omega_0 t)\left(\exp\left(i\frac{\omega}{C_\parallel}z\right) - \exp\left(-i\frac{\omega}{C_\parallel}z\right)\right) \\ &= \frac{2\pi\beta_3\sigma_0}{\omega_0\rho\varepsilon_0 C_\parallel}\exp(-i\omega_0 t)\sin\frac{\omega}{C_\parallel}z . \end{aligned} \tag{32.14}$$

We can see that the oscillating charge of the plane excites longitudinal acoustic waves in the piezocrystal, running in opposite directions away from the plane.

Chapter 5

Nonharmonic Waves in Nonequilibrium Media

33. Pulse propagation in Nonequilibrium Media

It has been established above that the propagation of a wave packet (pulse) in an equilibrium dispersive medium exhibits the following regular features: (i) in the quasi-harmonic approximation, a pulse propagates in space with the group velocity without distortion and (ii) on long time scales, the quasi-harmonic approximation is inapplicable because of the dispersion-induced widening (spreading) of the pulse.

The propagation of pulses in nonequilibrium media has its own specific features. In such media, the group velocities of the waves are generally complex. As a result, the pulses begin to spread out in the early stages, when the quasi-harmonic approximation is still applicable. In nonequilibrium media, the pulses spread out in the later stages as well. The main spreading mechanism in nonequilibrium media differs from that in equilibrium ones. Let us consider the dynamics of wave pulses under nonequilibrium conditions.

We assume that, in a certain nonequilibrium system, the eigenfrequency of a wave is a complex function of the real variable k: $\omega(k) = \omega'(k) + i\omega''(k)$, where $\omega''(k) > 0$. In deriving solution (29.18) in the quasi-harmonic approximation, we did not assume that the frequency $\omega(k)$ is real. Consequently, for a quasi-harmonic perturbation in a nonequilibrium medium, we can have a solution similar to that in (29.18):

$$A(t, z) = \Phi(z - V_g t)\exp[\omega''(k_0)t]\exp[-i\omega'(k_0)t + ik_0 z]\,. \qquad (33.1)$$

Here, the group velocity V_g is generally a complex quantity,

$$V_g = V_g' + iV_g'' = \left(\frac{d\omega'}{dk}\right)_{k=k_0} + i\left(\frac{d\omega''}{dk}\right)_{k=k_0}\,. \qquad (33.2)$$

Note that, when the frequency $\omega(k)$ is complex, the central point k_0 for expansion (29.11) should be taken to be the absolute maximum point of the function $\omega''(k)$, rather than the maximum point of spectral density (29.3) or the central point of its localization region. It is the spectral components from the vicinity of the absolute maximum point that grow with time to become dominant as the instability develops.

For a nonequilibrium system, the group velocity (33.2) is complex, so the argument of the envelope function $\Phi(z - V_g t)$ in formula (33.1) is also complex. Consequently, the function Φ does not vary along the complex line $z - V_g' t - iV_g'' t = \text{const}$. That is, in the complex plane $\zeta + i\eta$, formula (33.1) describes a wave packet that propagates without distortion along certain straight lines such that

$$\begin{aligned} \zeta &= z - V_g' t \,, \\ \eta &= V_g'' t \,. \end{aligned} \tag{33.3}$$

It is obvious that, being a function of the real variable ζ alone, the envelope of the packet is distorted. In order to investigate the dynamics of the envelope $\Phi(z - V_g t)$ of a wave packet with a complex group velocity V_g (33.2), we can use Eq. (29.33), which is also valid for such packets in the quasi-harmonic approximation. To do this, we introduce the notation

$$\begin{aligned} \Phi(z - V_g t) \equiv \Phi(z,t) &= \Phi_1(z,t) + i\Phi_2(z,t) \,, \\ \Phi(z,0) &= \Phi_{01}(z) + i\Phi_{02}(z) \,, \end{aligned} \tag{33.4}$$

where $\Phi_{1,2}$ and $\Phi_{01,2}$ are the real and imaginary parts of the corresponding complex functions and the functions $\Phi_{01,2}(z)$ describe the shape of the envelope of the initial perturbation.

Switching to variables (33.3) in Eq. (29.33) and separating the real and imaginary parts, we arrive at the familiar Cauchy–Riemann conditions for the functions $\Phi_{1,2}(\xi, \eta)$:

$$\begin{aligned} \frac{\partial \Phi_1}{\partial \eta} &= \frac{\partial \Phi_2}{\partial \zeta} \,, \\ \frac{\partial \Phi_1}{\partial \zeta} &= \frac{\partial \Phi_2}{\partial \eta} \,, \end{aligned} \tag{33.5}$$

We thus can see that each of the functions $\Phi_{1,2}(\xi, \eta)$ is a solution to the following problem for the Laplace equation:

$$\begin{aligned} \left(\frac{\partial^2}{\partial \zeta^2} + \frac{\partial^2}{\partial \eta^2} \right) \Phi_{1,2} &= 0 \,, \\ \Phi_{1,2}(\zeta, \eta)|_{\eta=0} &= \Phi_{01,2}(\zeta) \,, \end{aligned} \tag{33.6}$$

where the functions $\Phi_{01,2}(\zeta)$ are known from the initial conditions (see the second of relationships (33.4)). In the plane of the complex variable $\zeta + i\eta$, the solution to problem (33.6) is sought for in the half-plane determined by the condition $t \geq 0$ and by the sign of the imaginary part V_g'' of the group velocity. The solution to problem (33.6) that is regular at infinity is given by the Poisson integral,

$$\Phi_{1,2}(\zeta, \eta) = \frac{1}{\pi} \int_{-\infty}^{+\infty} \frac{\eta}{(z' - \zeta)^2 + \eta^2} \Phi_{01,2}(z') dz' \,. \tag{33.7}$$

With relationships (33.3), formula (33.7) describes the dynamics of the envelope of a wave packet in a nonequilibrium system in the quasi-harmonic approximation.

Let us consider as an example two model wave packets whose initial envelopes have the form

$$\Phi_{01}(z) = \frac{C}{\pi} \frac{\theta}{z^2 + \theta^2} \quad \text{and} \quad \Phi_{01}(z) = \frac{C}{\pi} \frac{\sin(\kappa z)}{z}, \tag{33.8}$$

where C, θ, and κ are constants. Note that, for a given function $\Phi_{01}(z)$, the function $\Phi_{02}(z)$ is not arbitrary but is determined from Cauchy–Riemann conditions (33.5) with the function $\Phi_1(\xi, \eta)$ found from formula (33.7).

Substituting the first of initial conditions (33.8) into formula (33.7) and performing integration, we obtain the following expression for the complex envelope of the wave packet:

$$\Phi_1(z,t) + i\Phi_2(z,t) = \frac{C}{\pi} \frac{(\theta + |V_g''|t) + i(z - V_g't)}{(z - V_g't)^2 + (\theta + |V_g''|t)^2}. \tag{33.9}$$

From this expression we can see that the perturbation propagates in space as a single entity with the velocity V_g' and the half-width of the envelope of the packet increases with time at a rate proportional to $|V_g''|$.

For the second of initial conditions (33.8), the solution to the problem is written as

$$\Phi_1(z,t) + i\Phi_2(z,t) = \frac{C}{\pi} \frac{\lfloor |V_g''|t + i(z - V_g't) \rfloor \cdot \lfloor 1 - \exp(-|V_g''|\kappa t + i\kappa(z - V_g't)) \rfloor}{(z - V_g't)^2 + (|V_g''|t)^2}. \tag{33.10}$$

This solution has the same properties as solution (33.9). For $\theta \to 0$ and $\kappa \to \infty$, formulas (33.9) and (33.10) describe the evolution of a delta-shaped initial perturbation.

Let us consider initial perturbations $\Phi_{01}(z)$ and $\Phi_{02}(z)$ that are large in absolute value on the closed interval $z \in [z_1^{(0)}, z_2^{(0)}]$ and are small outside the interval. For the initial stage of the process, i.e., when

$$|V_g''|t \ll z_2^{(0)} - z_1^{(0)}, \tag{33.11}$$

the integral in formula (33.7) can be calculated approximately. We take into account the fact that, under inequalities (33.11), the first factor in the integrand in formula (33.7) has a sharp maximum at $z = \zeta$. We also set the functions $\Phi_{01,2}(z)$ equal to zero outside the closed interval $[z_1^{(0)}, z_2^{(0)}]$. As a result, by approximate integration, we obtain the following expression for the envelope of the wave packet:

$$\Phi_{1,2}(z,t) = \frac{1}{\pi}\Phi_{01,2}(z - V_g't)\left(\operatorname{arctg}\frac{z_2^{(0)} - (z - V_g't)}{|V_g''|t} - \operatorname{arctg}\frac{z_1^{(0)} - (z - V_g't)}{|V_g''|t}\right). \tag{33.12}$$

This expression is valid for the early stage of the envelope evolution. We can readily see that the width of the localization region of function (33.12) increases with time at a rate proportional to $|V_g''|$ (or, more precisely, to $|V_g''|$) and that the localized perturbation itself propagates as a single entity with the velocity V_g', i.e.,

$$z \in [z_1(t), z_2(t)], \quad z_{1,2}(t) = z_{1,2}^{(0)} + V_g't \mp |V_g''|t, \quad z_2(t) - z_1(t) = 2|V_g''|t. \tag{33.13}$$

Hence, from approximate expression (33.12) and exact model solutions (33.9) and (33.10) we can conclude that, in a nonequilibrium system described in the quasi-harmonic approximation, a wave packet propagates in space with the velocity $V_g' = \mathrm{Re}(d\omega/dk)$ and spreads out at a rate proportional to $|V_g''| = |\mathrm{Im}(d\omega/dk)|$.

The same result can be immediately derived from the integral in solution (29.1). Since the frequency is complex, we can write the integral as

$$A(t,z) = \frac{1}{2\pi}\int_{-\infty}^{+\infty} A_0(k)\exp(\omega''(k)t)\exp(-i\omega'(k)t + ikz)dk\,. \tag{33.14}$$

Let the spectral density $A_0(k)$ be a finite function that is identically zero outside the closed interval $[k_1, k_2]$, and let the growth rate $\omega''(k)$ be a monotonic function inside the interval. For definiteness, we set $\omega''(k_2) > \omega''(k_1)$. In the quasi-harmonic approximation, under the inequality $k_2 - k_1 \ll |k_2 + k_1|/2$, the integral in formula (33.14) can be cast into the form

$$A(t,z) = \Phi(z,t)\cdot\exp[\omega''(k_2)t]\exp[-i\omega'(k_2)t + ik_2 z]\,, \tag{33.15}$$

where

$$\begin{gathered}\Phi(z,t) = \frac{1}{2\pi}\int_0^{k_2-k_1} A_0(k_2 - q)\exp[-iq(z - V_g't - iV_g''t)]dq\,,\\ V_g'' = \frac{\omega''(k_2) - \omega''(k_2)}{k_2 - k_1} \approx \frac{d\omega''}{dk} > 0\,, \quad V_g' = \frac{\omega'(k_2) - \omega'(k_1)}{k_2 - k_1} \approx \frac{d\omega'}{dk}\,.\end{gathered} \tag{33.16}$$

Expression (33.15) has the same structure as formula (33.1) with k_2 in place of k_0. It has already been mentioned that, when the frequency $\omega(k)$ is complex, the central point k_0 for expansion (29.11) in formula (33.1) should be taken to be the absolute maximum point of the function $\omega''(k)$. In accordance with the assumption made in deriving formulas (33.15) and (33.16), it is the point k_2 at which the growth rate achieves its maximum value. That is why formulas (33.15) and (33.1) coincide and the expression for the function $\Phi(z,t)$ from (33.16) is another form of expression (33.7) for the envelope in terms of the Poisson integral. Introducing the notation $A_0(k) = A_{01}(k) + iA_{02}(k)$ and $\Phi(z,t) = \Phi_1(\zeta,\eta) + i\Phi_2(\zeta,\eta)$ and separating the real and imaginary parts in (33.16), we obtain

$$\begin{aligned}&\Phi_{1,2}(\zeta,\eta)\\ &= \frac{1}{2\pi}\int_0^{k_2-k_1} dq\exp(-\eta q)\left[A_{01}(k_2 - 1)\begin{pmatrix}\cos q\zeta\\ -\sin q\zeta\end{pmatrix} + A_{02}(k_2 - q)\begin{pmatrix}\sin q\zeta\\ \cos q\zeta\end{pmatrix}\right].\end{aligned} \tag{33.17}$$

It can be easily verified by direct substitution that functions (33.17) satisfy Cauchy–Riemann conditions (33.5) and are solutions to problems (33.6).

Let us also consider the quadratic quantity $W(\zeta,\eta) = \Phi_1^2(\zeta,\eta) + \Phi_2^2(\zeta,\eta)$, which can be treated as an energy parameter of the perturbations in the system (see formula (29.15)). The task is to determine the velocity at which this quantity is transported in space. For simplicity, we restrict ourselves to an initial perturbation with a purely real spectral density, $A_{01}(k) = A_0(k)$ and $A_{02}(k) = 0$. Using formula (33.17), we readily find

$$W(\zeta,\eta) = (2\pi)^{-2} \int_0^{k_2-k_1} dq' \int_0^{k_2-k_1} dq'' A_0(k_2 - q')A_0(k_2 - q'')$$

$$\times \exp[-\eta(q' + q'')] \cos[\zeta(q' - q'')] , \tag{33.18}$$

The maximum of function (33.18) — which obviously coincides with the maximum of the function $\cos[\zeta(q' - q'')]$ — is achieved at $\zeta = 0$ and is therefore transported with the velocity $V_g' = \mathrm{Re}(d\omega/dk)$. This result provides a strong argument for treating the velocity V_g' as the perturbation energy transport velocity in nonequilibrium systems in the quasi-harmonic approximation. In addition, formulas (33.9), (33.10), and (33.13) imply that V_g' is the propagation velocity of the central point of the pulse envelope. Consequently, the real part of the complex group velocity in a nonequilibrium system is an analogue of the conventional group velocity in an equilibrium system. As for the absolute value of the imaginary part of the complex group velocity, $|V_g''| = |\mathrm{Im})(d\omega/\mathrm{dk})$, it characterizes the rate at which the width of the envelope of a wave pulse increases in space in the quasi-harmonic approximation.

It is an easy matter to understand the physical meaning of the spatial spreading of wave packets in nonequilibrium media in the quasi-harmonic approximation. When the imaginary part of the frequency, $\omega''(k)$, is nonzero, a substantial contribution to the integral in formula (33.14) comes from an increasingly narrowing interval $\Delta k(t)$ near the absolute maximum ω''_{max} of the function $\omega''(k)$. In other words, as time goes on, the spectral components from the interval $\Delta k(t)$ make an increasingly dominant contribution and the contribution from the remaining components is exponentially small, $\sim \exp[-(\omega''_{\mathrm{max}} - \omega''(k))t]$. Since the interval $\Delta k(t)$ narrows with time, the wave packet becomes wider, in accordance with the uncertainty principle (29.4).

In nonequilibrium media, quasi-harmonic wave packets can spread out in the manner just considered only when the following derivative is nonzero at the point k_0 of the maximum wave growth rate $\omega''(k)$:

$$\left(\frac{d\omega''}{dk}\right)_{k=k_0} = V_g''(k_0) \neq 0 \, . \tag{33.19}$$

Recall that, for a complex frequency $\omega(k)$, the central point k_0 for expansion (29.11) in formula (33.1) should be taken to be the absolute maximum point of the growth rate $\omega''(k)$. Consequently, condition (33.19) can be satisfied only when the function $\omega''(k)$ is monotonic within the localization region of the spectral density $A_0(k)$. Condition (33.19) is most easily satisfied by a finite spectral density function such that $A_0(k) \equiv 0$ for all k outside a closed interval $[k_1, k_2]$, $(k_1 < k_2)$. If the growth rate $\omega''(k)$ is monotonic on this interval, then condition (33.19) is certainly satisfied and the point k_0 is one of the end points of the interval (see formulas (33.15), (33.16)).

Now let the absolute maximum point k_0 of the growth rate $\omega''(k)$ be an interior point of the closed interval $[k_1, k_2]$ on which the spectral density is nonzero, $A_0(k) \neq 0$. In this case, derivative (33.19) vanishes and a pulse in a nonequilibrium system does not spread out in the quasi-harmonic approximation. In fact, taking into account the familiar relationship

$$\delta(z-\zeta) = \lim_{\eta \to 0} \frac{1}{\pi} \frac{\eta}{(z-\zeta)^2 + \eta^2} \tag{33.20}$$

and the corresponding property of the delta function, from the integral in formula (33.7) we obtain

$$\Phi_{1,2}(z,t) = \Phi_{01,2}(z - V_g' t)\,. \tag{33.21}$$

Solution (33.21) shows that the envelope of the initial perturbation, described by the functions $\Phi_{01,2}(z)$, propagates with the group velocity $V_g = V_g'$ without distortion. This case differs from the case of propagation of a pulse in an equilibrium system only in that the perturbation grows exponentially due to the instability according to the law $\exp[\omega''(k_0)t]$ (see formula (33.1)). The real group velocity calculated at the maximum point of the growth rate $\omega''(k)$ is called the convection velocity during instability.

34. Stationary-Phase Method for Complex Frequencies

Here, we consider how the integral

$$A(t,z) = \frac{1}{2\pi} \int_{k_1}^{k_2} A_0(k) \exp(\omega''(k)t) \exp(-i\omega'(k)t + ikz) dk \tag{34.1}$$

behaves asymptotically at $t \to \infty$ when the absolute maximum point k_0 of the growth rate $\omega''(k)$ is an interior point of the closed interval $[k_1, k_2]$ on which the spectral density is nonzero. Recall that, for equilibrium systems with $\omega''(k) \equiv 0$, this issue has been analyzed by the stationary-phase method. The stationary points have been defined as those at which Eq. (31.10) has solutions. For an equilibrium system, the existence of stationary points stems from the dispersion of the group velocity, with the result that the pulse components become increasingly monochromatic. The stationary-phase method for $\omega''(k) \neq 0$ is applied in its own way; moreover, the term "stationary-phase method for $\omega''(k) \neq 0$" is not a generally accepted one. Assume that the main contribution to the integral in formula (34.1) comes from the maximum point of the function $S(k) = A_0(k) \exp(\omega''(k)t)$. The maximum point is determined from the equation

$$\frac{d\omega''}{dk} = -\frac{1}{t} \frac{dA_0}{dk}\,. \tag{34.2}$$

As $t \to \infty$, the root of this equation approaches the absolute maximum point k_0 of the growth rate $\omega''(k)$. Moreover, the longer the time t, the sharper is the extremum of the function $S(k)$ around the point k_0. The assumption that, on long time scales, the main contribution to the integral in (34.1) comes from the vicinity of the point k_0 is justified only when the integration interval $[k_1, k_2]$ is sufficiently small. In the general case, the integral should be estimated by another method, as will be done in Chapter 6.

So let us expand the integrand in formula (34.1) in the vicinity of the point k_0 and retain terms up to second order. As a result, we obtain

$$A(t,z) = \frac{1}{2\pi} A_0(k_0) \exp(\omega''(k_0)t) \exp(-i\omega'(k_0)t + ik_0 z) \Im(\zeta, t), \qquad (34.3)$$

$$\Im(\zeta,t) = \int_{k-1-k_0}^{k_2-k_0} \exp(i\zeta x) \exp\left[\frac{1}{2}\left(\frac{d^2\omega''(k_0)}{dk^2} - i\frac{d^2\omega'(k_0)}{dk^2}\right) x^2 t\right] dx\,, \qquad (34.4)$$

where $\zeta = z - V_g' t$, with $V_g' = (d\omega'/dk)_{k=k_0}$. In deriving formula (34.4), we have taken into account the relationship $V_g'' = (d\omega''/dk)_{k=k_0} = 0$. Since, at the point k_0, the function $\omega''(k)$ reaches its maximum, we have $d^2\omega''(k_0)/dk^2 < 0$. For large t values, this inequality allows us to replace the limits of integration in formula (34.4) by $\mp\infty$. The resulting integral can then be calculated by using the well-known formula

$$\int_0^\infty \exp(-\beta x^2)\cos(bx)dx = \frac{1}{2}\sqrt{\frac{\pi}{\beta}} \exp\left(-\frac{b^2}{4\beta}\right), \quad \mathrm{Re}\,\beta > 0\,. \qquad (34.5)$$

Introducing the notation

$$\zeta = b\,, \quad \beta = \frac{1}{2} t \left(\left|\frac{d^2\omega''(k_0)}{dk^2}\right| + i\frac{d^2\omega'(k_0)}{dk^2}\right) \qquad (34.6)$$

and taking into account formula (34.5), we reduce the integral in formula (34.4) (with infinite limits of integration) to

$$\Im(\zeta,t) = \left[\sqrt{\frac{2\pi}{\rho t}} \exp\left(-\frac{\zeta^2}{2\rho^2\gamma_1^{-1}t}\right)\right] \exp\left(-i\frac{\varphi}{2} + i\frac{\zeta^2}{2\rho^2\gamma_2^{-1}t}\right), \qquad (34.7)$$

where

$$\gamma_1 = \left|\frac{d^2\omega''(k_0)}{dk^2}\right| \neq 0\,, \quad \gamma_2 = \frac{d^2\omega'(k_0)}{dk^2}\,, \quad \rho = \sqrt{\gamma_1^2 + \gamma_2^2}\,, \quad \varphi = \mathrm{arctg}\frac{\gamma_2}{\gamma_1}\,. \qquad (34.8)$$

Function (34.7) describes the envelope of the wave packet (34.3). The maximum of the envelope, which is attained at $\zeta = 0$, decreases with time as $t^{-1/2}$ and propagates in space with the velocity V_g'. Concurrently, the envelope becomes wider. The width of the envelope can be defined by the relationship

$$\zeta^2(2\rho^2\gamma_1^{-1}t)^{-1} = 1\,. \qquad (34.9)$$

Using this relationship, we arrive at the following formula for the localization region of perturbation (34.3):

$$V_g' t - \sqrt{2\frac{\rho^2}{\gamma_1}t} < z < V_g' t + \sqrt{2\frac{\rho^2}{\gamma_1}t}\,. \qquad (34.10)$$

That perturbation (34.3) spreads out stems from its harmonization: during instability, the main contribution to the integral in formula (34.1) comes from the spectral components with wavenumbers becoming increasingly close to k_0 — the maximum point of the growth rate $\omega''(k)$. In accordance with formula (34.7), the maximum of the perturbation amplitude (34.3) increases slower than in the initial stage (on short time scales, the rate of increase is $\exp[\omega''(k_0)t]$):

$$|A(t,z)|_{\max} \sim \frac{1}{\sqrt{t}} \exp[\omega''(k_0)t]\,. \tag{34.11}$$

This result also stems from the spreading of the perturbation and its harmonization.

35. Quasi-Harmonic Approximation in the Theory of Interaction of Electron Beams with Slowing-Down Media

In this section, we apply the above-described quasi-harmonic approximation to investigate some resonant instabilities occurring in the interaction of electron beams with slowing-down electromagnetic media (systems). By slowing-down systems we mean those in which there are waves with phase velocities that are close to the beam velocity and, accordingly, are less than the speed of light in vacuum. We begin with the basic equations of microwave electronics (see Eqs. (23.1)):

$$\begin{gathered}\left(\left(\frac{\partial}{\partial t} + U_{0b}\frac{\partial}{\partial z}\right)^2 + \omega_b^2\right) A_b = -\omega_b^2 S_b(\hat{\omega},\hat{k}) A_w\,, \\ D_w(\hat{\omega},\hat{k}) A_w = -\omega_w^2 S_2(\hat{\omega},\hat{k}) A_b\,.\end{gathered} \tag{35.1}$$

Here, $A_b(t,z)$ is the characteristic function of the state vector of the electron beam, $A_w(t,z)$ is the characteristic function of the state vector of the slowing-down medium, $D_w(\hat{\omega},\hat{k})$ is the dispersion operator of the medium, S_b and S_w are dimensionless operators, ω_w is a quantity having the dimension of frequency, and ω_b is the Langmuir frequency of the beam electrons.

Assume that the set of Eqs. (23.4), which determines the resonance point in the plane (k,ω), has a solution. We denote this solution by $(k = \chi_0,\ \omega = \Omega_0)$, a notation slightly differing from that in Secs. 22 and 23. Restricting ourselves to the resonant interaction between an electron beam and a slowing-down medium, we represent the solution to Eqs. (35.1) as

$$\begin{aligned} A_b(t,z) &= \tilde{A}_b(t,z)\exp(-i\Omega_0 t + i\chi_0 z)\,, \\ A_w(t,z) &= \tilde{A}_w(t,z)\exp(-i\Omega_0 t + i\chi_0 z)\,. \end{aligned} \tag{35.2}$$

Here, the slow amplitude of the eigenmode of the medium, $\tilde{A}_w(t,z)$, satisfies the inequalities

$$\left|\frac{\partial \tilde{A}_w}{\partial t}\right| \ll |\Omega_0 \tilde{A}_w|\,, \quad \left|\frac{\partial \tilde{A}_w}{\partial z}\right| \ll |\chi_0 \tilde{A}_w|\,. \tag{35.3}$$

As for the amplitude of the beam wave, $\tilde{A}_b(t,z)$, we do not now assume that it is slow.

We substitute representations (35.2) into Eqs. (35.1) and take into account the relationships $D_w(\Omega_0, \chi_0) = 0$ and $(\Omega_0 - \chi_0 U_{0b})^2 - \omega_b^2 = 0$, which hold at the resonance point. As a result, using inequalities (35.3) and formula (29.32), we arrive at the following set of equations for the amplitudes $\tilde{A}_b(t,z)$ and $\tilde{A}_w(t,z)$:

$$\begin{aligned} &\left(\frac{\partial}{\partial t} + U_{0b}\frac{\partial}{\partial z}\right)^2 \tilde{A}_b \mp 2i\omega_b \left(\frac{\partial}{\partial t} + U_{0b}\frac{\partial}{\partial z}\right) \tilde{A}_b = -\omega_b^2 S_b(\Omega_0, \chi_0)\tilde{A}_w\,, \\ &\left(\frac{\partial}{\partial t} + V_{0g}\frac{\partial}{\partial z}\right) \tilde{A}_w = -i\omega_w^2 S_w(\Omega_0, \chi_0) \left|\frac{\partial D_w(\Omega_0,\chi_0)}{\partial \omega}\right|^{-1} \tilde{A}_b\,. \end{aligned} \tag{35.4}$$

Here, V_{0g} is the group velocity of the wave in a slowing-down medium in the absence of perturbations introduced by an electron beam. Since Eqs. (35.4) are derived in largely the same way as Eqs. (23.12), we do not present here the details of the derivation.

Let us consider two limiting versions of general equations (35.4) that describe two limiting cases of interaction between an electron beam and the wave of the medium. In the collective regime, when the interaction is weak and when

$$\left|\left(\frac{\partial}{\partial t} + U_{0b}\frac{\partial}{\partial z}\right)\tilde{A}_b\right| \ll |\omega_b \tilde{A}_b|\,, \tag{35.5}$$

Eqs. (35.4) are written as (see Eqs. (23.16))

$$\begin{aligned} &\left(\frac{\partial}{\partial t} + U_{0b}\frac{\partial}{\partial z}\right) \tilde{A}_b = \mp\frac{1}{2} i\omega_b S_b(\Omega_0, \chi_0)\tilde{A}_w\,, \\ &\left(\frac{\partial}{\partial t} + V_{0g}\frac{\partial}{\partial z}\right) \tilde{A}_w = -i\omega_w^2 S_w(\Omega_0, \chi_0) \left|\frac{\partial D_w(\Omega_0,\chi_0)}{\partial \omega}\right|^{-1} \tilde{A}_b\,. \end{aligned} \tag{35.6}$$

Here, the upper sign is for the resonance between the wave of the medium and the fast beam wave and the lower sign is for the resonance with the slow beam wave. The set of quantities that completely determine the state of a physical system described by Eqs. (35.6) (the state vector of a "beam + medium" system) can be conveniently taken to be

$$\{\tilde{A}_b(t,z), \tilde{A}_w(t,z)\} \equiv A_{\text{coll.}}(t,z)\,. \tag{35.7}$$

In the single-particle regime, when the interaction is strong and when the inequality opposite to (35.5) is satisfied, Eqs. (35.4) reduce to

$$\begin{aligned} &\left(\frac{\partial}{\partial t} + U_{0b}\frac{\partial}{\partial z}\right)^2 \tilde{A}_b = -\omega_b^2 S_b(\Omega_0, \chi_0)\tilde{A}_w\,, \\ &\left(\frac{\partial}{\partial t} + V_{0g}\frac{\partial}{\partial z}\right) \tilde{A}_w = -i\omega_w^2 S_w(\Omega_0, \chi_0) \left|\frac{\partial D_w(\Omega_0,\chi_0)}{\partial \omega}\right|^{-1} \tilde{A}_b\,. \end{aligned} \tag{35.8}$$

The set of quantities that completely determine the state of a physical system described by Eqs. (35.8) in solving the initial-value problem can be chosen to be

$$\left\{\tilde{A}_b(t,z), \frac{\partial \tilde{A}_b(t,z)}{\partial t}, \tilde{A}_w(t,z)\right\} \equiv A_{sing}(t,z)\,. \tag{35.9}$$

The explicit form of inequality (35) and of the opposite inequality will be given later.

Let us first analyze Eqs. (35.6), describing the collective regime. Since the instability occurs only when there is a resonance between the wave of the medium and the slow beam wave, we choose the lower sign in the first of the equations. Inserting the functions $\tilde{A}_{b,w} = C_{b,w}\exp(-i\omega t + ikz)$ into Eqs. (35.6) and eliminating the constants C_b and C_w yields the dispersion relation

$$D(\omega,k) \equiv (\omega - kU)(\omega - kV) + a^2 = 0\,. \tag{35.10}$$

Here, in order to simplify the formulas to follow, we have introduced the notation

$$U = U_{0b}\,, \quad V = V_{0g}\,, \quad a^2 = \frac{1}{2}\omega_b\omega_2^2 S_b S_w \left|\frac{\partial D_w}{\partial \omega}\right|^{-1} = \frac{1}{2}\omega_b\tilde{\Omega}\,, \tag{35.11}$$

with $\tilde{\Omega}$ being the frequency introduced in dispersion relation (23.5). The quantity a^2 has the dimension of frequency squared. Note that the quantities ω and k in dispersion relation (35.10) are not actual frequency and wavenumber but rather are deviations from the resonant values Ω_0 and χ_0, as can be seen from formulas (35.2). In this section, unlike in Secs. 21–23, we do not introduce a new notation for these deviations in order to simplify the description. Assume that the beam velocity is positive, $U > 0$. As for the group velocity of the wave in the medium, V, it may be either positive or negative. The case $V > 0$ refers to the interaction of the beam with the forward wave, and the case $V < 0$, to the interaction with the backward wave.

In the initial-value problem, the solutions to dispersion relation (35.10) for the frequency ω are given by the formulas

$$\omega_{1,2}(k) = \frac{1}{2}k(U+V) \pm \sqrt{\frac{1}{4}k^2(U-V)^2 - a^2}\,. \tag{35.12}$$

Functions (35.12) determine two eigenmode branches in a "beam + medium" system. The instability occurs in the wavenumber range given by inequalities

$$-\frac{2a}{|U-V|} < k < \frac{2a}{|U-V|}\,. \tag{35.13}$$

The maximum of the instability growth rate, $\omega''_{\max} = a$, is reached at $k = 0$. Consequently, inequality (35.5) can be explicitly rewritten as $a \ll \omega_b$, a form that in fact coincides with inequality (23.10), including the notation.

The real and imaginary parts of the solutions to dispersion relation (35.10) are illustrated in Fig. 42, which shows the results calculated for a forward wave in a slowing-down medium with the following relative values of the parameters: $a = 1$,

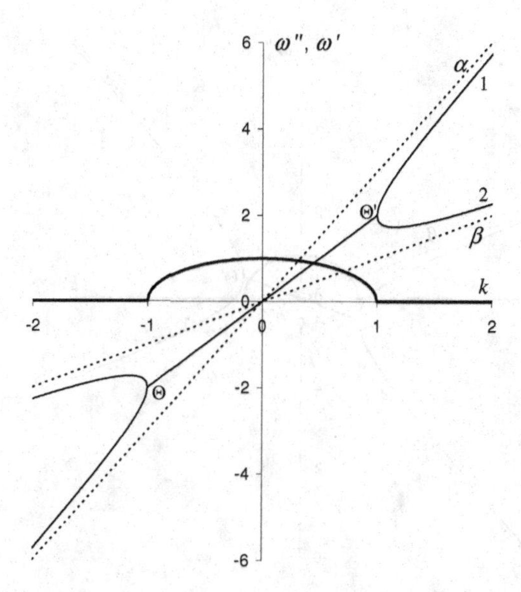

Fig. 42. Real ω' (lighter curves) and imaginary ω'' (heavier curve) parts of the frequency in the interaction of a beam with the forward wave in the collective regime.

$U = 3$ and $V = 1$. The absolute values of the parameters are obtained as follows. Take any real number σ. If the wavenumber k in the figure is in [cm^{-1}], then the frequency ω is in [σ rad s^{-1}] and the velocities U and V are in [σ cm s^{-1}]. In Fig. 42, the dotted line α is the asymptote $\omega = kU$ of the function $\omega_1(k)$, the dotted line β is the asymptote $\omega = kV$ of the function $\omega_2(k)$, curves *1* and *2* are for the real parts of the functions $\omega_1(k)$ and $\omega_2(k)$, and the heavy curve is for the nonnegative imaginary part of the frequency $\omega_1(k)$ (the instability growth rate). Note that, in Fig. 42 and subsequent figures, the branches $\omega_m(k)$ are numbered in accordance with their asymptotes at $k \to +\infty$, the first number being assigned to the frequency branch that has a positive imaginary part for real k values. One can easily see that the functions in dispersion relation (35.10) have the asymptotes $\omega_1 = kU - a^2k^{-1}(U-V)^{-1}$ and $\omega_2 = kV - a^2k^{-1}(U-V)^{-1}$.

Figure 43 shows the real and imaginary parts of the solutions to dispersion relation (35.10), calculated for a backward wave in a slowing-down medium with the following relative values of the parameters: $a = 1$, $U = 3$ and $V = -1$. The notation is the same as in Fig. 42.

Differentiating dispersion relation (35.10) with respect to k yields the following general expression for the group velocities of the eigenmodes of the system:

$$V_g = \frac{d\omega}{dk} = \frac{\omega(U+V) - 2kUV}{2\omega - k(U+V)}. \tag{35.14}$$

A particular value of the group velocity is calculated from expression (35.14) with ω_1 or ω_2 from (35.12) in place of the frequency ω. In the wavenumber range (35.10),

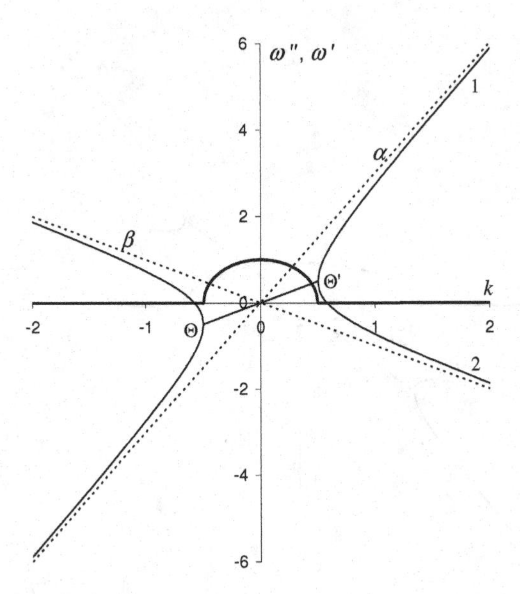

Fig. 43. Real ω' (lighter curves) and imaginary ω'' (heavier curve) parts of the frequency in the interaction of a beam with the backward wave in the collective regime.

the frequencies are complex, so the group velocities (35.4) are complex too. The only exception is the point $k = 0$, at which the group velocities of both waves are purely real. It has been shown above that, for the system under consideration, it is at the wavenumber $k = 0$ that the growth rate attains its maximum. Consequently, the group velocity (35.14) with $k = 0$ is the convective velocity during instability. We denote this velocity by U_g and from (35.14) obtain

$$U_g = \frac{1}{2}(U + V). \tag{35.15}$$

Formula (35.15) gives the velocity at which the fastest growing perturbations propagate during the collective instability. The propagation direction depends on the group velocity of the wave in the medium in the absence of perturbations introduced by the beam. For $V < -U < 0$, the propagation velocity (35.15) is negative. This means that the growing wave pulse propagates in space towards the electron beam. For $V > -U$, the pulse propagates in the same direction as the beam.

We also present the relationship between the amplitudes C_b and C_w of the wave perturbations of the electron beam and slowing-down medium:

$$C_w = \frac{w}{(\omega - kV)} C_b \,, \quad w = \omega_w^2 S_w |\partial D_w / \partial \omega|^{-1} \,. \tag{35.16}$$

This relationship, which will be used in the analysis to follow, is obtained by substituting the solution in the form $\tilde{A}_{b,w} = C_{b,w} \exp(-i\omega t + ikz)$ into Eqs. (35.6) when deriving dispersion relation (35.10). What is meant by the frequency ω for the eigenmodes in relationship (35.16) is one of the eigenfrequencies (35.12).

Let us consider the solution to the initial-value problem for the set of Eqs. (35.6). To do this, we utilize the eigenmode method. In accordance with general formula (28.9), the general solution to Eqs. (35.6) is given by the expressions

$$\tilde{A}_b(t,z) = \frac{1}{2\pi}\int_{-\infty}^{+\infty}\sum_{m=1}^{2}[C_{bm}(k)\exp(-i\omega_m(k)t + ikz)]dk\,,$$
$$\tilde{A}_w(t,z) = \frac{1}{2\pi}\int_{-\infty}^{+\infty}\sum_{m=1}^{2}\left[\frac{w}{\omega_m - kV}C_{bm}(k)\exp(-i\omega_m(k)t + ikz)\right]dk\,. \qquad (35.17)$$

In writing the second of these expressions, we have taken into account relationship (35.16). The unknown spectral densities $C_{bm}(k)$, $(m = 1, 2)$ are to be determined from the initial conditions.

For the system under consideration, the most general initial conditions have the form

$$\tilde{A}_{b0}(0,z) = A_{b0}(z)\,, \quad \tilde{A}_{w0}(0,z) = A_{w0}(z)\,. \qquad (35.18)$$

The right-hand sides of these conditions are known functions. Inserting solution (35.17) into initial conditions (35.18) yields the relationships

$$\frac{1}{2\pi}\int_{-\infty}^{+\infty}\sum_{m=1}^{2}[C_{bm}(k)\exp(ikz)]dk = A_{b0}(z)\,,$$
$$\frac{1}{2\pi}\int_{-\infty}^{+\infty}\sum_{m=1}^{2}\left[\frac{w}{\omega_m - kV}C_{bm}(k)\exp(ikz)\right]dk = A_{w0}(z)\,. \qquad (35.19)$$

We multiply both sides of relationships (35.19) by $\exp(-ik'z)$ and integrate over z from minus to plus infinity. Using integral representation (28.6) for the delta function and its main property, we arrive at the following set of algebraic equations:

$$C_{b1}(k) + C_{b2}(k) = A_{b0}(k)\,,$$
$$\frac{w}{\omega_1 - kV}C_{b1}(k) + \frac{w}{\omega_2 - kV}C_{b2}(k) = A_{w0}(k)\,, \qquad (35.20)$$

where we have introduced the spectral densities of the initial perturbations,

$$A_{b0}(k) = \int_{-\infty}^{+\infty} A_{b0}(z)\exp(-ikz)dz\,, \quad A_{w0}(k) = \int_{-\infty}^{+\infty} A_{w0}(z)\exp(-ikz)dz\,. \qquad (35.21)$$

The unknown quantities $C_{bm}(k)$, $(m = 1, 2)$ are determined from Eqs. (35.20).

Let us consider initial conditions that are simpler than conditions (35.18), specifically, $A_{b0}(z) \neq 0$, $A_{w0}(z) \equiv 0$. In this case, we have $A_{w0}(k) \equiv 0$ and from Eqs. (35.20) we obtain the coefficients

$$C_{b1}(k) = \frac{\omega_1(k) - kV}{\omega_1(k) - \omega_2(k)}A_{b0}(k)\,, \quad C_{b2}(k) = \frac{\omega_2(k) - kV}{\omega_2(k) - \omega_1(k)}A_{b0}(k)\,. \qquad (35.22)$$

Substituting coefficients (35.22) into expressions (35.17), we arrive at the sought-for solution to the problem. But we do not present the solution here because it is rather involved and poorly informative.

In order to make the problem easier to solve, we apply the quasi-harmonic approximation. It has been shown above that the growth rate is maximum at $k = 0$. Assume now that the spectral density of the perturbation, $A_{b0}(k)$, too, has a sharp and narrow maximum at $k = 0$. Under this assumption, we can expand the integrands in expressions (35.17) in powers of k and retain terms up to the first order. In addition, for $k = 0$, we have $\omega_1 = -\omega_2 = ia$ (see (35.12)) and formulas (35.22) can be substantially simplified:

$$C_{b1}(k) = C_{b2}(k) = \frac{1}{2}A_{b0}(k)\,. \tag{35.23}$$

As an example, let us transform the first term in the first of expressions (35.17):

$$\begin{aligned}&\frac{1}{2\pi}\int_{-\infty}^{+\infty} C_{b1}(k)\exp(-i\omega_1(k)t + ikz)dk\\ &= \frac{1}{2}\exp(-i\omega_1'(0)t + \omega_1''(0)t)\left\{\frac{1}{2\pi}\int_{-\infty}^{+\infty} A_{b0}(k)\exp\left(ikz - ik\frac{d\omega_1(0)}{dk}t\right)dk\right\}.\end{aligned} \tag{35.24}$$

In the case at hand, we have $\omega_1'(0) = 0$, $\omega_1''(0) = a$, and $d\omega_1(0)/dk = U_g$. Also, according to relationship (28.5), the expression in braces in (35.24) is the inverse Fourier transform of the spectral density $A_{b0}(k)$ in the argument $\xi = z - U_g t$. We thus arrive at the following final expression for integral (35.24):

$$\frac{1}{2\pi}\int_{-\infty}^{+\infty} C_{b1}(k)\exp(-i\omega_1(k)t + ikz)dk = \frac{1}{2}A_{b0}(z - U_g t)\exp(at)\,. \tag{35.25}$$

Performing similar manipulation, we find that, in the quasi-harmonic approximation, solution (35.17) to Eqs. (35.6) has the form

$$\begin{aligned}\tilde{A}_b(t,z) &= A_{b0}(z - U_g t)\mathrm{chat}\,,\\ \tilde{A}_w(t,z) &= A_{b0}(z - U_g t)\left(-i\frac{w}{a}\right)\mathrm{shat}\,,\end{aligned} \tag{35.26}$$

where the convective velocity during instability, U_g, is defined by formula (35.15).

It is also an easy matter to determine the asymptotic behavior of solution (35.17) on long time scales t. To do this, we must simply specify general expressions (34.3) and (34.7). In the case at hand, we can write

$$\frac{d^2\omega'(0)}{dk^2} = 0\,, \quad \frac{d^2\omega''(0)}{dk^2} = -\frac{(U-V)^2}{4a}\,. \tag{35.27}$$

As a result, from (34.3) and (34.7) we obtain the following asymptotic formula for the beam wave amplitude $\tilde{A}_b(t,z)$ during the collective instability of an electron beam in a medium:

$$\tilde{A}_b(t,z) = \frac{1}{2}(A_{b0}|_{k=0})\exp(at)\sqrt{\frac{2a}{\pi(U-V)^2 t}}\exp\left[-\frac{2a}{(U-V)^2 t}(z-U_g t)^2\right]. \tag{35.28}$$

The asymptotic behavior of the amplitude $\tilde{A}_w(t,z)$ is found in a similar way.

Let us now analyze Eqs. (35.8), which describe the single-particle regime. Substituting the functions $\tilde{A}_{b,w} = C_{b,w}\exp(-i\omega t + ikz)$ into these equations and eliminating the constants C_b and C_w yields the dispersion relation

$$D(\omega,k) \equiv (\omega - kU)^2(\omega - kV) - b^3 = 0, \tag{35.29}$$

where

$$U = U_{0b}, \quad V = V_{0g}, \quad b^3 = \omega_b^2\omega_w^2 S_b S_w \left|\frac{\partial D_w}{\partial \omega}\right|^{-1} = \omega_b^2\tilde{\Omega} > 0. \tag{35.30}$$

The quantity b^3 has the dimension of frequency cubed. The solutions to cubic equation (35.29) are expressed by means of the Cardano formulas. For the auxiliary equation

$$\delta^3 + k(U-V)\delta^2 - b^3 = 0, \tag{35.31}$$

these formulas are written as

$$q = \frac{1}{27}x^3 - \frac{1}{2}b^3, \quad p = -\frac{1}{9}x^2, \quad D = q^2 + p^3 = -\frac{1}{27}b^3\left(x^3 - \frac{27}{4}b^3\right),$$

$$X = \sqrt[3]{-q + \sqrt{D}}, \quad Y = \sqrt[3]{-q - \sqrt{D}},$$

$$\delta_1 = -\frac{1}{2}(X+Y) + i\frac{\sqrt{3}}{2}(X-Y), \quad \delta_2 = -\frac{1}{2}(X+Y) - i\frac{\sqrt{3}}{2}(X-Y), \quad \delta_3 = X+Y, \tag{35.32}$$

with $x = k(U-V)$. Formulas (35.32) can be conveniently used only for

$$D > 0 \quad \text{or} \quad x^3 < \frac{27}{4}b^3, \tag{35.33}$$

i.e., when the roots $\delta_{1,2,3}$ of Eq. (35.31) are complex. For $D < 0$, the roots of the equation are real and it is more convenient to use other formulas, which are not presented here. The solutions $\omega_{1,2,3}$ to dispersion relation (35.29) are expressed in terms of the solutions $\delta_{1,2,3}$ to auxiliary equation (35.31) by means of the relationships (for definiteness, we are assuming here that $V < U$)

$$\omega_{1,2,3}(k) = kU - x/3 + \delta_{1,2,3}(k). \tag{35.34}$$

Relationships (35.34) describe three branches of the eigenmodes of a "beam + slowing-down medium" system in the single-particle regime.

The growing eigenmode is that with the frequency ω_1, because, in the wavenumber range (see inequalities (35.33))

$$k < \frac{3}{\sqrt[3]{4}}\frac{b}{(U-V)}, \tag{35.35}$$

the roots of dispersion relation (35.29) are complex, showing that there is instability such that $\omega_1'' > 0$ (using inequality (35.35), one can easily find the positions of the points *1* and *2* in Fig. 19 — the boundaries of the instability region for the beam–plasma system considered in Sec. 19). In order to determine the maximum growth rate and the point at which this maximum is reached, we first calculate the wave group velocities. Differentiating dispersion relation (35.29) with respect to k yields

$$V_g = \frac{d\omega}{dk} = \frac{\omega(2U+V) - 3kUV}{3\omega - k(2V+U)}. \tag{35.36}$$

In the wavenumber range (35.35), where the frequencies are complex, the group velocities (35.36) are complex too. The only exception is at the point $k = 0$, at which the growth rate is maximum. Setting $k = 0$ in dispersion relation (35.29), we find $\omega''_{\max} = (\sqrt{3}/2)b$ and immediately arrive at the explicit form of an inequality opposite to (35.5), namely, $b \gg \omega_b$, which coincides with inequality (23.7) (including the notation). Setting $k = 0$ in expression (35.36), we then obtain the following formula for the convective velocity during instability:

$$U_g = \frac{2}{3}U + \frac{1}{3}V. \tag{35.37}$$

Formula (35.37) gives the velocity at which the fastest growing perturbations propagate during the single-particle instability. It is rather difficult to calculate this velocity, as well as the maximum growth rate, directly from formulas (35.34). But such a direct calculation from formulas (35.12) is a very easy task.

Figure 44 shows the real and imaginary parts of the solutions to dispersion relation (35.29) in the case of interaction of an electron beam with the forward wave in

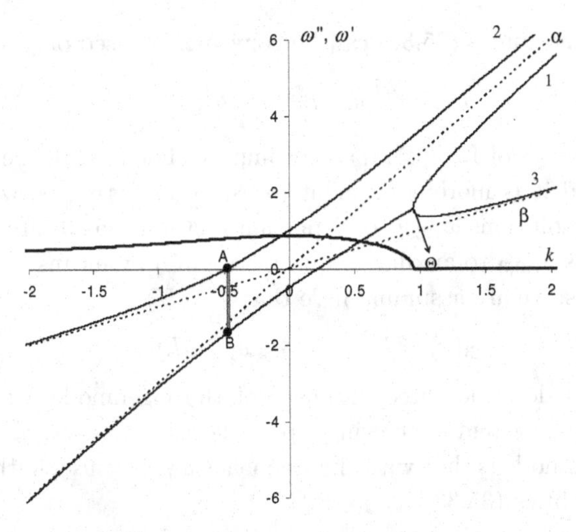

Fig. 44. Real ω' (lighter curves) and imaginary ω'' (heavier curve) parts of the frequency in the interaction of a beam with the forward wave in the single-particle regime.

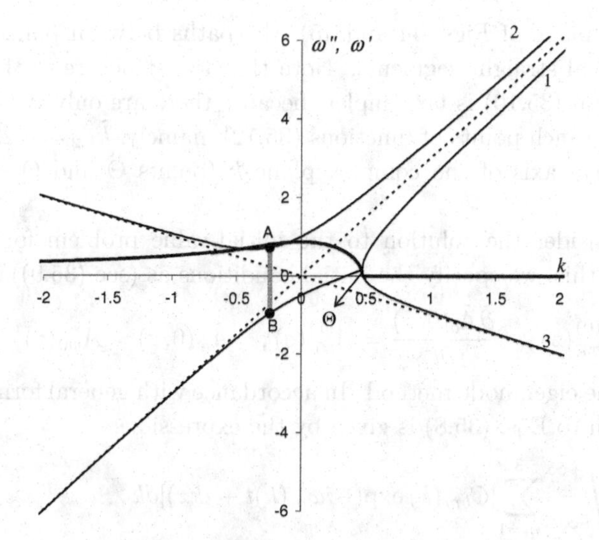

Fig. 45. Real ω' (lighter curves) and imaginary ω'' (heavier curve) parts of the frequency in the interaction of a beam with the backward wave in the single-particle regime.

a slowing-down medium. The calculations were carried out for the following relative values of the parameters: $b = 1$, $U = 3$ and $V = 1$. Curves *1*, *2*, and *3* in Fig. 44 are for the real parts of the frequencies $\omega_1(k)$, $\omega_2(k)$, and $\omega_3(k)$, respectively, and the heavy curve is for the imaginary part of the frequency $\omega_1(k)$. The asymptotes of the dispersion curves are the same as those in Fig. 42. The branches $\omega_m(k)$ of the dispersion curves are identified according to their asymptotes at $k \to +\infty$, specifically, $\omega_{1,2} = kU \mp b^{3/2}k^{-1/2}(U-V)^{-1/2}$ and $\omega_3 = kV + b^3k^{-2}(U-V)^{-2}$.

Figure 45 shows the real and imaginary parts of functions (35.34) in the case of interaction of a beam with the backward wave in a slowing-down medium. The calculations were carried out for the following relative values of the parameters: $b = 1$, $U = 3$, and $V = -1$. The notation is the same as in Fig. 44.

As functions of the wavenumber k, the solutions to cubic equation (35.29) have a fairly complicated structure. The functions $\omega_{1,2,3}(k)$ have three branch points, which are determined by simultaneously solving the dispersion relation $D = 0$ and the equation $\partial D/\partial\omega = 0$ (see (29.34)–(29.36)). For dispersion relation (35.29), the branch points $k = \tilde{k}$ are calculated from the formulas

$$\tilde{k}_{1,2,3} = (1, (-1+i\sqrt{3})/2\,, \quad (-1-i\sqrt{3})/2) \cdot 3b/(\sqrt[3]{4}(U-V))\,. \qquad (35.38)$$

One branch point, $\tilde{k}_1$, is on the real axis of the complex plane $k = k' + ik''$. In Figs. 44 and 45, this point is designated by the letter Θ. This is the merging point of the branches $\omega_1(k)$ and $\omega_3(k)$. The remaining two points, $\tilde{k}_{2,3}$, do not lie on the real axis of the complex plane k. These are the merging points of the branches $\omega_2(k)$ and $\omega_3(k)$. Thus, it is possible to pass from point A to point B through the branch points $\tilde{k}_{2,3}$, which do not lie in the plane of the figures. In the projection onto the

$k'' = 0$ plane (the plane of Figs. 44 and 45), the paths between points A and B are shown by vertical straight segments. Note that the structure of the solutions to dispersion relation (35.10) is far simpler, because there are only two eigenmode branches and two branch points of functions (35.12), namely, $\tilde{k}_{1,2} = \mp 2a/(U - V)$, which lie on the real axis of the complex plane k (points Θ and Θ' in Figs. 42 and 43).

Let us now consider the solution to the initial-value problem for the set of Eqs. (35.8). To do this, we specify the initial conditions as (see (35.9))

$$\tilde{A}_b(0,z) = A_{b0}^{(0)}(z)\,, \quad \frac{\partial \tilde{A}_b(0,z)}{\partial t} = A_{b0}^{(1)}(z)\,, \quad \tilde{A}_w(0,z) = A_{w0}(z) \tag{35.39}$$

and again utilize the eigenmode method. In accordance with general formula (28.9), the general solution to Eqs. (35.8) is given by the expressions

$$\begin{aligned} \tilde{A}_b(t,z) &= \frac{1}{2\pi}\int_{-\infty}^{+\infty}\sum_{m=1}^{3}[C_{bm}(k)\exp(-i\omega_m(k)t + ikz)]dk\,, \\ \tilde{A}_w(t,z) &= \frac{1}{2\pi}\int_{-\infty}^{+\infty}\sum_{m=1}^{3}\left[\frac{w}{\omega_m - kV}C_{bm}(k)\exp(-i\omega_m(k)t + ikz)\right]dk\,, \end{aligned} \tag{35.40}$$

where $\omega_m(k)$ are eigenfrequencies (35.34). In writing the second of expressions (35.40), we have taken into account relationship (35.16), which is also valid for the single-particle regime of the instability (the second solutions to the sets of Eqs. (35.6) and (35.8) are the same and it is from these equations that relationship (35.16) is derived). Substituting representation (35.40) into initial conditions (35.39) yields the relationships

$$\begin{aligned} \frac{1}{2\pi}\int_{-\infty}^{+\infty}\sum_{m=1}^{2}[C_{bm}(k)\exp(ikz)]dk &= A_{b0}^{(0)}(z)\,, \\ \frac{1}{2\pi}\int_{-\infty}^{+\infty}\sum_{m=1}^{2}[\omega_m C_{bm}(k)\exp(ikz)]dk &= iA_{b0}^{(1)}(z)\,, \\ \frac{1}{2\pi}\int_{-\infty}^{+\infty}\sum_{m=1}^{2}\left[\frac{w}{\omega_m - kV}C_{bm}(k)\exp(ikz)\right]dk &= A_{w0}(z)\,. \end{aligned} \tag{35.41}$$

Multiplying both sides of relationships (35.41) by $\exp(-ik'z)$ and integrating over z from minus to plus infinity, we arrive at the following set of algebraic equations for determining the unknown spectral densities $C_{bm}(k)$, $(m = 1, 2, 3)$:

$$\begin{aligned} C_{b1}(k) + C_{b2}(k) + C_{b3}(k) &= A_{b0}^{(0)}(k)\,, \\ \omega_1 C_{b1}(k) + \omega_2 C_{b2}(k) + \omega_3 C_{b3}(k) &= iA_{b0}^{(1)}(k)\,, \\ \frac{w}{\omega_1 - kV}C_{b1}(k) + \frac{w}{\omega_2 - kV}C_{b2}(k) + \frac{w}{\omega_3 - kV}C_{b2}(k) &= A_{w0}(k)\,. \end{aligned} \tag{35.42}$$

Here, we have introduced the spectral densities of the initial perturbations, $A_{b0}^{(0)}(k)$, $A_{b0}^{(1)}(k)$, and $A_{w0}(k)$ (see expressions (35.21)). Having found the spectral densities

$C_{bm}(k)$ from Eqs. (35.42) and substituting them into expressions (35.40), we finally obtain the general solution to the problem, which is not presented here because it is rather involved and poorly informative.

In order to simplify the general solution, we consider initial conditions that are simpler than conditions (35.39), specifically, $A_{b0}^{(0)}(z) \equiv A_{b0}(z) \neq 0$, $A_{b0}^{(1)}(z) \equiv 0$, and $A_{w0}(z) \equiv 0$. We also apply the quasi-harmonic approximation. It has been shown above that the maximum growth rate is attained at $k = 0$. Assume that the spectral density of the perturbation, $A_{b0}^{(0)}(k) \equiv A_{b0}(k)$, also has a sharp and narrow maximum at $k = 0$. In this case, we can expand the integrands in expressions (35.40) in powers of k and retain terms up to the first order. For $k = 0$, from formulas (35.32) and (35.34) we get

$$\omega_m = \delta_m b\,, \quad m = 1, 2, 3\,, \tag{35.43}$$

where δ_m are the cubic roots of unity (19.23). With relationships (35.43) and dispersion relation (35.29), the set of Eqs. (35.42) is substantially simplified to become

$$\begin{aligned} C_{b1}(k) + C_{b2}(k) + C_{b3}(k) &= A_{b0}(k)\,, \\ \delta_1 C_{b1}(k) + \delta_2 C_{b2}(k) + \delta_3 C_{b3}(k) &= 0\,, \\ \delta_1^2 C_{b1}(k) + \delta_2^2 C_{b2}(k) + \delta_3^2 C_{b2}(k) &= 0\,. \end{aligned} \tag{35.44}$$

These equations yield the relationships (see also the set of Eqs. (19.25))

$$C_{b1}(k) = C_{b2}(k) = C_{b3}(k) = \frac{1}{3} A_{b0}(k)\,. \tag{35.45}$$

As a result, with relationships (35.43) and (35.45), the first of expressions (35.40) in the quasi-harmonic approximation takes the form

$$\tilde{A}_b(t,z) = \frac{1}{3} \sum_{m=1}^{3} \left(\exp(-i\delta_m bt) \left\{ \frac{1}{2\pi} \int_{-\infty}^{+\infty} A_{b0}(k) \exp\left(ikz - ik \frac{d\omega_m(0)}{dk} t \right) dk \right\} \right). \tag{35.46}$$

In the case at hand, we have $d\omega_m(0)/dk = U_g$ $(m = 1, 2, 3)$, where U_g is the convective velocity during instability, given by formula (35.37). Consequently, according to relationship (28.5), the expression in braces in (35.46) is the inverse Fourier transform of the spectral density $A_{b0}(k)$ in the argument $\xi = z - U_g t$. Performing similar manipulation, we find that, in the quasi-harmonic approximation, solution (35.40) to Eqs. (35.8) has the form

$$\begin{aligned} \tilde{A}_b(t,z) &= \frac{1}{3} A_{b0}(z - U_g t) \sum_{m=1}^{3} \exp(-i\delta_m bt)\,, \\ \tilde{A}_w(t,z) &= \frac{1}{3} A_{b0}(z - U_g t) \left(\frac{w}{b} \right) \sum_{m=1}^{3} \delta_m^* \exp(-i\delta_m bt)\,, \end{aligned} \tag{35.47}$$

where the asterisk "*" stands for the complex conjugate and the convective velocity during instability, U_g, is defined by formula (35.37).

In order to deduce the asymptotic behavior of solution (35.40) on long time scales t, we specify general expressions (34.3) and (34.7). In this way, only the growing term proportional to $\exp(-i\omega_1 t)$ in the general solution (35.40) is relevant, so from Eq. (35.31) and relationships (35.34) and (35.43) we find

$$\frac{d^2\omega_1'(0)}{dk^2} = -\frac{(U-V)^2}{9b}, \quad \frac{d^2\omega_1''(0)}{dk^2} = -\frac{\sqrt{3}(U-V)^2}{9b}. \tag{35.48}$$

Substituting relationships (35.43) and (35.48) into general expressions (34.3) and (34.7) yields the following asymptotic formula for the beam wave amplitude $\tilde{A}_b(t,z)$ during the single-particle instability of an electron beam in a slowing-down medium:

$$\tilde{A}_b(t,z) = \sqrt{\frac{2}{\sqrt{3}}}(A_{b0}|_{k=0}) \exp\left(\frac{\sqrt{3}}{2}bt\right)\sqrt{\frac{9\sqrt{3}b}{\pi 8(U-V)^2 t}}$$
$$\times \exp\left[-\frac{9\sqrt{3}b}{8(U-V)^2 t}(z-U_g t)^2\right] S(t,z),$$
$$S(t,z) = \exp\left[\frac{1}{2}ibt - i\frac{9b}{8(U-V)^2 t}(z-U_g t)^2 + i\frac{\pi}{6}\right]. \tag{35.49}$$

The asymptotic behavior of the amplitude $\tilde{A}_w(t,z)$ is described by a similar formula. Asymptotic expression (35.49) has essentially the same structure as expression (35.28). The only difference is in the presence of the oscillating factor $S(t,z)$.

Although, in the limit $U-V \to 0$, asymptotic formulas (35.49) and (35.28) do not lose their meaning, they become inapplicable. The reason is that, for $d^2\omega''/dk^2 \to 0$, initial asymptotic representation (34.7) needs to be refined. But here we will not consider this question in general terms. As for the "beam + slowing-down medium" system, we can see from dispersion relations (35.10) and (35.29) that, for $V = U$, the frequency of the growing wave has the structure $\omega = kU + \beta$, with $\beta = \beta' + i\beta'' =$ const Given this structure of the frequency $\omega(k)$, the integral in formula (34.1) can be taken exactly:

$$A(t,z) = \text{const} \cdot \exp(\beta t)\Phi(z-Ut). \tag{35.50}$$

Formula (35.50) describes the characteristic function of the state vector of an unstable nondispersive system. By the way, in the limit $U-V \to 0$, asymptotic expressions (35.49) and (35.28) reduce to formula (35.50) with $\Phi(z-Ut) = \delta(z-Ut)$.

Chapter 6

Theory of Instabilities

36. Convective and Absolute Instabilities. First Criterion for the Type of Instability

The fact that the spectrum of an unstable system contains a frequency with a positive imaginary part does not yet imply that, at a certain fixed spatial point, an arbitrary perturbation will grow in time without bound. Any actual perturbation is a superposition of plane harmonic waves. In the unstable spectral range, each of the plane waves grows. But when the growing waves of different frequencies are added together, they may cancel each other. Mathematically, this means that, as $t \to \infty$, one of the factors, $\exp(\omega'' t)$, in the integrand in formula (34.1) increases without bound, while the other factor, $\exp(-i\omega' t)$, becomes an infinitely rapidly oscillating function. The result of integration depends on which of the factors turns out to dominate. That is why there are two different types of instabilities — convective and absolute.

If, for any $z =$ const and for $t \to \infty$, a perturbation $A(t, z)$ of an unstable system decreases to zero,

$$\lim_{\substack{t\to\infty \\ z=\text{const}}} A(t,z) = 0\,, \tag{36.1}$$

then the instability is called convective. In this case, a perturbation grows with time t, but simultaneously propagates in space so as to move away from any fixed point $z =$ const. And if, for a certain $z =$ const and for $t \to \infty$, a perturbation $A(t, z)$ of an unstable system grows without bound,

$$\lim_{\substack{t\to\infty \\ z=\text{const}}} A(t,z) = \infty\,, \tag{36.2}$$

then the instability is called absolute.

The above definitions obviously refer to spatially localized perturbations, in particular, those described by finite functions of the spatial coordinate z at any instant of time. In Sec. 29, it has been shown that finite perturbations cannot be described in the quasi-harmonic approximation. It also turns out that arbitrary perturbations cannot be described by the methods presented in Sec. 34, because these methods are based on expanding the integrand in formula (34.1) in powers of wavenumbers lying in a certain finite range. Hence, in order to determine the kind of instability, it is necessary to find out the asymptotic behavior (at $t \to \infty$) of the exact integral representation of solution (28.8) or (28.9) to the initial-value problem by applying methods other than those considered above.

It is convenient to use not solution (28.8) to homogeneous equation (3.11) in the initial-value problem, but rather solution (28.12) to inhomogeneous equation (3.12) with zero initial conditions. This can be done because the two solutions are equivalent. In fact, an arbitrary perturbation at any instant $t = t_0$ can be considered as a new initial condition. On the other hand, this same perturbation can be produced by a source operating during the time $t \leq t_0$. The asymptotic properties of solution (28.12) to the inhomogeneous equation are completely determined by Green's function (28.13),

$$G(t, z) = \frac{1}{(2\pi)^2} \int_{-\infty}^{+\infty} dk \int_{C(\omega)} \frac{1}{D(\omega, k)} \exp(-i\omega t + ikz) d\omega \,. \tag{36.3}$$

Hence, in order to determine the kind of instability, it is necessary to calculate the asymptotic behavior of function (36.3) at $t \to \infty$ for a certain fixed z value, say, $z = 0$. Note that Green's function (36.3) is obtained from the general solution (28.8) to the initial-value problem by setting $P_{n-1}(\omega, k) = 1$. It is for this reason that the Green's function can be used more conveniently than the solution to the initial-value problem: the properties of Green's function (36.3) depend only on the dispersion function $D(\omega, k)$, while the solution to the initial-value problem depends on $P_{n-1}(\omega, k)$ as well. We know that the dispersion function is governed exclusively by the physical system, while the polynomial $P_{n-1}(\omega, k)$ also contains information about the initial conditions, which are arbitrary.

Setting $z = 0$ in expression (36.3) yields the following integral representation of the Green's function:

$$G(t, 0) = \frac{1}{(2\pi)^2} \int_{-\infty}^{+\infty} dk \int_{C(\omega)} \frac{\exp(-i\omega t)}{D(\omega, k)} d\omega \,. \tag{36.4}$$

The task is to determine the asymptotic behavior of this representation. The contour of integration $C(\omega)$ is displayed in Fig. 1: this is a straight line that lies in the upper half of the complex plane ω and passes above all the singular points of the function $D^{-1}(\omega, k)$. The singular points are the roots $\omega_m(k)$ of the dispersion relation $D(\omega, k) = 0$. Assume that the number of roots is n and that there are no multiple roots. More precisely, multiple roots may exist, but they should be only at a finite number of branch points of the functions $\omega_m(k)$ in the complex plane $k = k' + ik''$. A detailed description of the branch points of the functions $\omega_m(k)$, as well as of their inverse functions $k_m(\omega)$, will be given later.

In the vicinity of the root, the dispersion function can be represented as

$$D(\omega,k)=(\omega-\omega_m)\frac{\partial D_m}{\partial\omega}+o(\omega-\omega_m)\,,\qquad \frac{\partial D_m}{\partial\omega}=\frac{\partial D(\omega_m)}{\partial\omega}\,,\qquad m=1,2,\ldots,n\,. \tag{36.5}$$

We integrate over ω in formula (36.4) and apply the methods of residue theory (for details, see Chapter 1, formulas (4.17)–(4.21)) to obtain

$$G(t,0)=-\frac{i}{2\pi}\sum_{m=1}^{n}\int_{-\infty}^{+\infty}\frac{\exp[-i\omega_m(k)t]}{\partial D_m/\partial\omega}dk\,, \tag{36.6}$$

where the integration is carried out along the real axis of the complex plane k.

Prior to determining the asymptotic behavior of the integral in formula (36.6) at $t\to\infty$, it is necessary to investigate the singular points of the integrand, i.e., the points where the derivative $\partial D_m/\partial\omega$ is zero. From (29.34)–(29.36) it follows that the derivative vanishes, $\partial D_m/\partial\omega=0$, at the branch points of the functions $\omega_m(k)$. Consequently, the integrand in formula (36.6) can have singularities only when at least one of the branch points of the functions $\omega_m(k)$ lies on the real axis of the complex plane k. Assume that such a point k_0 does indeed exist and that $\omega_1(k_0)=\omega_2(k_0)=\omega_0$ at this point. Using the representation $D=(\omega-\omega_1)(\omega-\omega_2)f(\omega)$ and taking into account formula (29.36), we can write the following relationships, which are valid in the vicinity of the branch point:

$$\omega_{1,2}=\omega_0\pm\varepsilon\,,\qquad \frac{\partial D_{1,2}}{\partial\omega}=\pm\varepsilon 2f(\omega_0)\,. \tag{36.7}$$

Here, $f(\omega_0)\neq 0$ and $\varepsilon\to 0$ as $k\to k_0$. We then find

$$\lim_{k\to k_0}\left(\frac{\exp[-i\omega_1(k)t]}{\partial D_1/\partial\omega}+\frac{\exp[-i\omega_2(k)t]}{\partial D_2/\partial\omega}\right)=-if^{-1}(\omega_0)t\exp(-i\omega_0 t)\,. \tag{36.8}$$

At the branch point, each of the terms on the left-hand side of relationship (36.8) approaches infinity, but when added together, the infinities cancel one another. Hence, in the integrand in formula (36.6), there are in fact no singularities associated with the branch points of the functions $\omega_m(k)$. Consequently, the integral in (36.6) can be calculated by ignoring the branch points of the functions $\omega_m(k)$.

Assume now that corresponding to certain real wavenumbers k, there are complex frequencies $\omega_m(k)$. The task is to determine which type of instability — convective or absolute — we deal with. For each of the terms in the sum over m in formula (36.6), we switch to its new own variable of integration:

$$\omega=\omega_m(k)\,,\qquad d\omega=\frac{d\omega_m}{dk}dk\,. \tag{36.9}$$

Using relationship (29.32), we can convert the integral in formula (36.6) to

$$G(t,0)=\frac{i}{2\pi}\sum_{m=1}^{n}\int_{\Omega_m}\frac{\exp(-i\omega t)}{\partial D_m/\partial k}d\omega\,. \tag{36.10}$$

Here, $D_m=D[\omega,k_m(\omega)]$, with $k_m(\omega)$ being the function inverse to the function $\omega_m(k)$, and Ω_m is the contour in the complex plane $\omega=\omega'+i\omega''$ along which the point $\omega = \omega_m(k)$ passes as k goes from minus to plus infinity. From formula (36.10) we can see that, in the integrand, there are no any singularities associated with the zeros of the derivative $\partial D_m/\partial\omega$, but there are singular points associated with the zeros of the derivative $\partial D_m/\partial k$. Relationships (29.31) and (29.32) imply that, at these points in the complex plane ω, the complex group velocity $d\omega_m/dk$ vanishes. It is the positions of these singular points that determine the asymptotic behavior of the Green's function at $t \to \infty$.

If there are no points at which $\partial D_m/\partial k = 0$ between the integration contour Ω_m and the real axis in the complex plane ω, then the contour Ω_m can be deformed into the real axis. In this case, the corresponding term in formula (36.10) is written as

$$\int_{-\infty}^{+\infty} \frac{\exp(-i\omega t)}{\partial D_m/\partial k} d\omega \,, \tag{36.11}$$

where the integration is carried out along the real axis and the function $(\partial D_m/\partial k)^{-1}$ does not have singularities. The Riemann–Lebesgue lemma (see (29.19)) then implies that integral (36.11) approaches zero as $t \to \infty$. Hence, we can formulate the first criterion for convective instability: the instability is convective when, for any $m = 1, 2, \ldots, n$, there are no points at which $\partial D_m/\partial k = 0$ between the integration contour Ω_m and the real axis in the complex plane ω.

Assume now that, for a certain m value, there is a point ω_0 at which $\partial D_m/\partial k = 0$ between the integration contour Ω_m and the real axis in the upper half of the complex plane ω. The type of the point has to be clarified. To do this, we expand the dispersion function in the relation $D(\omega, k) = 0$ in a power series in the vicinity of the values ω_0 and k_0 related by the equation $D(\omega_0, k_0) = 0$:

$$D(\omega, k) = \frac{\partial D}{\partial \omega}(\omega - \omega_0) + \frac{\partial D}{\partial k}(k - k_0) + \frac{1}{2}\frac{\partial^2 D}{\partial k^2}(k - k_0)^2 + \cdots = 0 \,. \tag{36.12}$$

Setting $\partial D/\partial k = 0$ leads to the following representation of the function $k(\omega)$ around the point ω_0 in question:

$$k = k_0 \pm \left(-2\frac{\partial D}{\partial \omega} \Big/ \frac{\partial^2 D}{\partial k^2}\right)^{1/2} \sqrt{\omega - \omega_0} \,. \tag{36.13}$$

We thus see that the point ω_0 at which $\partial D/\partial k = 0$ is a branch point of the function $k(\omega)$. This is the merging point of the two solutions to the dispersion relation $D(\omega, k) = 0$ regarded as an equation in the wavenumber k. The branch points of the function $k(\omega)$ are found by solving the following set of two equations:

$$D(\omega, k) = 0 \,, \qquad \frac{\partial D(\omega, k)}{\partial k} = 0 \,. \tag{36.14}$$

Dispersion relation (36.12) and representation (36.13) imply that, in the vicinity of the branch point ω_0, the derivative of the dispersion function with respect to k has the form

$$\frac{\partial D(\omega, k)}{\partial k} = \pm\left(-2\frac{\partial D}{\partial \omega}\frac{\partial^2 D}{\partial k^2}\right)^{1/2} \sqrt{\omega - \omega_0} = \pm\sigma\sqrt{\omega - \omega_0} \,, \tag{36.15}$$

where σ is a certain quantity, which is to be calculated from the derivatives of the dispersion function at the point (ω_0, k_0). The choice of the sign in expression (36.15) is unimportant now. The asymptotic behavior of the integral in formula (36.10) is determined by the vicinity of the branch point in the upper half-plane, provided of course that such a point is present. Therefore, taking into account expression (36.15), we can write

$$G(t,0) = \frac{i}{2\pi\sigma}\int_{\Omega_m}\frac{\exp(-i\omega t)}{\sqrt{\omega-\omega_0}}d\omega\,. \tag{36.16}$$

Recall that the point ω_0 lies between the contour Ω_m and the real axis in the upper half of the complex plane ω. The asymptotic behavior of the integral in formula (36.16) is known, so, using this familiar mathematical result, we finally obtain

$$G(t,0) = \frac{\text{Const}}{\sigma\sqrt{\pi}}\frac{1}{\sqrt{t}}\exp(-i\omega_0 t)\,, \tag{36.17}$$

where Const is a constant, which is not specified here. Since the imaginary part of ω_0 is positive, we have $G(t,0) \to \infty$ as $t \to \infty$. Consequently, the instability is absolute.

Hence, we can formulate the first criterion for absolute instability: the instability is absolute when, for some m, there is a branch point of the functions $k_m(\omega)$ at which $\partial D_m/\partial k = 0$ between the contour Ω_m and the real axis in the upper half of the complex plane ω. At the branch point of the functions $k_m(\omega)$, the complex group velocity vanishes. Accordingly, we can give a more illustrative formulation of the first criterion for absolute instability: the instability is absolute if, for some m, there is a point at which the complex group velocity of an unstable eigenmode vanishes between the contour Ω_m and the real axis in the upper half of the complex plane ω.

Let us apply the above criteria to clarify the kind of beam instability in a slowing-down medium. We begin with the case in which the instability is described by dispersion relation (35.10). Differentiating this relation with respect to k, we arrive at the following form of the second of Eq. (36.14) for finding the branch points of the function $k(\omega)$:

$$\omega(U+V) - 2kUV = 0\,. \tag{36.18}$$

The branch points of the function $k(\omega)$ are obtained by eliminating the wavenumber in dispersion relation (35.10) and Eq. (36.18):

$$\omega^2 = 4a^2\frac{UV}{(U-V)^2}\,. \tag{36.19}$$

For $V > 0$, the branch points lie on the real axis:

$$\omega_0 = \pm 2a\frac{\sqrt{UV}}{|U-V|}\,. \tag{36.20}$$

Consequently, in the case of collective interaction of a beam with the forward wave in a slowing-down medium, the instability is convective. Figure 46 shows the integration contour Ω_1 (a contour passing from $-\infty$ to $+\infty$ through the points A, B,

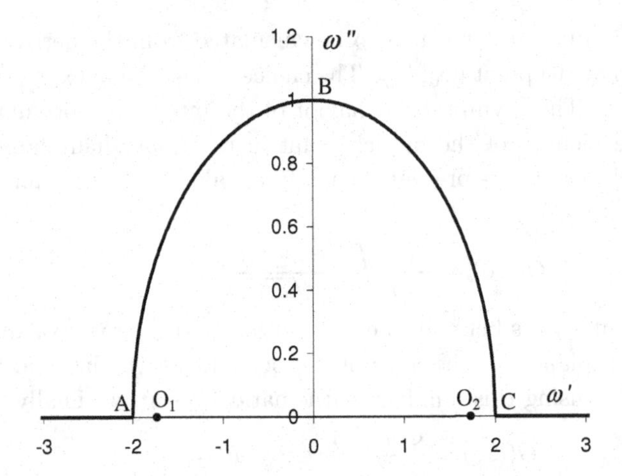

Fig. 46. Integration contour Ω_1 and branch points in the complex plane ω in the case of collective interaction of a beam with the forward wave.

C) and the branch points $O_{1,2}$ in the complex plane ω. The contour Ω_1 is plotted from the first of formulas (35.12). The real and imaginary parts of the function $\omega_1(k)$ are a parameterized form of the contour Ω_1, the parameter being k. The parameters of Fig. 46 are the same as those of Fig. 42.

For $V < 0$, the branch points lie on the imaginary axis:

$$\omega_0 = \pm i2a\frac{\sqrt{U|V|}}{U+|V|}\,. \tag{36.21}$$

We can readily see that $|\omega_0| \leq a$, the equality holding only for $|V| = U$. The uppermost point of the contour Ω_1 also lies on the imaginary axis at $\omega'' = a$, which is above the upper of the branch points (36.21). Consequently, in the case of collective interaction of a beam with the backward wave in a slowing-down medium, the instability is absolute. Figure 47 shows the integration contour Ω_1 (a contour passing from $-\infty$ to $+\infty$ through the points A, B, C) in the complex plane ω and the branch point O in the upper half-plane. The parameters of the system in Fig. 47 are the same as those in Fig. 43.

Let us now consider the case in which the instability is described by dispersion relation (35.29). Differentiating this relation with respect to k, we arrive at another equation for determining the branch points of the function $k(\omega)$:

$$\omega(2U+V) - 3kUV = 0\,. \tag{36.22}$$

The branch points are obtained by eliminating the wavenumber in dispersion relation (35.29) and Eq. (36.22):

$$\omega^3 = \frac{27}{4}b^3\frac{x^2}{(1-x)^3}\,, \tag{36.23}$$

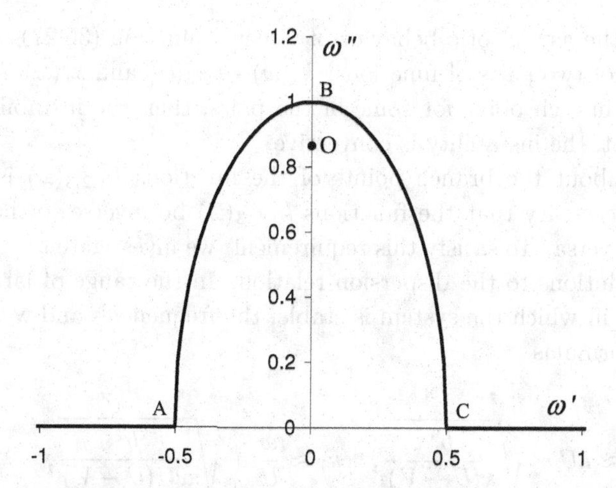

Fig. 47. Integration contour Ω_1 and branch point in the complex plane ω in the case of collective interaction of a beam with the backward wave.

where $x = V/U$. Under the above assumption $V < U$, we have $x < 1$. From Eq. (36.26) we find the branch point in the upper half of the complex plane ω:

$$\omega_0 = \frac{-1 + i\sqrt{3}}{2} bY(x) , \quad Y(x) = \frac{3}{\sqrt[3]{4}} \frac{x^{2/3}}{(1-x)} . \tag{36.24}$$

Further analysis of the kind of instability in a system described by dispersion relation (35.29) is far more complicated than in the previous case.

From the numbering adopted in formulas (35.32) and (35.34) we can see that, for real wavenumbers k, the frequency $\omega_1(k)$ has a positive imaginary part, the frequency $\omega_2(k)$ has a negative imaginary part, and the imaginary part of the frequency $\omega_3(k)$ is zero. We denote by $k_{1,2,3}(\omega)$ the functions inverse to the functions $\omega_{1,2,3}(k)$. Since the functions $k_{1,2,3}(\omega)$ are solutions to the dispersion relation, the latter can be written as

$$D(\omega, k) = C(k - k_1(\omega))(k - k_2(\omega))(k - k_3(\omega)) , \tag{36.25}$$

where C is a constant. Consequently, the following relationship is satisfied:

$$\frac{\partial D_1}{\partial k} = C(k_1(\omega) - k_2(\omega))(k_1(\omega) - k_3(\omega)) . \tag{36.26}$$

Now, we are going to estimate the integral in formula (36.10). In this formula, only the first term is relevant because it is the only term that contains the integral over the contour in the upper half of the complex plane ω. At $t \to \infty$, the second and third terms will certainly vanish. We omit the second and third terms in formula (36.10) and take into account relationship (36.26) to obtain

$$G(t, 0) = \frac{i}{2\pi C} \int_{\Omega_1} \frac{\exp(-i\omega t)}{(k_1(\omega) - k_2(\omega))(k_1(\omega) - k_3(\omega))} d\omega . \tag{36.27}$$

We thus see that the asymptotic behavior of Green's function (36.27) depends on the branch points of two pairs of functions: $k_1(\omega) \leftrightarrow k_2(\omega)$ and $k_1(\omega) \leftrightarrow k_3(\omega)$. If point (36.24) is a branch point for some of the pairs, then the instability can be absolute, and if not, the instability is convective.

The question about the branch points of the functions $k_{1,2,3}(\omega)$ is answered as follows. It is necessary that the functions $k_{1,2,3}(\omega)$ be inverse to the solutions $\omega_{1,2,3}(k)$, and vice versa. To satisfy this requirement, we investigate the asymptotic behavior of the solutions to the dispersion relation. In the range of large positive values of k and ω, in which the system is stable, the frequencies and wavenumbers are given by the formulas

$$\begin{aligned} \omega_1 &\approx kU - \sqrt{\frac{b^3}{k(U-V)}}, & k_1 &\approx \frac{\omega}{U} + \sqrt{\frac{b^3}{\omega U(U-V)}}, \\ \omega_2 &\approx kU + \sqrt{\frac{b^3}{k(U-V)}}, & k_2 &\approx \frac{\omega}{U} - \sqrt{\frac{b^3}{\omega U(U-V)}}, \\ \omega_3 &\approx kV + \frac{b^3}{k^2(U-V)^2}, & k_3 &\approx \frac{\omega}{V} - \frac{b^3 V}{\omega^2(U-V)^2}. \end{aligned} \tag{36.28}$$

It is obvious that, in these formulas, the functions in the pairs $\omega_1 \leftrightarrow k_1$, $\omega_2 \leftrightarrow k_2$, and $\omega_3 \leftrightarrow k_3$ are mutually inverse. This circumstance allows the solutions $k_m(\omega)$ to be identified in accordance with the functions $\omega_m(k)$.

Figure 48 shows the real and imaginary parts of the functions $k_{1,2,3}(\omega)$, numbered in accordance with asymptotic formulas (36.28), in the case of single-particle interaction of an electron beam with the forward wave in a slowing-down medium. The calculations were carried out for the following relative values of the parameters: $b = 1$, $U = 3$, and $V = 1$. Dotted curve α is for the asymptote $k = \omega/U$ of the functions $k_{1,2}(\omega)$, and dotted curve β is for the asymptote $k = \omega/V$ of the function $k_3(\omega)$. Curves *1*, *2*, and *3* are for the real parts of the functions $k_1(\omega)$, $k_2(\omega)$, and $k_3(\omega)$, and the heavy curve is for the imaginary part of the function $k_1(\omega)$. The branch point (36.24) is the merging point of the branches $k_2(\omega)$ and $k_3(\omega)$. Thus, it is possible to pass from point A to point B through the branch point ω_0, which does not lie in the plane of Fig. 48. In the projection onto the $\omega'' = 0$ plane (the plane of the figure), the path between points A and B is shown by a vertical straight segment.

Hence, in the case of interaction of an electron beam with the forward wave, point (36.24) is a branch point of the functions $k_2(\omega)$ and $k_3(\omega)$. This point does not contribute to the integral in formula (36.27). That is why the instability in the single-particle interaction with the forward wave is convective.

Next, Fig. 49 gives an additional, more detailed illustration of the position of the branches $k_{1,2,3}(\omega)$ and their merging point, corresponding to the frequency branch

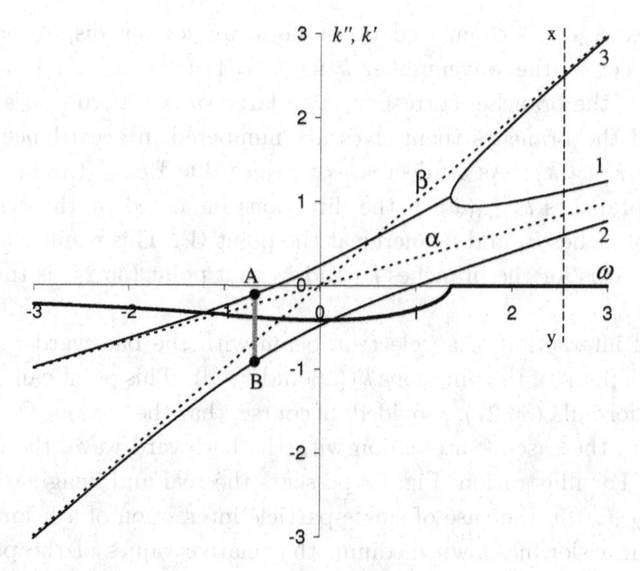

Fig. 48. Real parts k' (lighter curves) and imaginary part k'' (heavier curve) of the wavenumber in the case of single-particle interaction of a beam with the forward wave.

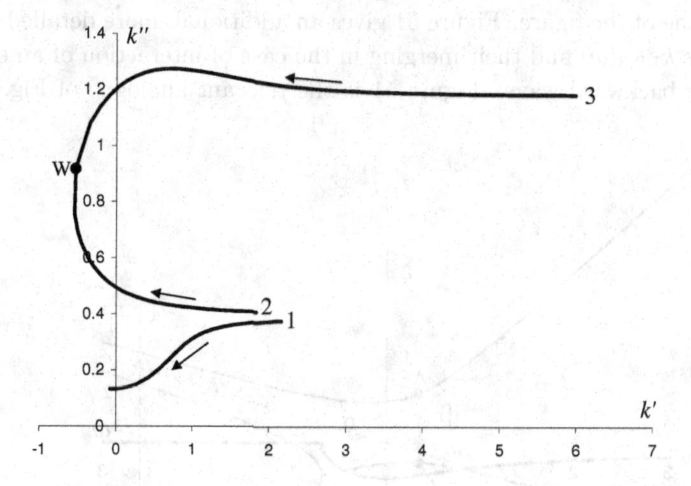

Fig. 49. Branches $k_{1,2,3}(\omega)$ and branch point W for convective instability.

point ω_0 (36.24), in the complex plane $k' + ik''$. In the complex plane $\omega' + i\omega''$, we move from infinity along a contour given by the relationships $\omega'' = \operatorname{Im}\omega_0$ and $\omega' \in [\operatorname{Re}\omega_0, +\infty)$. For large positive ω' (such that $\omega' \to +\infty$ or $\omega' \gg |\omega_0|$), the branches $k_{1,2,3}(\omega)$ are uniquely identified (see formulas (36.28)). In Fig. 48, the vertical dashed line x–y is drawn corresponding to a sufficiently large value $\omega' > 0$. This line intersects all the three wavenumber branches, enabling their real parts to be arranged in decreasing order, $k_3' > k_1' > k_2'$. Figure 49 displays the three complex branches $k_{1,2,3}(\omega)$ calculated by numerically solving dispersion relation (35.29) with respect to the wavenumber k for $b = 1$, $U = 3$, and $V = 1$. The rightmost points of the branches correspond to a large value $\omega' > 0$ (more precisely, to $\omega' = 10$), and the branches themselves are numbered in accordance with the inequalities $k_3' > k_1' > k_2'$. As ω' decreases to the value $\mathrm{Re}\,\omega_0$, the image points move along the branches $k_{1,2,3}(\omega)$ in the directions indicated by the arrows. For $\omega' = \mathrm{Re}\,\omega_0$, the branches k_2 and k_3 merge at the point W. This result confirms the above conclusion: it is for the branches k_2 and k_3 that point (36.24) is the merging point.

In the case of interaction of an electron beam with the backward wave, point (36.24) is a branch point of the functions $k_1(\omega)$ and $k_3(\omega)$. This point can contribute to the integral in formula (36.27), provided, of course, that the contour Ω_1 lies above it. That is why, in the case of interaction with the backward wave, the instability can be absolute. For illustration, Fig. 50 presents the real and imaginary parts of the functions $k_{1,2,3}(\omega)$ in the case of single-particle interaction of a beam with the backward wave in a slowing-down medium, the relative values of the parameters being $b = 1$, $U = 3$, and $V = -1$. The notation is the same as in Fig. 48. What distinguishes Fig. 50 is that the vertical straight segment AB connects the branches $k_1(\omega)$ and $k_3(\omega)$, showing that they merge at the branch point (36.24), which is outside the plane of the figure. Figure 51 gives an additional, more detailed picture of the branches $k_{1,2,3}(\omega)$ and their merging in the case of interaction of an electron beam with the backward wave. Figure 51 is the relevant analogue of Fig. 49. In

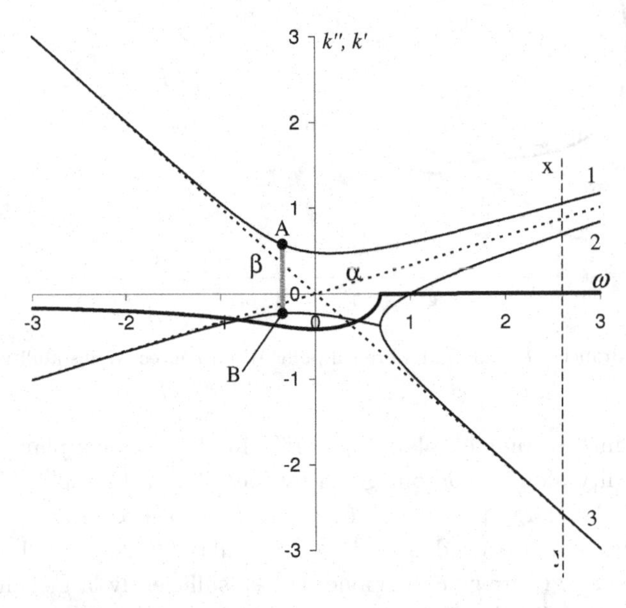

Fig. 50. Real parts k' (lighter curves) and imaginary part k'' (heavier curve) of the wavenumber in the case of single-particle interaction of a beam with the backward wave.

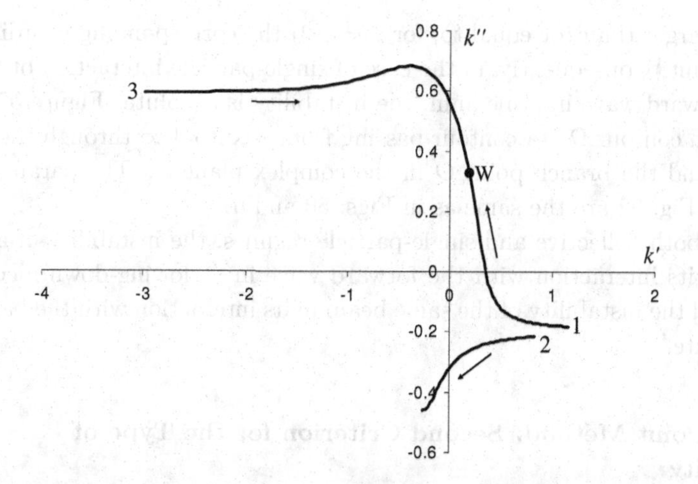

Fig. 51. Branches $k_{1,2,3}(\omega)$ and branch point W for absolute instability.

Fig. 50, the vertical dashed line x–y intersects the three wavenumber branches in such a way as to yield the inequalities $k_1' > k_2' > k_3'$, with $k_3' < 0$. We thus immediately see that the numbering of the branches $k_{1,2,3}(\omega)$ in Fig. 51 is correct. Consequently, the point W is the merging point of the branches k_1 and k_3.

In order to finally answer the question about the kind of instability, it is necessary to examine the relative positions of the contour Ω_1 and point (36.24). In the case of interaction of a beam with the backward wave, we have $V < 0$ and therefore $x < 0$. In the region $x < 0$, the maximum of the function $Y(x)$ is equal to one and is reached at $x = -2$ ($V = -2U$). We therefore have $\omega''_{0\,\max} = (\sqrt{3}/2)b$. The imaginary coordinate of the uppermost point of the contour Ω_1 is $\omega'' = (\sqrt{3}/2)b$, a

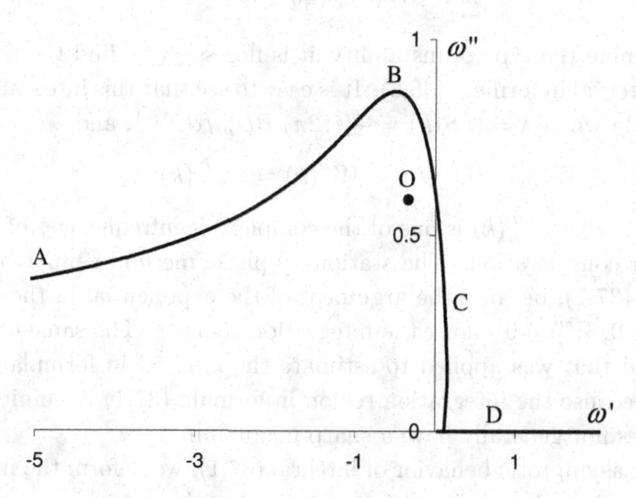

Fig. 52. Integration contour Ω_1 and branch point in the complex plane ω in the case of single-particle interaction of a beam with the backward wave.

value that is larger than (or equal to, for $x = -2$) the corresponding coordinate of the branch point. Consequently, in the case of single-particle interaction of a beam with the backward wave in a medium, the instability is absolute. Figure 52 shows the integration contour Ω_1 (a contour passing from $-\infty$ to $+\infty$ through the points A, B, C, D) and the branch point O in the complex plane ω. The parameters of the system in Fig. 52 are the same as in Figs. 50 and 51.

Hence, in both collective and single-particle regimes, the instability of an electron beam in its interaction with the forward wave in a slowing-down medium is convective and the instability of the same beam in its interaction with the backward wave is absolute.

37. Saddle-Point Method. Second Criterion for the Type of Instability

In this section, we give another formulation of the criteria for absolute and convective instabilities. We begin by considering a purely mathematical problem of calculating the asymptotic behavior of the integral

$$Y(\lambda) = \int_{-\infty}^{+\infty} S(k) \exp[\lambda W(k)] dk \tag{37.1}$$

for large positive values of the real argument λ ($\lambda \to +\infty$). Here, $S(k)$ and $W(k)$ are analytic functions of the complex variable $k = k' + ik''$ and the integration is carried out along the real axis $k'' = 0$ of the complex plane k. We assume that, for real values of the variable k, the function $W(k) = W'(k) + iW''(k)$ has a nonzero real part $W'(k)$ such that

$$\lim_{k' \to \pm\infty} W'(k)|_{k''=0} \to 0\,. \tag{37.2}$$

In order to determine the type of instability, it is necessary to find the asymptotic behavior of the integral in formula (36.6). It is easy to see that this integral coincides with integral (37.1) when $\lambda = t$, $S(k) = -(i/2\pi)(\partial D_m/\partial\omega)^{-1}$, and

$$W'(k) = \omega''_m(k)\,, \quad W''(k) = -\omega'_m(k)\,, \tag{37.3}$$

where $\omega_m(k) = \omega'_m(k) + i\omega''_m(k)$ is one of the complex eigenfrequencies of the physical system under consideration. The stationary-phase method cannot be used to estimate integral (37.1), because the argument of the exponential in the integrand is complex ($W' \neq 0$, $W'' \neq 0$) along the integration contour. The same can be said about the method that was applied to estimate the integral in formula (34.1): it cannot be used because the integration region in formula (37.1) is infinite and the function $S(k)$ does not generally have a sharp maximum.

To deduce the asymptotic behavior of integral (37.1), we deform the integration contour so that its initial and end points are fixed at $k' = \mp\infty$ on the real axis of the complex plane k. The integrand is an analytic function; consequently, such a deformation of the integration contour does not change the value of the integral (by Cauchy's theorem). We must also take into account the fact that, since the physical system is unstable, there would necessarily be an interval on the real axis of the complex plane k in which $W' = \omega''_m > 0$. If there exist a deformed integration contour $C(k)$ along which

$$W'(k) = \omega''_m(k) \leq 0 \tag{37.4}$$

everywhere, then, for $\lambda \to \infty$, integral (37.1) obviously vanishes. If such an integration contour does not exist, then, for $\lambda \to \infty$, integral (37.1) can be nonzero. In this case, the asymptotic behavior of integral (37.1) is usually constructed by using the saddle-point method, also known as the method of steepest descent.

The essence of the saddle-point method consists in choosing such an integration contour $C(k)$ that, along its certain small portion, the function $W'(k)$ increases sharply to its maximum value and then decreases abruptly and the function $W''(k)$ remains constant (in order that there be no rapid oscillations in the integrand, which lower the value of the integral). It can thus be expected that, for large λ values, the main contribution to the integral will come just from this small portion of the contour $C(k)$.

The analytic nature of the function $W(k)$ implies that its real and imaginary parts, $W'(k)$ and $W''(k)$, are harmonic functions. In accordance with the maximum principle for harmonic functions, the real part $W'(k)$ cannot reach its absolute maximum at the interior points of the region where it is analytic; i.e., inside this region, there are no points away from which the function $W'(k)$ decreases in all directions. Consequently, the surface of the function $W'(k) = W'(k', k'')$ can have only saddle points. If there are no saddle points, then the integration contour can be deformed so as to satisfy condition (37.4), in which case integral (37.1) vanishes for $\lambda \to \infty$.

We assume that there is only one saddle point $k_0 = k'_0 + ik''_0$ on the surface $W'(k', k'') = \omega''_m(k', k'')$ and introduce the notation $W'(k_0) = W'_0$. If $W'_0 < 0$, then the integration contour can again be chosen so as to satisfy condition (37.4), in which case integral (37.1) again vanishes for $\lambda \to \infty$. Hence, the only case when the integral can be nonzero for $\lambda \to \infty$ is that in which $W'_0 > 0$, where W'_0 is the value of the real part of the function $W(k) = W'(k) + iW''(k)$ at the saddle point $k_0 = k'_0 + ik''_0$. Now, we assume that $W'_0 > 0$ and consider the lines of constant values of the function $W'(k) = W'(k_0) = W'_0$. By virtue of the maximum principle for harmonic functions, these lines cannot be closed curves because they go to infinity in the complex plane k. The lines of constant values of $W'(k) = W'(k_0) = w'_0$ divide the complex plane K into sectors, within which the function $W'(k)$ is either less than or greater than W'_0. The sectors in which $W'(k) < W'_0$ are called negative, and the sectors in which $W'(k) > W'_0$ are called positive. Let there be only four sectors — two negative and two positive. Since the lines of constant values of $W'(k) = W'(k_0) = W'_0$ emerge from and go to infinity and since conditions (37.2) and $W'_0 > 0$ are satisfied, the points $k' = \mp\infty$ of the real axis of the complex plane k

lie in the opposite negative sectors. Consequently, the integration contour $C(k)$ can be chosen such that it emerges from infinity in one negative sector, passes through the saddle point k_0, and goes to infinity in the opposite negative sector. This is the case to which the saddle-point method applies.

The saddle point k_0 is the intersection point of two lines, which we denote by L_{max} and L_{min} and along which the function $W'(k)$ varies especially sharply: along the line L_{max}, the function $W'(k)$ decreases in both directions away from the saddle point and, along the line L_{min}, this function increases in both directions away from k_0 (see Figs. 53, 54 below). Along the same lines, the imaginary part

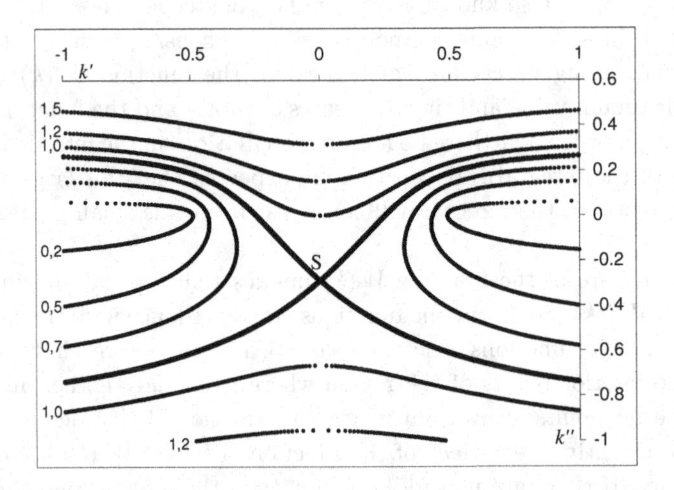

Fig. 53. Lines of constant values of the imaginary part of the frequency, $\omega''(k)$, in the complex plane $k = k' + ik''$ in the case of instability in the collective interaction with the backward wave. The saddle point is designated by S.

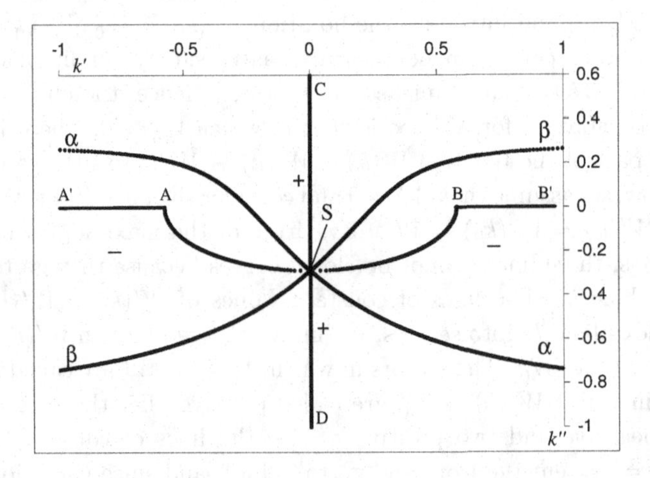

Fig. 54. Saddle point S, negative and positive sectors, and steepest descent line AB.

$W''(k)$ is constant — an assertion that follows from the analytic nature of the function $W(k)$, or, more precisely, from the Cauchy–Riemann conditions (see below for details on how to apply these conditions). The analytic nature of the function $W(k)$ also implies that its derivative vanishes at the saddle point, $dW(k_0)/dk = 0$. The integration contour $C(k)$ should pass through the saddle point along the line $L_{\max}$, i.e., in the direction of steepest descent from the saddle of the surface $W'(k', k'') = \omega''_m(k', k'')$ — a requirement that has evidently given the name to the method.

Let us determine the direction of the tangent to the line $L_{\max}$ at the saddle point k_0. To do this, we introduce the notation

$$k - k_0 = \rho \exp(i\varphi) \tag{37.5}$$

(with $\rho > 0$). Near the saddle point, we can write

$$W(k) - W(k_0) = \frac{1}{2}\frac{d^2W(k_0)}{dk^2}\rho^2 \exp(2i\varphi) + o(\rho^2)\,, \tag{37.6}$$

where we have taken into account the fact that the first derivative of the function $W(k)$ vanishes at the saddle point. We also introduce the notation

$$\frac{d^2W(k_0)}{dk^2} = q\exp(i\mu) \quad (q > 0) \tag{37.7}$$

and, ignoring terms on the order of $o(\rho^2)$, write out the imaginary component of relationship (37.6):

$$W''(k) - W''(k_0) = \frac{1}{2}q\rho^2 \sin(2\varphi + \mu)\,. \tag{37.8}$$

Since $W''(k) = \text{const}$ along the lines $L_{\max}$ and $L_{\min}$, from relationship (37.8) we find the following equation for determining the directions of the tangents to these lines at the saddle point:

$$\sin(2\varphi + \mu) = 0\,. \tag{37.9}$$

Equation (37.9) yields two different directions in the complex plane k:

$$\varphi = \varphi_1 = -\frac{\mu}{2}\,, \quad \varphi = \varphi_2 = -\frac{\mu}{2} + \frac{\pi}{2}\,. \tag{37.10}$$

These are precisely the slope angles of the sought-for tangents to the lines $L_{\max}$ and $L_{\min}$ at the saddle point. In order to identify the tangent to the line $L_{\max}$, we consider the real component of relationship (37.6):

$$W'(k) - W'(k_0) = \frac{1}{2}q\rho^2 \cos(2\varphi + \mu) = \pm\frac{1}{2}q\rho^2\,, \tag{37.11}$$

where the plus "+" and minus "−" signs refer to the angles $\varphi = \varphi_1$ and $\varphi = \varphi_2$, respectively.

Let us now take into account the fact that the main contribution to integral (37.1) along the contour $C(k) = L_{\max}$ comes from the vicinity of the saddle point (away from this point, the integrand is exponentially small in comparison with the function $\exp(\lambda W_0'))$. That is why we can switch from integration along the line $L_{\max}$ to integration along the tangent to $L_{\max}$ at the saddle point k_0. By virtue of relationships (37.6)–(37.11), the following expansion is valid on the tangent to $L_{\max}$ in the vicinity of the point k_0:

$$W(k) = W(k_0) - \frac{1}{2}qs^2 . \tag{37.12}$$

Here, s is the distance from the point k_0 along the tangent. If we treat s as a natural parameter, then, to highest order in s, we can represent the equation of the tangent as

$$k = k_0 + s\exp(i\varphi_2) , \tag{37.13}$$

where the angle φ_2 is defined in (37.10). Proceeding in the same way as with the stationary-phase method and taking into account relationships (37.12) and (37.13), we can transform integral (37.1) at large positive λ values to

$$Y(\lambda) \approx S(k_0)\exp[\lambda W(k_0) + i\varphi_2] \int_{-\sigma}^{+\sigma} \exp\left(-\frac{\lambda}{2}qs^2\right) ds , \tag{37.14}$$

where the real number σ determines the size of the vicinity of the saddle point that provides a correct estimate of the integral. Since, for a large λ value, the integrand in formula (37.14) decreases sharply with increasing s^2, we can set $\sigma = \infty$. Using formula (34.5), we then arrive at the following final asymptotic estimate of integral (37.1):

$$Y(\lambda) \approx S(k_0)\sqrt{\frac{2\pi}{\lambda q}}\exp[\lambda W(k_0) + i\varphi_2] . \tag{37.15}$$

Here, k_0 is the complex coordinate of the saddle point of the function $W'(k)$ and

$$q = \left|\frac{d^2W(k_0)}{dk^2}\right| , \quad \varphi_2 = \frac{\pi}{2} - \frac{1}{2}\arg\left(\frac{d^2W(k_0)}{dk^2}\right) . \tag{37.16}$$

For $W'(k_0) > 0$, the quantity (37.15) approaches infinity as $\lambda \to \infty$. In the opposite case, we have $Y(\lambda) \to 0$. If the function $W'(k)$ does not have a saddle point, then the initial integral (37.1) also vanishes for large λ values.

Taking into account the relationships

$$\begin{aligned} W(k) &= W'(k) + iW''(k) = -i\omega_m(k) , \\ W'(k) &= W_m''(k) , \quad W''(k) = -\omega_m'(k) , \end{aligned} \tag{37.17}$$

which are valid by virtue of (37.3), we can give a formulation of the criteria for absolute and convective instabilities that differs from the formulation presented in Sec. 36. The second criterion for convective instability reads: if, for any m, the imaginary parts of the frequency — the functions $\omega_m''(k)$ — have no saddle points, then the instability is convective; and, if the functions $\omega_m''(k)$ have saddle points k_0 in each of which $\omega_m''(k_0) \leq 0$, then the instability is convective, too. The second criterion for absolute instability reads: if, at least at one saddle point, the imaginary part of at least one of the eigenfrequencies is positive, $\omega''_m(k_0) > 0$, then the instability is absolute. From estimate (37.15), we can see that, for the absolute instability, the asymptotic behavior of Green's function (36.6) is described by the formula

$$G(0,t) = \text{Const}\frac{1}{\sqrt{t}}\exp(-i\omega_0 t)\,, \tag{37.18}$$

where $\omega_0 = \omega'_m(k_0) + i\omega''_m(k_0)$ is the value of the complex frequency at the saddle point k_0. Formula (37.18) has the same structure as formula (36.17).

At the saddle point of the function $\omega''_m(k)$, the derivative of the analytic function $\omega_m(k) = \omega'_m(k) + i\omega''_m(k)$ of the complex wavenumber vanishes, as can be seen from the Cauchy–Riemann conditions

$$\frac{\partial\omega'_m(k',k'')}{\partial k'} = \frac{\partial\omega''_m(k',k'')}{\partial k''}\,, \quad \frac{\partial\omega'_m(k',k'')}{\partial k''} = -\frac{\partial\omega''_m(k',k'')}{\partial k'} \tag{37.19}$$

and from the fact that the derivative of an analytic function is independent of the direction of differentiation. Therefore, at the saddle point k_0, we have

$$\frac{d\omega_m(k_0)}{dk} = 0\,. \tag{37.20}$$

This equality means that the complex group velocity vanishes at the saddle point. Accordingly, the second criterion for absolute instability, formulated in this section in terms of the notion of a saddle point, can be reformulated as follows: if, in the complex plane k, there is a point at which the group velocity (37.20) vanishes and the complex frequency has a positive imaginary part, then the instability is absolute. If this is not the case, then the instability is convective.

The second criterion for absolute instability, which has been constructed based on the saddle point method, agrees with the first criterion, established above in the language of the saddle point of the functions $k_m(\omega)$. Note that, at the branch point ω_0 of the functions $k_m(\omega)$, two of their branches coincide and the functions are equal to a certain complex value k_0; on the other hand, in the complex plane k, this same k_0 value determines a point at which the complex group velocity of one of the eigenmodes vanishes, i.e., equality (37.20) holds. Of course, at the point k_0, the frequency of this eigenmode is ω_0.

Nevertheless, the formulations of the second and first criteria for absolute instability are somewhat contradictory. Thus, the first criterion says that the instability is absolute if, in the upper half of the complex plane ω, there is a point "between the contour Ω_m and the real axis" at which the complex group velocity of the unstable eigenmode vanishes. Except for the words in quotation marks, the first and second criteria are identical. In order to remove the contradiction, let us consider the properties of the surface $\omega''_m(k', k'')$. The saddle point k_0 of this surface is the intersection point of the lines $L_{\max}$ and $L_{\min}$ (at this point, the lines are perpendicular). At the saddle point, the imaginary part of the frequency is $\omega''_m(k_0) = \omega''_0$. Along the line $L_{\max}$, the function $\omega''_m(k)$ decreases in both directions away from the saddle point, and, along the line $L_{\min}$, the function $\omega''_m(k)$ increases in both directions away from k_0. The maximum instability growth rate $\omega''_{\max}$ is attained at the intersection point of the line $L_{\min}$ with the real axis of the complex plane k. In the complex plane ω, this same maximum growth rate $\omega''_{\max}$ determines the uppermost point of the contour Ω_m. Since the points $\omega''_m(k_0)$ and $\omega''_{\max}$ lie on the line $L_{\min}$, we have $\omega''_m(k_0) \le \omega''_{\max}$, the equality holding only when the saddle point lies on the real axis $k''_0 = 0$ of the complex plane k. Consequently, in the complex plane ω, the saddle point ω_0 (which is the branch point of the functions $k_m(\omega)$) is closer to the real axis than the uppermost point of the contour Ω_m. Hence, the words in quotation marks — "between the contour Ω_m and the real axis" — are unnecessary because they repeat the condition that is certainly satisfied.

Let us apply the above criteria to analyze the type of instability of an electron beam in a slowing-down medium. We begin with the collective regime. In this case, the functions $\omega_m(k)$ are solutions (35.12) to dispersion relation (35.10). In the wavenumber range determined by inequalities (35.13), the imaginary part of the frequency $\omega_1(k)$ is positive for real k values. Figure 53 shows the lines of constant values of the function $\omega''_1(k', k'')$ in the complex plane $k = k' + ik''$ in the case of collective interaction of a beam with the backward wave in a slowing-down medium. The relative values of the parameters, $a = 1$, $U = 3$, and $V = -1$ are the same as in Fig. 43. In Fig. 53, the saddle point is designated by S. At this point, the imaginary part of the complex frequency is $\omega''_0 = \sqrt{3}/2 \approx 0.87$. The saddle point is at $k' = 0$ and $k'' = -1/(2\sqrt{3}) \approx -0.29$ (these coordinates are found by using relationships (36.18) and (36.19)). For the above parameter values, we have $\omega''_{\max} = 1 > \omega''_0$.

Next, Fig. 54 shows the saddle point (point S), the boundaries of the sectors (curves α, β), the line $L_{\max}$ (curve AB), and the line $L_{\min}$ (curve CD). The negative and positive sectors are designated by the minus "−" and plus "+" symbols, respectively. Since the saddle point S is present and, at this point, we have $\omega''_0 > 0$, the second criterion, obtained by the saddle point method, implies that, in the collective interaction of a beam with the backward wave, the instability is absolute. This result is in complete agreement with what follows from the first criterion, obtained based on an analysis of the branch point of the functions $k_{1,3}(\omega)$. Moreover, these two criteria are equivalent. Indeed, in the case at hand, we have $\omega''_0 < \omega''_{\max}$, so the branch point is between the contour Ω_1 and the real axis of the complex plane ω.

Substituting formula (36.21) — the value of the frequency at the saddle point S — into formula (37.18), we obtain the following asymptotic expression for the Green's function of a "beam + slowing-down medium" system in the case of instability in the collective interaction with the backward wave:

$$G(0,t) = \text{Const}\frac{1}{\sqrt{t}}\exp\left(2a\frac{\sqrt{U|V|}}{U+|V|}t\right), \quad t \to \infty\,. \tag{37.21}$$

In the case of collective interaction of a beam with the forward wave, the situation is different. Figure 55 shows the relevant lines of constant values of the function $\omega''_1(k', k'')$ in the complex plane $k = k' + ik''$. The relative values of the parameters,

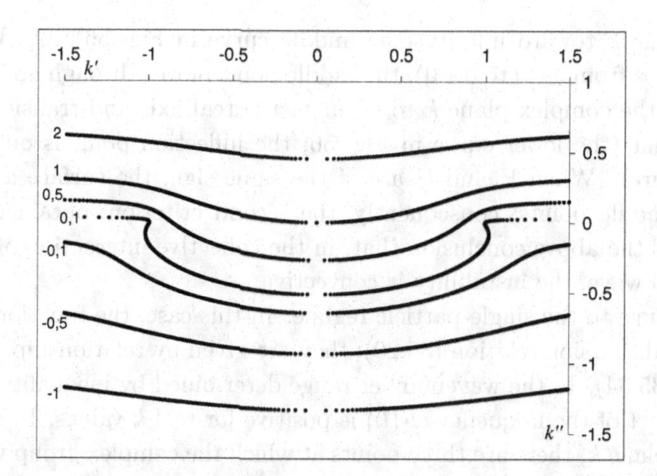

Fig. 55. Lines of constant values of the imaginary part of the frequency, $\omega''(k)$, in the complex plane $k = k' + ik''$ in the case of instability in the collective interaction with the forward wave.

$a = 1$, $U = 3$, and $V = 1$, are the same as in Fig. 53. We can see that the topology of the surface of the function differs from that shown in Fig. 53. In particular, the saddle point S, at which the function ω_1'' is positive, disappears. The points at which equality (37.20) holds are also present for V and U of the same sign, but now we have $\omega_1'' = 0$ at these points (see formula (36.20)).

The disappearance of the saddle point of the surface $\omega_1''(k', k'')$ is illustrated in Fig. 56, which shows the function $\omega_1''(k' = 0, k'')$ for $U = 3$ and for different values of the velocity V. For $V = -1$, the saddle point is at the imaginary axis of the complex plane k (it is the lowermost point of the upper curve). As V increases from a negative value to zero, the saddle point moves along the imaginary axis of

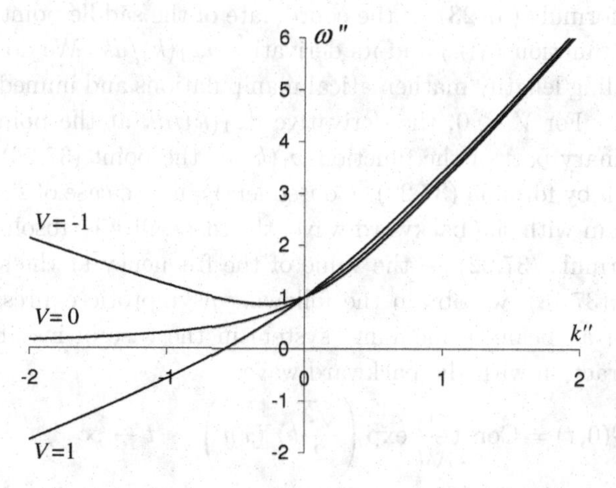

Fig. 56. Illustration of the disappearance of the saddle point as the group velocity of a wave interacting with a beam changes sign.

the complex plane k toward infinity (the middle curve in Fig. 56). As V passes through zero (rises from $0-0$ to $0+0$), the saddle point moves through an infinitely remote point in the complex plane k to occur at the real axis and transforms into an inflection point (the lower curve in Fig. 56; the inflection point is outside the plane of the figure). When V and U are of the same sign, the surface $\omega_1''(k', k'')$ does not have saddle points; consequently, the second criterion, obtained in this section, confirms the above conclusion that, in the collective interaction of a beam with the forward wave, the instability is convective.

Now, let us turn to the single-particle regime. In this case, the functions $\omega_m(k)$ are solutions to dispersion relation (35.29); they are given by relationships and formulas (35.31)–(35.34). In the wavenumber range determined by inequality (35.35), the imaginary part of the frequency $\omega_1(k)$ is positive for real k values. In addition, in the complex plane k, there are three points at which the complex group velocities (35.36) vanish. From formula (36.24), we can see that, at one of these points, the imaginary part of the frequency is positive and is given by the formula

$$\omega_0'' = \frac{\sqrt{3}}{2} bY(x) , \tag{37.22}$$

where $x = V/U < 1$. The point in the complex plane k at which the group velocity vanishes and equality (37.22) holds is determined from relationships (36.22)–(36.24):

$$k_0 = \frac{-1 + i\sqrt{3}}{2} \frac{2U + V}{3UV} bY(x) . \tag{37.23}$$

If point (37.23) is a saddle point of the surface $\omega_1''(k)$ and if the frequency ω_1'' at this point is given by formula (37.22), then the second criterion implies that the instability is absolute. In any other case, the instability is convective. Although it is possible to check directly whether the second criterion holds or not, but doing so requires laborious manipulations. Specifically, it is necessary to obtain an expression for the derivative $d\omega_1(k)/dk$ from relationships and formulas (35.31)–(35.34) and then to substitute formula (37.23) — the coordinate of the saddle point — into the expressions for the function $\omega_1(k)$ and its derivative $d\omega_1(k)/dk$. We do not present here the corresponding lengthy mathematical manipulations and immediately write out the final result. For $V < 0$, the derivative $d\omega_1(k)/dk$ at the point (37.23) is zero and the imaginary part of the function $\omega_1(k)$ at the point (37.23) is positive, its value being given by formula (37.22). Consequently, in the case of single-particle interaction of a beam with the backward wave, the instability is absolute.

Substituting formula (37.22) — the value of the frequency at the saddle point S — into formula (37.18), we obtain the following asymptotic expression for the Green's function of a "beam + medium" system in the case of instability in the single-particle interaction with the backward wave:

$$G(0, t) = \text{Const} \frac{1}{\sqrt{t}} \exp\left(\frac{\sqrt{3}}{2} bY(x)t\right) , \quad t \to \infty . \tag{37.24}$$

That the instability in the single-particle interaction of a beam with the backward wave ($V > 0$) is convective can be verified in the same way.

38. Third Criterion for the Type of Instability

Of course, we speak not of three different criteria for absolute and convective instabilities but of three formulations of one and the same criterion or, more precisely, of different approaches to estimating the asymptotic value of the integral in formula (36.4). In considering the first criterion in Sec. 36, we have first integrated over frequency ω in formula (36.4) and then converted the result into formula (36.6). In this section, instead of doing so, we first integrate over wavenumber k (along the real axis $k'' = 0$) in formula (36.4). So, we represent the dispersion function $D(\omega, k)$ in the form (see (36.25))

$$D(\omega, k) = C \prod_{m=1}^{N} (k - k_m(\omega)), \tag{38.1}$$

where C is a constant, N is the order of the dispersion function in the variable k, and $k_m(\omega)$ $(m = 1, 2, \ldots, N)$ are solutions to the dispersion relation $D(\omega, k) = 0$ regarded as an equation in the wavenumber. The order N can generally differ from the order n of the dispersion function in the frequency ω. We assume for the moment that the dispersion function is a polynomial of degree two or higher in k, i.e., that $N \geq 2$.

The function $D^{-1}(\omega, k)$ is analytic over the entire complex plane k except at a finite number of isolated singular points $k = k_m(\omega)$, and, at infinity, it decreases to zero according to the law $|k|^{-2}$ or more sharply. Consequently, in formula (36.4), we can integrate over k by using one of the following formulas:

$$J(\omega) \equiv \int_{-\infty}^{+\infty} \frac{1}{D(\omega, k)} dk = \pm 2\pi i \sum_{m=1}^{N^{(\pm)}} \text{Res}[D^{-1}(\omega, k), k_m^{(\pm)}(\omega)]. \tag{38.2}$$

Here, $k_m^{(\pm)}(\omega)$ are the roots of dispersion relation (38.1) in the upper and lower halves of the complex plane k, respectively, and $N^{(\pm)}$ are the numbers of these roots in the corresponding half-planes. Formulas (38.2) are obtained from residue theory by closing the integration path $-\infty < k < +\infty$ by a semicircle of infinite radius in the upper (the plus sign) or lower (the minus sign) half of the complex plane k. Function (38.2) being calculated, the problem of determining the kind of instability, in accordance with formula (36.4), reduces to that of finding the asymptotic behavior of the Green's function

$$G(t, 0) = \frac{1}{(2\pi)^2} \int_{C(\omega)} J(\omega) \exp(-i\omega t) d\omega. \tag{38.3}$$

It is obvious that the integral in formula (38.3) should not depend on which semicircle closes the integration path $-\infty < k < +\infty$ in calculating the integral. But it is also obvious that the value of the integral in formula (38.2) depends strongly on the positions of the poles $k_m^{(\pm)}(\omega)$ in the complex plane k. In turn, the positions of the poles $k_m^{(\pm)}(\omega)$ depend on the frequency ω, which varies along the integration contour $C(\omega)$ in formula (38.3) (see Fig. 1). The contour $C(\omega)$ is the straight line $\omega'' = \sigma = \text{const}$, which lies above all the singular points of the function $D^{-1}(\omega, k)$. Consequently, in the complex plane ω, this contour can be displaced an arbitrarily large distance upward, $\omega'' = \sigma \to +\infty$. Assume that, for $|\omega| \to \infty$, the functions $k_m(\omega)$ can be represented as

$$k_m(\omega) = \frac{\omega}{v_m} + \alpha_m + \varphi_m(\omega)\,, \tag{38.4}$$

where v_m and α_m are constants and $\varphi_m(\omega)$ is a function that goes to zero as $|\omega| \to \infty$. Note that representation (38.4) is valid for solutions to dispersion relations (35.10) and (35.29) regarded as equations in the wavenumber k (see, e.g., formulas (36.28)). If relationships (38.4) are valid for all $m = 1, 2, \ldots, N$, then, for $\omega'' \to +\infty$, the question of to which of the half-planes — upper and lower — in the complex plane k the poles $k_m(\omega)$ belong is uniquely answered for the entire integration contour $C(\omega)$: for $v_m > 0$, the pole $k_m = k_m^{(+)}$ is in the upper half-plane and, for $v_m < 0$, the pole $k_m = k_m^{(-)}$ is in the lower half-plane.

Let us show that if the integration contour $C(\omega)$ lies above all the singularities of the function $D^{-1}(\omega, k)$, then the poles $k_m^{(\pm)}(\omega)$ cannot move from one half-plane into the other in the complex plane k. Since the physical system is unstable, the dispersion relation $D(\omega, k) = 0$ for real k values has at least one solution $\omega(k)$ with a positive imaginary part. Consider the solution $\omega = \omega_p(k)$ with the largest imaginary part. Let the absolute maximum $\operatorname{Im}\omega_p(k) = \sigma_0 > 0$ be reached at a certain real wavenumber $k = k_0$. The integration contour $C(\omega)$ is described by the relationship $\omega'' = \sigma > \sigma_0$. We assume that, as the frequency ω varies along the contour, one of the poles $k = k_q(\omega)$ in the complex plane k moves from one half-plane into the other. Let this pole cross the real axis of the complex plane k at a certain frequency $\tilde{\omega} = \omega' + i\sigma$, so, at the axis, we have $k_q(\tilde{\omega}) = \tilde{k} = k' + i0$. Since $k_q(\omega)$ is a solution to the dispersion relation $D(\omega, k) = 0$, we should also have $D(\tilde{\omega}, \tilde{k}) = D(\omega' + i\sigma,\ k' + i0) = 0$, where $\sigma > \sigma_0$. But the above assumption implies that, for a real k value, the frequency ω with this imaginary part cannot satisfy the dispersion relation. This contradiction proves the assertion that, in the complex plane k, the poles $k_m^{(\pm)}(\omega)$ cannot move from one half-plane into the other if the integration contour $C(\omega)$ passes above all the singularities of the function $D^{-1}(\omega, k)$.

Regardless of which of the formulas (38.2) is used to calculate the integral in formula (38.3), the result will be the same since the positions of the poles $k_m^{(\pm)}(\omega)$ in the complex plane k are independent of frequency.

Assume first that all the zeros $k_m(\omega)$ $(m = 1, 2, \ldots, N)$ of dispersion function (38.1) are in the upper (lower) half of the complex plane k, i.e., $N^{(-)} = 0$ $(N^{(+)} = 0)$. In accordance with the second (first) of formulas (38.2), we have $J(\omega) = 0$, which implies that Green's function (38.3) is zero. In turn, the fact that all the zeros $k_m(\omega)$ of dispersion function (38.1) belong to the upper (lower) half of the complex plane k implies, in accordance with formulas (38.4), that, for $\omega'' \to +\infty$, all the imaginary parts $\operatorname{Im} k_m(\omega) = k''_m(\omega)$ are positive (negative). Hence, the third criterion for convective instability can be formulated as follows: if, for $\omega'' \to +\infty$, all the solutions $k_m(\omega)$ $(m = 1, 2, \ldots, N)$ to the dispersion relation $D(\omega, k) = 0$ belong to one and the same half-plane — upper or lower — of the complex plane k, then the instability is convective.

Now let one of roots — say, the root $k_N(\omega) = k_N^{(-)}(\omega)$ — belong to the lower half of the complex plane k and let the remaining roots $k_m(\omega) = k_m^{(+)}(\omega)$ $(m = 1, 2, \ldots, N-1)$ lie in the upper half-plane. Using the second of formulas (38.2), we find

$$J(\omega) = -\frac{2\pi i}{C} \prod_{m=1}^{N-1} (k_N^{(-)}(\omega) - k_m^{(+)}(\omega))^{-1} . \tag{38.5}$$

The result of a further calculation of Green's function (38.3) depends on the singularities of expression (38.5). These singularities are determined by the zeros of the relationships

$$k_N^{(-)}(\omega) - k_m^{(+)}(\omega) = 0 , \quad m = 1, 2, \ldots, N-1 . \tag{38.6}$$

It is obvious that equalities (38.6) hold at the branch points of the roots $k_m(\omega)$ of the dispersion relation; moreover, they hold not for all of the roots, but only for those that, at $\omega'' \to +\infty$, belong to different half-planes — upper and lower — of the complex plane k. An increasing contribution to the integral in formula (38.3) comes only from the branch points that lie in the upper half of the complex plane ω.

And finally, let the roots $k_m^{(+)}(\omega)$, $m = 1, 2, \ldots, N^{(+)}$ be in the upper half of the complex plane k and let the roots $k_s^{(-)}(\omega)$, $s = 1, 2, \ldots, N^{(-)}$ $(N^{(+)} + N^{(-)} = N)$ be in the lower half-plane. Using the second of formulas (38.2), we find

$$J(\omega) = \sum_{s=1}^{N^{(-)}} \left[P_s(\omega) \prod_{m=1}^{N^{(+)}} (k_s^{(-)}(\omega) - k_m^{(+)}(\omega))^{-1} \right] , \tag{38.7}$$

where $P_s(\omega)$ are analytic functions over the entire complex plane ω. The singularities of expression (38.7) are determined by the zeros of the differences

$$k_s^{(-)}(\omega) - k_m^{(+)}(\omega) = 0 , \quad m = 1, 2, \ldots, N^{(+)} , \quad s = 1, 2, \ldots, N^{(-)} . \tag{38.8}$$

Like (38.6), equalities (38.8) hold at the branch points of those roots $k_m(\omega)$ of the dispersion relation that belong to different half-planes — upper and lower — of the complex plane k. An increasing contribution to the integral in formula (38.3) comes only from the branch points that lie in the upper half of the complex plane ω. Hence, the third criterion for absolute instability can be formulated as follows: if, in the upper half of the complex plane ω, there is a branch point of those solutions $k_m(\omega)$ $(m = 1, 2, \ldots, N)$ to the dispersion relation $D(\omega, k) = 0$ that, for $\omega'' \to +\infty$, belong to different half-planes — upper and lower — of the complex plane k, then the instability is absolute.

The third criterion for absolute instability is consistent with the second criterion, obtained by invoking the notion of a saddle point and by using the saddle point method. Indeed, if there is a branch point of the functions $k_m(\omega)$ in the upper half of the complex plane ω, then the complex group velocity at this point vanishes. And the branch point is a saddle point since it is the merging point of the branches of the functions $k_m(\omega)$ that are in the different halves of the complex plane k for $\omega'' \to +\infty$. That this is the case is clearly seen from the upper curve (for $V = -1$) in Fig. 56.

Let us consider how the third criterion can be applied to analyze the type of instability of an electron beam in a slowing-down medium. We begin with the collective regime, for which, in accordance with dispersion relation (35.10), relationships (38.4) take the form

$$k_1(\omega) = \frac{\omega}{U} + \frac{a^2}{\omega(U-V)}, \quad k_2(\omega) = \frac{\omega}{V} + \frac{a^2}{\omega(U-V)}. \tag{38.9}$$

For $U > 0$ and $V > 0$, both of the solutions $k_{1,2}(\omega)$ are in the upper half of the complex plane k. Consequently, in this case, the instability is convective. For $U > 0$ and $V < 0$, the root $k_1(\omega)$ is in the upper half-plane and the root $k_2(\omega)$ is in the lower half-plane. Further, if the velocities U and V have opposite signs, then one of the branch points (36.21) of the functions $k_{1,2}(\omega)$ is in the upper half of the complex plane ω. Consequently, in this case, the instability is absolute.

For the single-particle regime of the instability, relationships (38.4) are written as (see formulas (36.28))

$$k_1(\omega) = \frac{\omega}{U} + \sqrt{\frac{b^3}{\omega U(U-V)}}, \quad k_2(\omega) = \frac{\omega}{U} + \sqrt{\frac{b^3}{\omega U(U-V)}}, \quad k_3(\omega) = \frac{\omega}{V} + \frac{b^3 V}{\omega^2 (U-V)^2}. \tag{38.10}$$

For $U > 0$ and $V > 0$, all the three solutions $k_{1,2,3}(\omega)$ are in the upper half of the complex plane k. Consequently, in this case, the instability is convective. For $U > 0$ and $V < 0$, the roots $k_{1,2}(\omega)$ are in the upper half-plane and the root $k_3(\omega)$ is in the lower half-plane. One of the branch points of the functions $k_{1,2,3}(\omega)$ (see formula (36.24)) lies in the upper half of the complex plane ω. In Sec. 36, we have shown that, at this point, the following equality is satisfied:

$$k_k^{(-)}(\omega) - k_1^{(+)}(\omega) = 0\,. \tag{38.11}$$

Consequently, in this case, the instability is absolute, in accordance with equalities (38.8). In what follows, we will show that, if the functions $k_m(\omega)$ belong to different halves of the complex plane k, then, in a physical system, the eigenmodes, characterized by these functions, propagate in different directions. Accordingly, the third criterion for absolute instability can be reformulated as follows: if, in the upper half of the complex plane ω, there is a branch point of the functions $k_m(\omega)$ that correspond to waves propagating in opposite directions, then the instability is absolute. We can say that, during the absolute instability, each point of a physical system acts as a "generator" of growing backward waves. From a practical standpoint, the third criterion for the type of instability seems to be most convenient.

39. Type of Beam Instability in the Interaction with a Slowed Wave of Zero Group Velocity in a Medium

In some cases, the above criteria for the type of instability are inapplicable. In particular, when one of the velocities v_m in asymptotic formulas (38.4) is zero, the third criterion cannot be applied directly. The results from the first and second criteria may also turn out to be inadequate (see below). Of course, the case $v_m = 0$ is an exclusive one. The situation can easily be made standard by setting $v_m = 0 \pm \varepsilon$, where ε is a small quantity. But from the theoretical point of view, it is worthwhile to consider the case $v_m = 0$ separately — just what we are going to do in this section.

We begin with the beam instability in the collective interaction with a wave of zero group velocity in a medium. Setting $V = 0$ in dispersion relation (35.10), we obtain the following expressions for the complex frequencies:

$$\omega_{1,2}(k) = \frac{1}{2}(kU \pm \sqrt{k^2U^2 - 4a^2}). \tag{39.1}$$

Note that, for $V = 0$, the order of dispersion relation (35.10) regarded as an equation in the frequency ω is higher than the order of this relation regarded as an equation in the wavenumber k. We substitute frequencies (39.1) into formula (36.6) and retain only the term that increases with time. As a result, we arrive at the following formula for the Green's function at the point $z = 0$:

$$G(t,0) = -\frac{i}{2\pi}\int_{-\infty}^{+\infty}\frac{\exp[-i\omega_1(k)t]}{\sqrt{k^2U^2 - a^2}}dk\,. \tag{39.2}$$

The integral in formula (39.2) can be estimated by the same method as that used to formulate the first criterion for the type of instability. Introducing the new variable of integration

$$\Omega = \omega_1(k)\,, \quad d\Omega = \frac{\Omega U}{\sqrt{k^2U^2 - 4a^2}}dk\,, \tag{39.3}$$

we convert formula (39.2) into the form

$$G(t,0) = -\frac{i}{2\pi U}\int_{\Omega_1}\frac{\exp(-i\Omega t)}{\Omega}d\Omega\,. \tag{39.4}$$

Here, Ω_1 is the contour in the complex plane $\omega = \omega' + i\omega''$ along which the point $\omega = \omega_1(k)$ passes as k runs from minus to plus infinity along the real axis. This contour has the same shape as that shown in Figs. 46 and 47. Calculating the integral in formula (39.4) gives $G(0,t) = U^{-1}$. We can see that, at the point $z = 0$, the Green's function does not increase with time, on the one hand, and does not vanish, on the other. That is why the type of instability in the system needs to be refined. To do this, it is sufficient to find the Green's function at an arbitrary point z.

To begin, we formulate the general problem of determining the Green's function for beam instabilities in slowing-down media. Using the notation (35.11) and (35.30), we write the following two sets of inhomogeneous differential equations describing beam instabilities in the collective and the single-particle interaction in slowing-down media (see sets of Eqs. (35.6) and (35.8)):

$$\begin{aligned} &\left(\frac{\partial}{\partial t} + U\frac{\partial}{\partial z}\right) A_b - ia^2 A_w = \delta(z) f(t)\,, \\ &\left(\frac{\partial}{\partial t} + V\frac{\partial}{\partial z}\right) A_w + iA_b = 0 \end{aligned} \tag{39.5a}$$

and

$$\begin{aligned} &\left(\frac{\partial}{\partial t} + U\frac{\partial}{\partial z}\right)^2 A_b + b^3 A_w = \delta(z) f(t)\,, \\ &\left(\frac{\partial}{\partial t} + V\frac{\partial}{\partial z}\right) A_w + iA_b = 0\,. \end{aligned} \tag{39.5b}$$

Here, $f(t)$ is a function of time, $U > 0$, and either $V \geq 0$ or $V \leq 0$. The sets of Eqs. (39.5a) and (39.5b) are solved for $t > 0$ in the region $z > 0$. The initial conditions are $A_b(0,z) = A_w(0,z) = 0$. For the set of Eqs. (39.5b), the additional initial condition at $t = 0$ is $\partial A_b/\partial t = 0$. It is assumed that, for $t < t_0 \to 0$, the right-hand sides are identically zero, i.e., $f(t) \equiv 0$. The Green's functions are to be constructed by setting $f(t) = \delta(t)$ (this will be done later).

Applying the Laplace transformation in the time t and the Fourier transformation in the coordinate z, from the sets of Eqs. (39.5a) and (39.5b), respectively, we find the following expressions for the transformed functions $A_b(t,z)$ and $A_w(t,z)$:

$$A_b(\omega,k) = i\frac{\omega - kV}{D_1(\omega,k)} f(\omega)\,, \quad A_w(\omega,k) = i\frac{A_b(\omega,k)}{\omega - kV} \tag{39.6a}$$

and

$$A_b(\omega,k) = -\frac{\omega - kV}{D_2(\omega,k)} f(\omega)\,, \quad A_w(\omega,k) = \frac{A_b(\omega,k)}{\omega - kV}\,. \tag{39.6b}$$

Here,

$$D_1(\omega,k) = (\omega - kU)(\omega - kV) + a^2 \tag{39.7a}$$

and

$$D_2(\omega,k) = (\omega - kU)^2(\omega - kV) - b^3 \tag{39.7b}$$

are the dispersion functions for the collective and the single-particle interaction (they coincide with the left-hand sides of dispersion relations (35.10) and (35.29)), and

$$f(\omega) = \int_0^\infty f(t)\exp(i\omega t)dt\,. \tag{39.8}$$

For $f(t) = \delta(t)$, we have $f(\omega) = 1$.

Applying the inverse transformations, from expressions (39.6) we find

$$\begin{aligned} A_b(t,z) &= \frac{C_\alpha}{(2\pi)^2}\int_{-\infty}^{+\infty} dk \int_{C(\omega)} \frac{(\omega - kV)}{D_\alpha(\omega,k)} f(\omega)\exp(-i\omega t + ikz)d\omega\,, \\ & \qquad\qquad\qquad\qquad\qquad\qquad\qquad\qquad \alpha = 1,2\,. \\ A_w(t,z) &= \frac{C_\alpha}{(2\pi)^2}\int_{-\infty}^{+\infty} dk \int_{C(\omega)} \frac{1}{D_\alpha(\omega,k)} f(\omega)\exp(-i\omega t + ikz)d\omega\,, \end{aligned} \tag{39.9}$$

Here, C_α are constants such that $|C_\alpha| = 1$, with the subscripts $\alpha = 1$ and $\alpha = 2$ referring to the collective and the single-particle interaction, respectively. For $f(\omega) = 1$, the second expressions in (39.9) coincide to within factors on the order of unity in absolute value with the solutions to the equations

$$D_\alpha(\hat\omega, \hat k)G(t,z) = \delta(z)\delta(t)\,, \quad \alpha = 1,2\,, \tag{39.10}$$

thereby being the sought-for Green's functions.

In expressions (39.9), we will first integrate over k and will estimate the resulting integrals over ω by using the following formulas of the saddle-point method (see Sec. 37):

$$\begin{aligned} &\int_{C(\omega)} S(\omega)\exp(-it\psi(\omega))d\omega \xrightarrow[t\to\infty]{} S(\omega_0)\sqrt{\frac{2\pi}{tq}}\exp(-it\psi(\omega_0) + i\varphi)\,, \\ &\qquad q = \left|\frac{d^2\psi}{d\omega^2}(\omega_0)\right|, \quad \varphi = \arg\frac{d^2\psi}{d\omega^2}(\omega_0)\,. \end{aligned} \tag{39.11}$$

Here, ω_0 is a saddle-point determined from the equation $d\psi/d\omega = 0$. More insight into the integration over k along the real axis in expressions (39.9) is required. Recall that the solution is sought for in the region $z > 0$. Jordan's lemma (see Sec. 4) implies that the integration can be performed by closing the real axis $k'' = 0$ by a semicircle of infinite radius in the upper half of the complex plane $k = k' + ik''$. In this case, the contribution to the integral comes only from those roots $k_m(\omega)$ of the dispersion relations $D_\alpha(\omega, k) = 0$ that lie precisely in the upper half of the complex plane k. Since, in the complex plane $\omega = \omega' + i\omega''$, the contour of integration over ω in expressions (39.9) lies above all the singular points of the integrand, it can be displaced an arbitrarily large distance upward, $C(\omega) = (-\infty + i\sigma,\ +\infty + i\sigma)$, $\sigma \to +\infty$. This makes it possible to uniquely determine to which of the half-planes — upper and lower — in the complex plane k the roots $k_m(\omega)$ belong. If, for $\operatorname{Im}\omega = \sigma \to \infty$, the imaginary part of the root $k_m(\omega)$ of the dispersion relation is positive, $\operatorname{Im} k_m(\omega) > 0$, then the root belongs to the upper half of the complex plane k and contributes to the integral determining the solution in the region $z > 0$. In the opposite case, the contribution of the root is zero. We will regard the waves for which

$$\operatorname{Im} k_m(\omega) > 0 \quad \text{at} \quad \operatorname{Im} \omega \to \infty \tag{39.12}$$

as propagating rightward (into the region $z > 0$) and will distinguish them by the plus superscript, $k_m(\omega) \equiv k_m^{(+)}(\omega)$, as in formulas (38.2).

This same analysis can be applied completely to construct solutions in the region $z < 0$, the only difference being that, in the integrals over the variable k in expressions (39.9), the integration path along the real axis is closed by a semicircle of infinite radius in the lower half of the complex plane $k = k' + ik''$. That is why we will regard the waves for which

$$\operatorname{Im} k_m(\omega) < 0 \quad \text{at} \quad \operatorname{Im} \omega \to \infty \tag{39.13}$$

as propagating leftward (into the region $z < 0$) and will distinguish them by the minus superscript, $k_m(\omega) \equiv k_m^{(-)}(\omega)$, as in formulas (38.2).

Now, we consider Green's functions in particular cases. Let the dispersion function have the form

$$D_1(\omega, k) = (\omega - kU)\omega + a^2 = -U\omega(k - k_0(\omega)), \tag{39.14}$$

which corresponds to the collective interaction between an electron beam and a wave of zero group velocity, $V = 0$. Here,

$$k_0(\omega) = \frac{\omega}{U} + \frac{a^2}{U\omega}. \tag{39.15}$$

The solutions to dispersion relation (39.14) considered as an equation in the variable ω are given by formulas (39.1). Substituting dispersion function (39.14) into expressions (39.9) and integrating over k yields

$$G(t, z) = -\frac{i}{2\pi U} \int_{C(\omega)} \frac{d\omega}{\omega} \exp(-it\psi(\omega)). \tag{39.16}$$

In deriving expression (39.16), we have integrated over k by closing the real axis $k'' = 0$ by a semicircle of infinite radius in the upper half of the complex plane k. In fact, the wavenumber (35.15) satisfies condition (39.12) and as such refers to a wave propagating rightward along the z axis. In the example at hand, there are no leftward propagating waves. Consequently, we have $G(t, z < 0) \equiv 0$. In expression (39.16), we use the following notation:

$$\psi(\omega) = \omega(1 - \xi) - \frac{a^2}{\omega}\xi, \quad \xi = \frac{z}{Ut}. \tag{39.17}$$

For $\xi > 1$, we have $\sigma(1 - \xi) < 0$ on the contour $C(\omega) = (-\infty + i\sigma, \ +\infty + i\sigma)$. Therefore, for $\xi > 1$, the integral in expression (39.16) is identically zero ($\sigma \to +\infty$). The reason is that, in the system under consideration, perturbations cannot propagate with velocities higher than U. That is why we set $\xi < 1$.

From the equation

$$\frac{d\psi}{d\omega} = (1-\xi) + \frac{a^2}{\omega^2}\,\xi = 0 \tag{39.18}$$

we find the saddle point:

$$\omega_0 = ia\sqrt{\frac{\xi}{1-\xi}}\,. \tag{39.19}$$

Then, we have

$$\psi(\omega_0) = 2ia\sqrt{\xi(1-\xi)}\,,\quad q = 2\frac{(1-\xi)^{3/2}}{a\xi^{1/2}}\,,\quad \varphi = \frac{\pi}{2}\,,\quad S(\omega_0) = \frac{1}{\omega_0} = -i\frac{(1-\xi)^{1/2}}{a\xi^{1/2}}\,. \tag{39.20}$$

Substituting relationships (39.20) into formulas (39.11) and taking into account expression (39.16), we arrive at the asymptotic formula for the Green's function:

$$G(t,z) = \frac{i}{2U}\left(\pi a t\sqrt{\xi(1-\xi)}\right)^{-1/2} \exp(2at\sqrt{\xi(1-\xi)})\,. \tag{39.21}$$

For estimates, we can use the following main asymptotic formula:

$$G(t,z) \sim \frac{1}{\sqrt{at}} \exp(2at\sqrt{\xi(1-\xi)})\,. \tag{39.22}$$

Figure 57(a) shows how function (39.21) depends on z at different times t for $a = 1$ and $U = 1$. Figure 57(b) compares function (39.21) (curve *1*) and function (39.22) (curve *2*). The maximum of function (39.22) is at $\xi = 1/2$; i.e., it moves according to the law $z = Ut/2$ — a result that also follows from Fig. 57(a). Note that the solutions obtained in the quasi-harmonic approximation agree very well with asymptotic solutions (39.21) and (39.22), although, for long time scales, such a good agreement is not necessary at all. In the quasi-harmonic approximation, the maximum of a perturbation described by solution (35.26) propagates with the convective velocity (35.15) during instability; for $V = 0$, this velocity is $U/2$. As for solution (35.28), which qualitatively accounts for the deviation of the perturbation from being quasi-harmonic, it coincides to graphical accuracy with function (39.22), although differs in structure from the latter.

From formulas (39.21) and (39.22), as well as from Fig. 57, we can see that, in the collective interaction of an electron beam with a wave having a zero group velocity, the perturbation propagates downstream along the beam, like during the convective instability. But at each fixed point $z = \text{const}$, the perturbation grows without bound (as $\exp(2aU^{-1/2}t^{1/2}z^{1/2})$ for $z \ll Ut$), so the instability resembles absolute instability.

Now, let the dispersion function have the form

$$D_2(\omega,k) = (\omega - kU)^2\omega - b^3 = U^2\omega(k - k_1(\omega))(k - k_2(\omega))\,, \tag{39.23}$$

which corresponds to the single-particle interaction between an electron beam and a wave of zero group velocity, $V = 0$. Here,

$$k_{1,2}(\omega) = \frac{\omega}{U} \mp \frac{1}{U}\sqrt{\frac{b^3}{\omega}}\,. \tag{39.24}$$

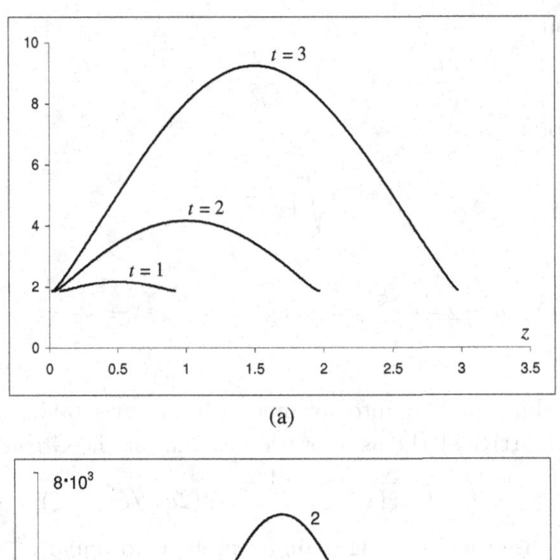

(a)

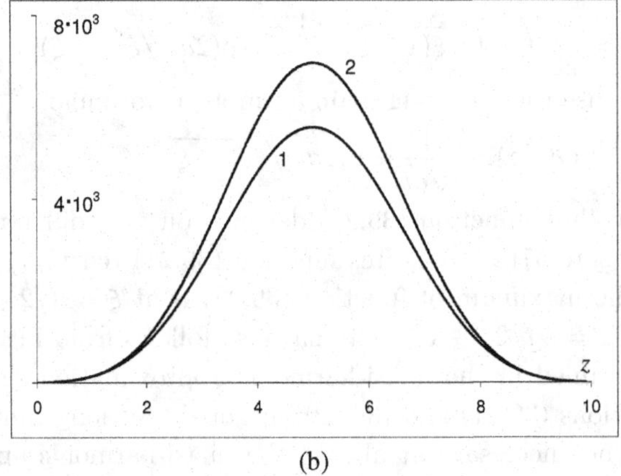

(b)

Fig. 57. (a) Green's function in the case of collective interaction of an electron beam with a wave of zero group velocity and (b) Green's functions calculated for $t = 10$ from formula (39.21) (curve *1*) and formula (39.22) (curve *2*).

The wavenumbers (39.24) satisfy condition (39.12), so they refer to rightward propagating perturbations. Consequently, in the region $z > 0$, the Green's function is described by the formula

$$G(t,z) = -\frac{i}{4\pi u b^{3/2}} \int_{C(\omega)} \frac{d\omega}{\sqrt{\omega}} [\exp(-it\psi_1(\omega)) - \exp(-it\psi_2(\omega))] \,. \tag{39.25}$$

For $z < 0$, the Green's function is identically zero. In formula (39.25), we use the following notation:

$$\begin{aligned} \psi_1(\omega) &= \omega(1-\xi) - \sqrt{\frac{b^3}{\omega}}\xi \,, \\ \psi_2(\omega) &= \omega(1-\xi) + \sqrt{\frac{b^3}{\omega}}\xi \,. \end{aligned} \tag{39.26}$$

From the equations

$$\frac{d\psi_1}{d\omega} = (1-\xi) + \frac{1}{2}b^{3/2}\xi\omega^{-3/2} = 0\,, \quad \frac{d\psi_2}{d\omega} = (1-\xi) - \frac{1}{2}b^{3/2}\xi\omega^{-3/2} = 0\,. \quad (39.27)$$

we find the saddle points:

$$\omega_1 = \omega_0 \exp(i2\pi/3) = \frac{-1+i\sqrt{3}}{2}\omega_0\,,$$
$$\omega_2 = \omega_0 \exp(i4\pi/3) = \frac{-1-i\sqrt{3}}{2}\omega_0\,, \quad (39.28)$$

where

$$\omega_0 = \frac{b}{\sqrt[3]{4}}\left(\frac{\xi}{1-\xi}\right)^{2/3}. \quad (39.29)$$

Taking into account only the saddle point that lies in the upper half-plane, we reduce formula (39.25) for the Green's function to

$$G(t,z) = -\frac{i}{4\pi u b^{3/2}} \int_{C(\omega)} \frac{d\omega}{\sqrt{\omega}} \exp(-it\psi_1(\omega))\,. \quad (39.30)$$

We also have

$$\psi_1(\omega_1) = \exp(i2\pi/3)\frac{3}{\sqrt[3]{4}}b\xi^{2/3}(1-\xi)^{1/3}\,, \quad q = \frac{3}{\sqrt[3]{2}}b^{-1}\xi^{-2/3}(1-\xi)^{5/3}\,,$$
$$\psi = \frac{4\pi}{3}\,, \quad S(\omega_1) = \exp(-i\pi/3)\sqrt[3]{2b}^{-1/2}(1-\xi)^{1/3}\,. \quad (39.31)$$

Inserting relationships (39.31) into formulas (39.11) and taking into account formula (39.30) yields the following asymptotic expression for the Green's function:

$$G(t,z) = \frac{i}{2\sqrt{3\pi}Ub}(bt(1-\xi))^{-1/2} \exp\left[\left(\frac{1}{2}i + \frac{\sqrt{3}}{2}\right)\frac{3}{\sqrt[3]{4}}bt\xi^{2/3}(1-\xi)^{1/3}\right]. \quad (39.32)$$

For estimates, we can use the following main asymptotic formula:

$$G(t,z) \sim \frac{1}{\sqrt{bt}} \exp\left[\frac{3\sqrt{3}}{2\sqrt[3]{4}}bt\xi^{2/3}(1-\xi)^{1/3}\right]. \quad (39.33)$$

Figure 58 shows how function (39.33), which gives the main asymptotic formula for estimating Green's function (39.32), depends on z at different times t for $b = 1$ and $U = 1$. The maximum of function (39.33) is at $\xi = 2/3$; i.e., it moves according to the law $z = 2Ut/3$ (as in the quasi-harmonic approximation, see (35.47) and (35.37)) — a result that can also be inferred from Fig. 58. From formulas (39.32) and (39.33), as well as from Fig. 58, we can see that, in the single-particle interaction of an electron beam with a wave having a zero group velocity, the perturbation propagates downstream along the beam, like during the convective instability. But at each fixed point $z = \text{const}$, the perturbation grows without bound (as $\exp(\beta bU^{-2/3}t^{1/3}z^{2/3})$, with $\beta \approx 1.6$, for $z \ll Ut$), so the instability resembles absolute instability.

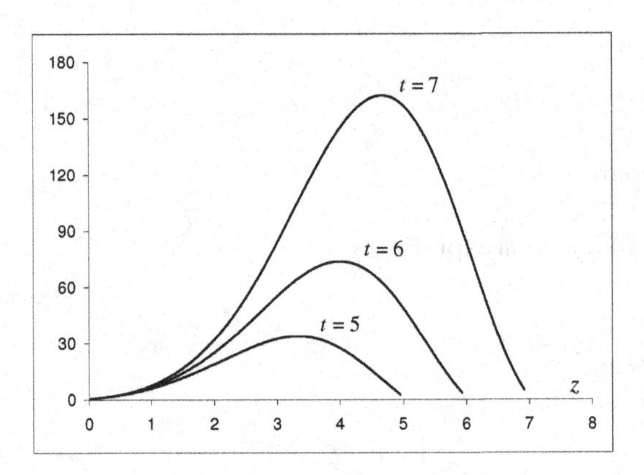

Fig. 58. Results of estimating the Green's function from the main asymptotic formula in the case of single-particle interaction of an electron beam with a wave of zero group velocity.

It is clear that the case $V = 0$ is intermediate between the cases of absolute and convective instability. But this is not the only reason why we have considered this case separately. Another reason has been to show that, using formulas (39.21), (39.22), (39.32), and (39.33), one can easily obtain asymptotic formulas for the Green's functions in the most general case.

40. Calculation of the Green's Functions of Unstable Systems

As an example, let us calculate Green's functions for beam instabilities in slowing-down media for an arbitrary ratio between the velocities U and V. This example, being of interest in itself, provides a clear illustration of the dynamics of convective and absolute instabilities, as well as of the propagation of the fronts of localized perturbations. For definiteness, we assume that $V < U$.

We begin, however, with the particular case $V = -U$, in which dispersion relation for collective interaction has the form

$$D_1(\omega, k) = \omega^2 - k^2U^2 + a^2 = -U^2(k - k_1(\omega))(k - k_2(\omega)), \tag{40.1}$$

where

$$k_{1,2}(\omega) = \pm\frac{1}{U}\sqrt{\omega^2 + a^2}. \tag{40.2}$$

Since condition (39.12) is satisfied by the wavenumber $k_1(\omega)$, the Green's function in the region $z > 0$ is described by the formula

$$G(t, z) = -\frac{i}{4\pi U}\int_{C(\omega)} \frac{d\omega}{\sqrt{\omega^2 + a^2}} \exp(-it\psi(\omega)), \tag{40.3}$$

with

$$\psi(\omega) = \omega - \sqrt{\omega^2 + a^2}\xi. \tag{40.4}$$

From the equation

$$\frac{d\psi}{d\omega} = 1 - \frac{\omega}{\sqrt{\omega^2 + a^2}}\xi = 0 \tag{40.5}$$

we find the saddle point:

$$\omega_0 = \frac{ia}{\sqrt{1-\xi^2}}. \tag{40.6}$$

Then, we have

$$\psi(\omega_0) = ia\sqrt{1-\xi^2}, \quad q = \frac{(1-\xi^2)^{3/2}}{a\xi^2}, \quad \varphi = \frac{\pi}{2}, \quad S(\omega_0) = -i\frac{(1-\xi^2)^{1/2}}{a\xi}. \tag{40.7}$$

Substituting relationships (40.7) into formulas (39.11) and taking into account expression (40.3), we arrive at the following asymptotic formula for the Green's function:

$$G(t,z) = \frac{i}{2\sqrt{2}U}\left(\pi at\sqrt{1-\xi^2}\right)^{-1/2} \exp(at\sqrt{1-\xi^2}). \tag{40.8}$$

For estimates, we can use the main asymptotic formula

$$G(t,z) \sim \frac{1}{\sqrt{at}} \exp(at\sqrt{1-\xi^2}). \tag{40.9}$$

It is easy to see that this same formula describes the Green's function in the region $z < 0$.

Figure 59 shows how function (40.8) depends on z at different times t for $a = 1$ and $U = -V = 1$. The maximum of function (40.9) is at $\xi = 0$; i.e., it remains at rest at the point $z = 0$, as is clearly seen in Fig. 59. Note that, in the quasi-harmonic approximation, the propagation velocity (35.15) of the perturbation — the convective velocity during instability in the collective interaction — is also zero for $V = -U$.

An analysis of the general case of an arbitrary ratio between the velocities U and V requires lengthy calculations. This especially concerns the single-particle interaction, because, in this case, dispersion function (39.7b) is a polynomial of degree three in both ω and k. But there is no need for us to perform such calculations because we can use the results of the previous section. Indeed, in a moving coordinate system, the type of instability can change from absolute to convective, and vice versa. It is also possible to choose a coordinate system in which the velocity of one of the waves is zero. This is what we are going to do in order to calculate the Green's function in the general case.

In Eqs. (39.5), we change the variables,

$$z' = z + Wt, \quad t' = t, \tag{40.10}$$

where W is a constant having the dimension of velocity. Under this change of variables, the differential operators are transformed as

$$\begin{aligned} \frac{\partial}{\partial t} + U\frac{\partial}{\partial z} &\to \frac{\partial}{\partial t'} + (U+W)\frac{\partial}{\partial z'}, \\ \frac{\partial}{\partial t} + V\frac{\partial}{\partial z} &\to \frac{\partial}{\partial t'} + (V+W)\frac{\partial}{\partial z'}. \end{aligned} \tag{40.11}$$

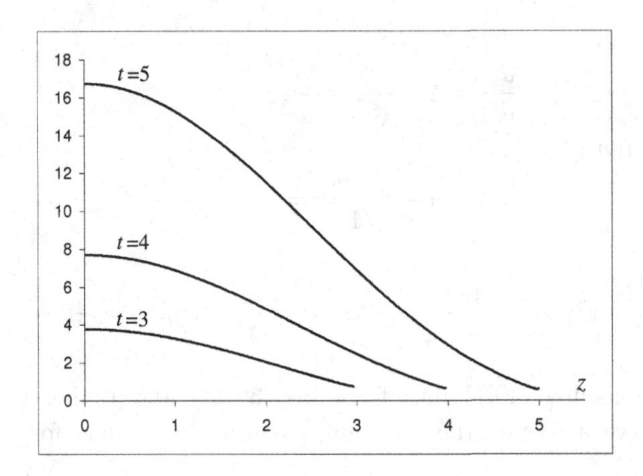

Fig. 59. Green's function in the case of collective interaction of an electron beam with a wave having the group velocity $V = -U$.

Since $\delta(z' - Wt')\delta(t') = \delta(z')\delta(t')$, switching to variables (40.10) leaves the right-hand side of Eqs. (39.5) unchanged. We set $W = -V$. In this case, in the moving coordinate system, the velocity of the slowed wave in a medium is zero and the beam velocity is

$$U' = U - V. \tag{40.12}$$

We also use the primed coordinate,

$$z' = z - Vt, \tag{40.13}$$

and omit the prime for the time t. For a wave of zero velocity in a medium, the Green's functions have been constructed in the previous section. Accordingly, the Green's functions in which we are interested here can be obtained by designating the velocity U and coordinate z ($\xi = z/Ut$) in formulas (39.21), (39.22), (39.32), and (39.33) by a prime and then by switching to the unprimed quantities in accordance with relationships (40.12) and (40.13).

Hence, for collective interaction, the Green's function has the form

$$G(t,z) = \frac{i \exp\left[2at\sqrt{\dfrac{(Ut-z)(z-Vt)}{(U-V)^2t^2}}\right]}{2\sqrt{\pi a(U-V)\sqrt{(Ut-z)(z-Vt)}}}. \tag{40.14}$$

For $V = -U$, formula (40.14) goes over to formula (40.8). For estimates, we can use the following main asymptotic formula:

$$G(t,z) \sim \frac{1}{\sqrt{at}} \exp\left[2at\sqrt{\frac{(Ut-z)(z-Vt)}{(U-V)^2t^2}}\right]. \tag{40.15}$$

Formulas (40.14) and (40.15) are valid only in the region given by the inequalities

$$Vt < z < Ut, \tag{40.16}$$

which determine the positions of the leading and trailing edges of an initial point perturbation. Recall that we have set $U > V$. As the perturbation propagates, it spreads out with the velocity (40.12). The propagation velocity can be found from the maximum of function (40.15):

$$z = U_g t = \frac{U+V}{2} t \equiv z_{\max}. \tag{40.17}$$

Consequently, as in the quasi-harmonic approximation, the propagation velocity of the perturbation is the convective velocity (35.15) during instability. For $V < 0$, formulas (40.14) and (40.15) are also valid for negative z values, but only those that lie in the region (40.16). Substituting relationship (40.17) into formula (40.14), we find that, at the point (40.17), the amplitude of the perturbation described by formula (40.14) grows according to the law

$$G(t, z_{\max}) = \frac{i \exp(at)}{(U-V)\sqrt{\pi a t}}. \tag{40.18}$$

This is the fastest growth over the entire perturbation.

Figure 60 shows how function (40.15) depends on z at different times t for $a = 1$, $U = 2$, and $V = 1$ in the case of convective instability. Figure 61 shows the same function for $a = 1$, $U = 2$ and $V = -1$ in the case of absolute instability. These two figures provide quite a clear illustration of such notions as convective and absolute instabilities.

Analogously, making replacements (40.12) and (40.13) in formulas (39.32) and (39.33), we obtain the Green's functions for beam instabilities in the single-particle

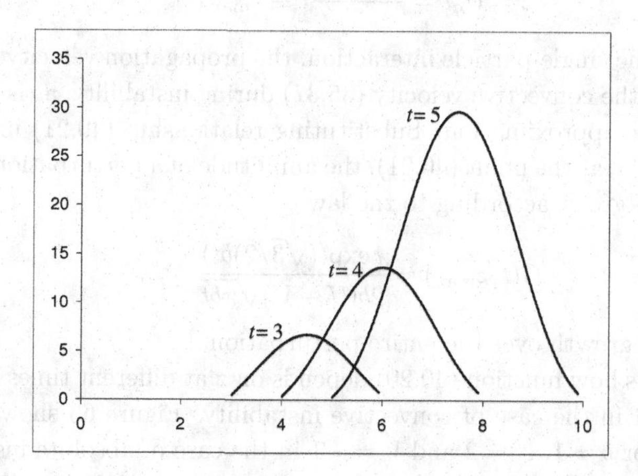

Fig. 60. General form of the Green's function in the case of convective instability in the collective interaction.

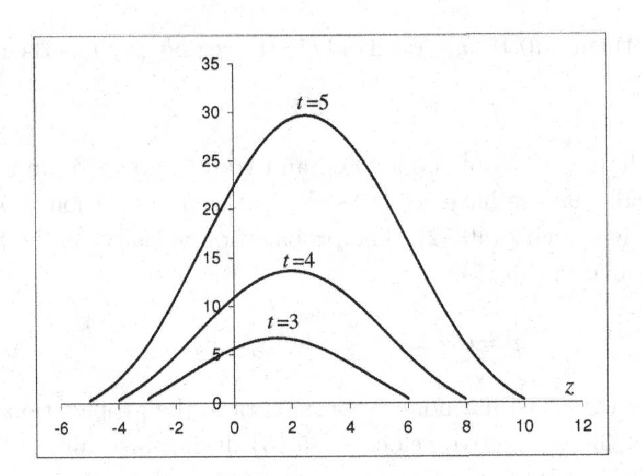

Fig. 61. General form of the Green's function in the case of absolute instability in the collective interaction.

interaction:

$$G(t,z) = i\frac{\exp\left[\left(\frac{1}{2}i + \frac{\sqrt{3}}{2}\right)\frac{3}{\sqrt[3]{4}}b\frac{(z-Vt)^{2/3}(Ut-z)^{1/3}}{U-V}\right]}{2\sqrt{3\pi}b\sqrt{b(U-V)(Ut-z)}}, \tag{40.19}$$

$$G(t,z) \sim \frac{1}{\sqrt{bt}}\exp\left[\frac{3\sqrt{3}}{2\sqrt[3]{4}}b\frac{(z-Vt)^{2/3}(Ut-z)^{1/3}}{U-V}\right]. \tag{40.20}$$

Formulas (40.19) and (40.20), too, are valid only in the region given by inequalities (40.16), which determine the velocities of the leading and trailing edges of the perturbation. The maximum of function (40.19) propagates according to the law

$$z = U_g t = \frac{2U+V}{3}t \equiv z_{\max}. \tag{40.21}$$

Consequently, in the single-particle interaction, the propagation velocity of the perturbation is again the convective velocity (35.37) during instability, as is the case in the quasi-harmonic approximation. Substituting relationship (40.21) into formula (40.19), we find that, at the point (40.21), the amplitude of a perturbation described by formula (40.19) grows according to the law

$$G(t, z_{\max}) = \frac{i\exp((\sqrt{3}/2)bt)}{2b(U-V)\sqrt{\pi bt}}. \tag{40.22}$$

This is the fastest growth over the entire perturbation.

Figure 62 shows how function (40.20) depends on z at different times t for $b = 1$, $U = 2$, and $V = 1$ in the case of convective instability. Figure 63 shows the same function (40.20) for $b = 1$, $U = 2$ and $V = -1$ in the case of absolute instability.

For $V \to U$, formulas (40.14), (40.15), (40.18), (40.19), (40.20), and (40.22) are inapplicable, although they are valid for a small, but nevertheless finite, velocity

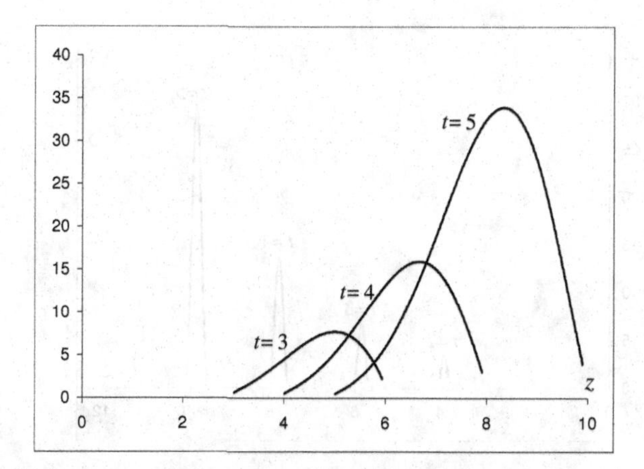

Fig. 62. General form of the Green's function in the case of convective instability in the single-particle interaction.

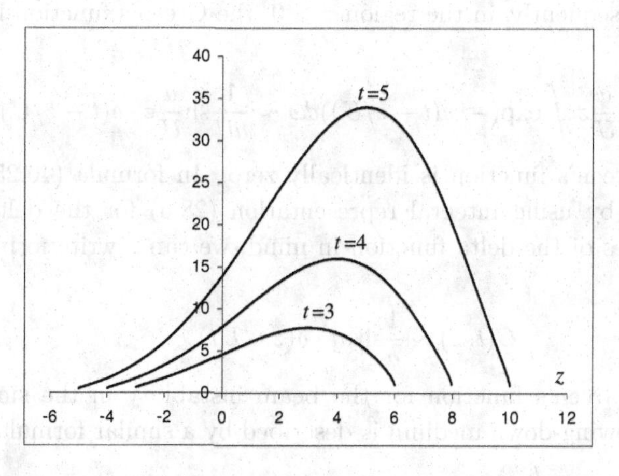

Fig. 63. General form of the Green's function in the case of absolute instability in the single-particle interaction.

difference, $U - V \ll U$. Figure 64 shows how function (40.15) for instability in the collective interaction depends on z at different times t for $a = 1$, $U = 2$, and $V = 1.9$. Green's function (40.20) for instability in the single-particle interaction has approximately the same form. The case in which the velocities are exactly the same, $V = U$, requires a separate analysis. We examine this case for the collective interaction. The relevant dispersion relation is written as

$$D_1(\omega, k) = (\omega - kU)^2 + a^2 = U^2(k - k_1(\omega))(k - k_2(\omega)), \tag{40.23}$$

where

$$k_{1,2}(\omega) = \frac{\omega}{U} \mp i\frac{a}{U}. \tag{40.24}$$

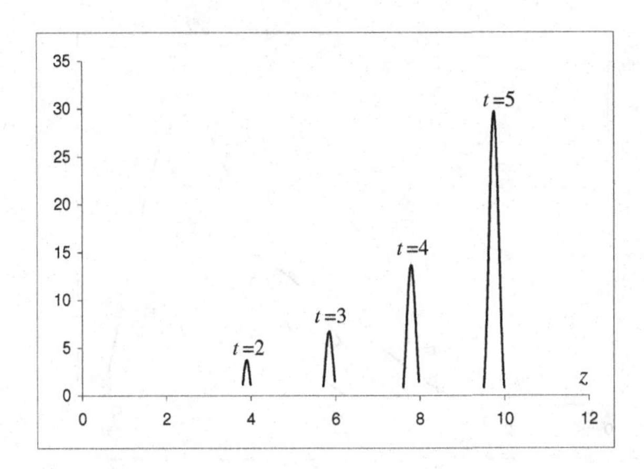

Fig. 64. Green's function for close values of the velocities U and V.

According to condition (39.12), the wavenumbers (40.24) refer to rightward propagating waves. Consequently, in the region $z > 0$, the Green's function is described by the formula

$$G(t, z) \sim \frac{1}{aU} \text{sh} \frac{a}{U} z \int \exp(-i\omega(t - z/U)) d\omega \sim \frac{1}{aU} \text{sh} \frac{a}{U} z \cdot \delta(t - z/U) \,. \quad (40.25)$$

For $z < 0$, the Green's function is identically zero. In formula (40.25), we have integrated over ω by using integral representation (28.6) for the delta function. With the properties of the delta function in mind, we can rewrite formula (40.25) as

$$G(t, z) \sim \frac{1}{a} \text{sh}\, at \cdot \delta(z - Ut) \,. \quad (40.26)$$

For $U = V$, the Green's function for the beam instability in the single-particle interaction in a slowing-down medium is described by a similar formula:

$$G(t, z) \sim \exp\left((\sqrt{3}/2) \frac{b}{U} z\right) \cdot \delta(t - z/U) = U \exp((\sqrt{3}/2) bt) \cdot \delta(z - Ut) \,. \quad (40.27)$$

It is clear that formulas (40.25)–(40.27) describe the evolution of a delta-shaped initial perturbation in unstable nondispersive media. Although the monograph begins and ends with nondispersive media, an attentive reader must have noticed that its main body is devoted to media and systems with dispersion.

Chapter 7

Hamiltonian Method in the Theory of Electromagnetic Radiation in Dispersive Media

41. Equations for the Excitation of Transverse Electromagnetic Field Oscillators

An important branch of the theory of wave phenomena deals with problems of the excitation (radiation) of waves by bodies executing a given motion. In this section, we consider electromagnetic problems of this type. We also present the theory of radiation of transverse electromagnetic waves in a medium by a moving point charge (e.g., a relativistic electron). Note that the problem of the emission of longitudinal (electrostatic) waves by a charge moving along a straight line in a dispersive medium has been considered in Sec. 32. The radiation from an electron executing a given motion, unperturbed by this radiation, is called spontaneous emission. The spontaneous emission theory makes use of two equivalent approaches: the first is to calculate the work done by an electromagnetic field on an electron in its unperturbed motion and the second is to consider the excitation of electromagnetic field oscillators by an unperturbed electron. In Sec. 32, we have utilized the first approach. In the present section, we apply the second one: being fairly general, it provides a detailed study of the properties of the emitted radiation, in particular, the conditions for its excitation, as well as its spectral content, angular distribution, and power.

The problem of the excitation of field oscillators reduces to that of solving inhomogeneous differential equations (3.2) for the state vector of the electromagnetic field. The right-hand sides of these equations contain known functions determined by the law of motion of an electron. The charge exciting the field is a point charge; consequently, instead of solving equations like (3.2), it is necessary to solve a set of more general equations containing derivatives with respect to all three spatial coordinates. In Sec. 32, this circumstance has been taken into account in writing Eq. (32.4). We have solved Eq. (32.4) by expanding the electrostatic potential in Fourier integrals in spatial coordinates and time (see formulas (32.5)) under the assumption that there are no perturbations in the infinite past (at $t \to -\infty$). It is sometimes more convenient to expand the electromagnetic field not in Fourier integrals but in infinite series of plane waves of the form $\exp(i\mathbf{kr})$. The time-dependent coefficients of the series expansions are treated as generalized field coordinates. The electromagnetic field is thereby represented as an infinite set of harmonic oscillators. An approach based on this representation came to be called the Hamiltonian method.

Prior to applying the Hamiltonian method in the theory of spontaneous emission from an electron, it is worthwhile to refine the formulation of the problem and to put the set of electromagnetic field equations into a convenient form. Particular versions of the electromagnetic field equations for different media have been presented in Chapter 2. In their general form, the equations can be written as

$$\begin{aligned}
[\boldsymbol{\nabla}\cdot\mathbf{B}] &= \frac{1}{c}\frac{\partial\mathbf{D}}{\partial t} + \frac{4\pi}{c}\mathbf{j}_0\,,\\
(\boldsymbol{\nabla}\cdot\mathbf{D}) &= 4\pi\rho_0\,,\\
[\boldsymbol{\nabla}\cdot\mathbf{E}] &= -\frac{1}{c}\frac{\partial\mathbf{B}}{\partial t}\,,\\
(\boldsymbol{\nabla}\cdot\mathbf{B}) &= 0\,.
\end{aligned} \tag{41.1}$$

Here, $\mathbf{j}_0(t,\mathbf{r})$ and $\rho_0(t,\mathbf{r})$ are the current and charge densities of the external sources and $\mathbf{D}(t,\mathbf{r})$ is the electric field induction. For a linear homogeneous isotropic medium, we have the relationship

$$\mathbf{D}(t,\mathbf{r}) = \varepsilon(\hat{\omega})\mathbf{E}(t,\mathbf{r})\,, \tag{41.2}$$

where $\varepsilon(\hat{\omega})$ is the permittivity and $\hat{\omega}$ is the frequency operator defined in (3.7). For a medium without frequency dispersion, we have $\varepsilon(\hat{\omega}) = \varepsilon = \text{const}$. We assume for the moment that the frequency dispersion is zero and will take it into account later.

Introducing the vector and scalar potentials, $\mathbf{A}(t,\mathbf{r})$ and $\Phi(t,\mathbf{r})$, in a conventional manner,

$$\mathbf{E} = -\frac{1}{c}\frac{\partial\mathbf{A}}{\partial t} - \boldsymbol{\nabla}\Phi\,, \quad \mathbf{B} = [\boldsymbol{\nabla}\cdot\mathbf{A}]\,, \tag{41.3}$$

and using the Coulomb gauge $(\boldsymbol{\nabla}\cdot\mathbf{A}) = 0$, from Eqs. (41.1) we obtain

$$\begin{aligned}
\Delta\mathbf{A} - \frac{\varepsilon}{c^2}\frac{\partial^2\mathbf{A}}{\partial t^2} &= -\frac{4\pi}{c}\mathbf{j}_0 + \frac{\varepsilon}{c}\boldsymbol{\nabla}\frac{\partial\Phi}{\partial t}\,,\\
\varepsilon\Delta\Phi &= -4\pi\rho_0\,.
\end{aligned} \tag{41.4}$$

In the Coulomb gauge for the potential, the fields are described by the formulas

$$\mathbf{E} = \mathbf{E}_{tr} + \mathbf{E}_l\,, \quad (\boldsymbol{\nabla}\cdot\mathbf{E}_{tr}) = 0\,, \quad \mathbf{E}_{tr} = -\frac{1}{c}\frac{\partial\mathbf{A}}{\partial t}\,, \quad \mathbf{E}_l = -\boldsymbol{\nabla}\Phi\,, \quad \mathbf{B} = [\boldsymbol{\nabla}\cdot\mathbf{A}]\,, \tag{41.5}$$

where $\mathbf{E}_{tr}$ is the transverse (vortex) component of the electric field and $\mathbf{E}_l$ is the longitudinal component. The Coulomb gauge is advantageous in that the scalar potential Φ satisfies Poisson's equation, as in electrostatics (see also Eq. (32.4)). Therefore, the scalar potential Φ determines the Coulomb field of a source — a field that is important only in the near region. The field that is radiated in the form of electromagnetic waves away from the source is mainly determined by the vector potential $\mathbf{A}$. But the fact that the scalar potential Φ enters the first of Eqs. (41.4) makes impossible a complete and mathematically strict separation of the fields into those radiated away from the charge and those confined to the charge. Nevertheless, at large distances from the source — in the far (or wave) region — the term with Φ in the first of Eqs. (41.4) can be ignored. In this case, the scalar potential is not at all required in describing the radiation field.

Hence, in the far region, the emitted field has the structure of transverse waves, whose description requires only the vector potential $\mathbf{A}(t, \mathbf{r})$ satisfying the wave equation

$$\Delta\mathbf{A} - \frac{\varepsilon}{c^2}\frac{\partial^2\mathbf{A}}{\partial t^2} = -\frac{4\pi}{c}\mathbf{j}_0\,. \tag{41.6}$$

We can also use the following relationships:

$$\mathbf{E} = -\frac{1}{c}\frac{\partial\mathbf{A}}{\partial t}\,, \quad \mathbf{B} = [\boldsymbol{\nabla}\cdot\mathbf{A}]\,, \quad W = \int\frac{\varepsilon E^2 + B^2}{8\pi}dV\,, \tag{41.7}$$

where W is the electromagnetic field energy, $\mathbf{E} \equiv \mathbf{E}_{tr}$, and $\mathbf{E}_l = 0$. If the current is produced by an electron executing a given motion, then we have

$$\mathbf{j}_0(t, \mathbf{r}) = e\mathbf{v}_0(t)\delta(\mathbf{r} - \mathbf{r}_0(t))\,, \quad \mathbf{r}_0'(t) = \mathbf{v}_0(t)\,, \tag{41.8}$$

where the coordinate $\mathbf{r}_0(t)$ and velocity $\mathbf{v}_0(t)$ are known functions.

In the Hamiltonian method, it is assumed that the field occupies a region of sufficiently large volume L^3 and that it can be expanded in plane waves with the wavenumbers

$$\mathbf{k}_\lambda = \left\{\frac{2\pi}{L}n_x,\ \frac{2\pi}{L}n_y,\ \frac{2\pi}{L}n_z\right\}. \tag{41.9}$$

Here, λ denotes a set (n_x, n_y, n_z) of negative integers and zeros, except for $n_x = n_y = n_z = 0$. The expansion of the vector potential in plane waves can be written as

$$\begin{gathered}\mathbf{A}(t, \mathbf{r}) = \sqrt{4\pi}\frac{c}{n}\sum_\lambda(\mathbf{q}_\lambda(t)\exp(i\mathbf{k}_\lambda\mathbf{r}) + \mathbf{q}_\lambda^*\exp(-i\mathbf{k}_\lambda\mathbf{r}))\,,\\ \mathbf{q}_\lambda\mathbf{k}_\lambda = \mathbf{q}_\lambda^*\mathbf{k}_\lambda = 0\,.\end{gathered} \tag{41.10}$$

Here, $n = \sqrt{\varepsilon}$ is the refractive index, which for the moment is assumed to be constant (λ-independent). The relationship $\mathbf{q}_\lambda\mathbf{k}_\lambda = 0$ follows from the transverse nature of the electromagnetic field, i.e., from the second of relationships (41.5). Substituting expansion (41.10) into Eq. (41.6) leads to the following inhomogeneous equations describing the excitation of harmonic oscillators of the electromagnetic field:

$$\mathbf{q}_\lambda'' + \omega_\lambda^2\mathbf{q}_\lambda = \sqrt{4\pi}\frac{e}{n}\mathbf{v}_0\exp(-i\mathbf{k}_\lambda\mathbf{r}_0)\,, \quad \omega_\lambda^2 = \frac{c^2}{n^2}k_\lambda^2\,. \tag{41.11}$$

Since the vectors $\mathbf{q}_\lambda$ and $\mathbf{k}_\lambda$ are orthogonal, the electromagnetic field energy, defined in (41.7), has the form

$$W = \sum_\lambda (\mathbf{q}'_\lambda \mathbf{q}'^*_\lambda + \omega_\lambda^2 \mathbf{q}_\lambda \mathbf{q}^*_\lambda) . \qquad (41.12)$$

In Eqs. (41.11) and formula (41.12), the volume of the region occupied by the electromagnetic field has been set equal to unity, $L^3 = 1$.

Equations (41.11) imply that the vectors $\mathbf{q}_\lambda$ and $\mathbf{v}_0$ are collinear and, accordingly, that the vectors $\mathbf{v}_0$ and $\mathbf{k}_\lambda$ are orthogonal; but this is not generally the case. The contradiction arises from the fact that the field is not purely transverse in the region near the charge where it is produced. That is why Eqs. (41.11) can only be used to determine the projections of the vectors $\mathbf{q}_\lambda$ onto the direction of the electron velocity. Let us introduce the expansion

$$\mathbf{q}_\lambda = q_{\lambda 1}\mathbf{e}_{\lambda 1} + q_{\lambda 2}\mathbf{e}_{\lambda 2} , \qquad (41.13)$$

where the unit polarization vectors of the field, $\mathbf{e}_{\lambda 1,2}$, satisfy the orthogonality conditions

$$\mathbf{e}_{\lambda 1}\mathbf{e}_{\lambda 2} = 0 , \quad \mathbf{e}_{\lambda 1}\mathbf{k}_\lambda = 0 , \quad \mathbf{e}_{\lambda 2}\mathbf{k}_\lambda = 0 . \qquad (41.14)$$

The last two of conditions (41.14) are consequences of the transverse nature of the field. Equations (41.11) and expansion (41.13) yield the following equations for the vector components $q_{\lambda 1,2}$:

$$q''_{\lambda 1,2} + \omega_\lambda^2 q_{\lambda 1,2} = \sqrt{4\pi}\frac{e}{n}(\mathbf{e}_{\lambda 1,2}\mathbf{v}_0)\exp(-i\mathbf{k}_\lambda \mathbf{r}_0) . \qquad (41.15)$$

Let the law of motion of an electron be described by the formulas

$$\begin{aligned} \mathbf{r}_0(t) &= \mathbf{u}t + \mathbf{a}_0 \sin\omega_0 t , \\ \mathbf{v}_0(t) &= \mathbf{u}t + \mathbf{a}_0\omega_0 \cos\omega_0 t , \end{aligned} \qquad (41.16)$$

where $\mathbf{u}$ and $\mathbf{a}_0$ are constant vectors and ω_0 is a certain frequency. Let also the following inequality be satisfied:

$$|\mathbf{k}_\lambda \mathbf{a}_0| \ll 1 . \qquad (41.17)$$

Substituting formulas (41.16) into Eqs. (41.15) and taking into account inequality (41.17), we arrive at the following equations describing the excitation of field oscillators by an electron:

$$\begin{aligned} q''_{\lambda 1,2} + \omega_\lambda^2 q_{\lambda 1,2} = \sqrt{4\pi}\frac{e}{n}((\mathbf{e}_{\lambda 1,2}\mathbf{u}) + \omega_0(\mathbf{e}_{\lambda 1,2}\mathbf{a}_0)\cos\omega_0 t \\ - i(\mathbf{e}_{\lambda 1,2}\mathbf{u})(\mathbf{k}_\lambda\mathbf{a}_0)\sin\omega_0 t)\exp(-i\mathbf{k}_\lambda\mathbf{u}t) , \end{aligned} \qquad (41.18a)$$

or

$$\begin{aligned} q''_{\lambda 1,2} + \omega_\lambda^2 q_{\lambda 1,2} &= a_{\lambda 1,2}\exp(-i\mathbf{k}_\lambda\mathbf{u}t) + \frac{1}{2}b^{(+)}_{\lambda 1,2}\exp(-i(\mathbf{k}_\lambda\mathbf{u} + \omega_0)t) \\ &\quad + \frac{1}{2}b^{(-)}_{\lambda 1,2}\exp(-i(\mathbf{k}_\lambda\mathbf{u} - \omega_0)t) , \\ a_{\lambda 1,2} &= \sqrt{4\pi}\frac{e}{n}(\mathbf{e}_{\lambda 1,2}\mathbf{u}) , \quad b^{(+)}_{\lambda 1,2} = b_{\lambda 1,2} \pm c_{\lambda 1,2} , \\ b_{\lambda 1,2} &= \sqrt{4\pi}\frac{e}{n}\omega_0(\mathbf{e}_{\lambda 1,2}\mathbf{a}_0) , \quad c_{\lambda 1,2} = \sqrt{4\pi}\frac{e}{n}(\mathbf{e}_{\lambda 1,2}\mathbf{u})(\mathbf{k}_\lambda\mathbf{a}_0) . \end{aligned} \qquad (41.18b)$$

Before solving Eqs. (41.18), we present an auxiliary result: the initial-value problem for a second-order equation like Eqs. (41.18a) or (41.18b),

$$\begin{aligned} x'' + \omega^2 x &= C \exp(-i\Omega t)\,, \\ x(0) &= x'(0) = 0\,, \end{aligned} \tag{41.19}$$

has the solution

$$x(t) = \frac{C}{\omega^2 - \Omega^2}\left[\exp(-i\Omega t) - \frac{1}{2}\left(1 - \frac{\Omega}{\omega}\right)\exp(i\omega t) - \frac{1}{2}\left(1 + \frac{\Omega}{\omega}\right)\exp(-i\omega t)\right]. \tag{41.20}$$

We also present the relationship

$$\begin{aligned} S &\equiv x'x'^* + \omega^2 x x^* \\ &= \frac{C^2}{(\omega^2 - \Omega^2)^2}[2(\omega^2 + \Omega^2) - (\omega - \Omega)^2 \cos(\omega + \Omega)t - (\omega + \Omega)^2 \cos(\omega - \Omega)t]\,. \end{aligned}$$

Since we are going to consider resonances such that $\omega - \Omega \approx 0$, we can use the following approximate expression for S:

$$S = C^2 \frac{1 - \cos(\omega - \Omega)t}{(\omega - \Omega)^2}\,. \tag{41.21}$$

In what follows, expression (41.21) will be used to write the electromagnetic field energy in different particular cases.

42. Dipole Radiation

Let us consider conventional dipole radiation that occurs for $\mathbf{u} = 0$ (see formulas (41.16)). We choose a spherical coordinate system (r, θ, φ) with the z axis directed along the vector $\mathbf{a}_0$. Here, $\varphi \in [0, 2\pi]$ is the azimuthal angle in a plane perpendicular to the z axis and $\theta \in [0, \pi]$ is the angle measured from the positive direction of the z axis. The vector $\mathbf{e}_{\lambda 1}$ lies in the plane $\varphi =$ const, and the vector $\mathbf{e}_{\lambda 2}$ is perpendicular to this plane, so we can write $\mathbf{e}_{\lambda 2}\mathbf{a}_0 = 0$ and $\mathbf{e}_{\lambda 1}\mathbf{a}_0 = a_0 \sin\theta_\lambda$, where θ_λ is the angle between the vectors $\mathbf{a}_0$ and $\mathbf{k}_\lambda$. Accordingly, for the right-hand sides of Eqs. (41.18b), we have $a_{\lambda 1,2} = c_{\lambda 1,2} = 0$, $b_{\lambda 2} = 0$, and $b_{\lambda 1} = \sqrt{4\pi}(e/n)a_0\omega_0 \sin\theta_\lambda$. Using formula (41.21), we can then write the field energy (41.12) as

$$W = \pi \frac{e^2 a_0^2 \omega_0^2}{n^2}\left(\sum_\lambda \frac{1 - \cos(\omega_\lambda - \omega_0)t}{(\omega_\lambda - \omega_0)^2} \sin^2\theta_\lambda\right). \tag{42.1}$$

Expression (42.1) has a complicated structure because both the frequency ω_λ and the angle θ_λ depend on the index of summation λ. In fact, we have

$$\omega_\lambda = \frac{c}{n}k_\lambda = \frac{2\pi c}{Ln} n_\perp \sqrt{1 + \frac{n_z^2}{n_\perp^2}}\,, \quad \sin^2\theta_\lambda = \left(1 + \frac{n_z^2}{n_\perp^2}\right)^{-1}, \tag{42.2}$$

where $n_\perp^2 = n_x^2 + n_y^2$. Taking a sum over two indices is a difficult task. On the other hand, the final result should be independent of the order of summation. That is why we first take the sum over $n_\perp$ from zero to ∞ and then over the angles. The first summation yields the energy emitted by a dipole at a certain angle $\theta_\lambda = \theta$ into a solid angle element $do = \sin\theta\, d\theta\, d\varphi$. It is this physical quantity that is of primary interest. Accordingly, in expression (42.1), we can factor the sine function out of the summation sign to rewrite this expression as

$$W = \pi \frac{e^2 a_0^2 \omega_0^2}{n^2} \left(\sum_\lambda \frac{1 - \cos(\omega_\lambda - \omega_0)t}{(\omega_\lambda - \omega_0)^2} \right) \sin^2\theta\,. \tag{42.3}$$

In expression (42.3), we switch from summation over λ to integration over the frequencies ω_λ. Note that, according to the first of relationships (42.2), the range of values of the frequencies is determined by the inequality $\omega_\lambda \geq 0$. Let us formulate a general rule for switching from summation over λ to integration over ω. Since $\omega = kc/n$, integration over ω can be reduced to that over k. In k space, the volume is $V_k = (4/3)\pi k^3$. In accordance with (41.9), one electromagnetic field oscillator occupies the volume $(\Delta k)^3 = (2\pi)^3 L^{-3} = (2\pi)^3$ (where we again set $L^3 = 1$). Therefore, the number of oscillators in the volume V_k is $N_k = V_k/(\Delta k)^3 = (1/6)\pi^{-2}k^3$, so we have $dN_k = (1/2)\pi^{-2}k^2 dk$. Multiplying dN_k by $do/4\pi$ gives the number of oscillators emitted into a solid angle element do, specifically,

$$dN(\omega, \theta, \varphi) = \frac{k^2 dk}{(2\pi)^3} do = \frac{n^3 \omega^2 d\omega}{(2\pi c)^3} do\,, \tag{42.4}$$

where we have used the equality $\omega = kc/n$. Thus, the rule for switching from summation to integration can obviously be written as

$$\sum_\lambda \cdots \to \int \cdots dN(\omega, \theta, \varphi) = \frac{1}{(2\pi c)^3} \int \cdots n^3(\omega)\omega^2 d\omega do\,. \tag{42.5}$$

It is in writing rule (42.5) that we have taken into account a possible frequency dispersion of the medium, because we have used the frequency-dependent refractive index, $n = n(\omega)$, and have incorporated it into the integrand.

With rule (42.5), we transform expression (42.3) into

$$W = \frac{e^2 a_0^2 \omega_0^2}{8\pi^2 c^3} \left(\int n(\omega)\omega^2 \frac{1 - \cos(\omega - \omega_0)t}{(\omega - \omega_0)^2} d\omega \right) \sin^2\theta\, do\,. \tag{42.6}$$

Differentiating the field energy (42.6) with respect to time yields the radiation power emitted at an angle θ into a solid angle element do:

$$P(\theta) \equiv \frac{dW}{dt} = \frac{e^2 a_0^2 \omega_0^2}{8\pi^2 c^3} \left(\int n(\omega)\omega^2 \frac{\sin(\omega - \omega_0)t}{(\omega - \omega_0)} d\omega \right) \sin^2\theta\, do\,. \tag{42.7}$$

The function that describes the frequency–angular radiation spectrum has the form

$$I(\omega, \theta) = n(\omega)\omega^2 \frac{\sin(\omega - \omega_0)t}{(\omega - \omega_0)} \sin^2\theta \xrightarrow[t\to\infty]{} \pi n(\omega)\omega^2 \delta(\omega - \omega_0) \sin^2\theta\,. \tag{42.8}$$

Calculating the integral in formula (42.7) at $n(\omega) = n = \text{const}$ for long time scales gives the radiation power emitted per unit time:

$$P(\theta) \equiv \frac{dW}{dt} = \frac{W}{t} = \frac{e^2 a_0^2 \omega_0^4 n}{8\pi c^3} \sin^2\theta \, do\,. \tag{42.9}$$

And finally, integrating expression (42.9) over the solid angle, we find the total power radiated by a dipole per unit time:

$$P_0 = \frac{dW_0}{dt} = \frac{W_0}{t} = \frac{e^2 a_0^2 \omega_0^4 n}{3c^3}\,. \tag{42.10}$$

This is a familiar result of the field theory. In fact, from classical electrodynamics it is known that the total power of the dipole radiation is given by the formula

$$P_d = \frac{2}{3c^3} n (d'')^2\,. \tag{42.11}$$

Substituting $d = ea_0 \sin\omega_0 t$ into formula (42.11) and averaging over the period $2\pi/\omega_0$, we arrive at expression (42.10).

43. Radiation from a Moving Dipole — Undulator Radiation

Here, we consider radiation from a moving dipole — undulator radiation. We first assume that $\mathbf{a}_0 \perp \mathbf{u}$; such a dipole is called "transverse." We work in the same spherical coordinate system as that used to consider an immobile dipole. For the moment, on the right-hand side of Eqs. (41.18b), we discard the term describing Cherenkov radiation (this term will be considered later) and also ignore the third term. In other words, we take into account only resonances $\omega_\lambda \approx \mathbf{k}_\lambda \mathbf{u} + \omega_0$ (for more detail on resonances $\omega_\lambda \approx \mathbf{k}_\lambda \mathbf{u} - \omega_0$, see Sec. 45). For the second term, we have

$$\mathbf{u} \neq 0\,, \quad \mathbf{k}_\lambda \mathbf{u} = \omega_\lambda \beta n \sin\theta_\lambda \cos\varphi_\lambda\,, \quad b_{\lambda 1} = \sqrt{4\pi}(e/n) a_0 \omega_0 \sin\theta_\lambda\,, \quad b_{\lambda 2} = 0\,,$$
$$c_{\lambda 1} = \sqrt{4\pi}(e/n) a_0 \omega_\lambda \beta n \cos\varphi_\lambda \cos^2\theta_\lambda\,, \quad c_{\lambda 2} = \sqrt{4\pi}(e/n) a_0 \omega_\lambda \beta n \sin\varphi_\lambda \cos\theta_\lambda\,, \tag{43.1}$$

where $\beta = u/c$ and φ_λ is the angle measured from the positive direction of the vector $\mathbf{u}$, lying in the $z = 0$ plane. With formula (41.21), the field energy (41.12) can then be written as

$$W = \pi \frac{e^2 a_0^2 \omega_0^2}{n^2} \left(\sum_\lambda \frac{1 - \cos(\omega_\lambda - \omega_\lambda \beta n \sin\theta_\lambda \cos\varphi_\lambda - \omega_0)t}{(\omega_\lambda - \omega_\lambda \beta n \sin\theta_\lambda \cos\varphi_\lambda - \omega_0)^2} F_\perp(\omega_\lambda, \varphi_\lambda, \theta_\lambda) \right),$$
$$F_\perp(\omega_\lambda, \varphi_\lambda, \theta_\lambda) = \left(\sin\theta_\lambda + \frac{\omega_\lambda}{\omega_0} \beta n \cos\varphi_\lambda \cos^2\theta_\lambda \right)^2 + \left(\frac{\omega_\lambda}{\omega_0} \beta n \sin\varphi_\lambda \cos\theta_\lambda \right)^2. \tag{43.2}$$

We transform expression (43.2) in the same manner as we did in deriving the field energy (42.6) from formula (42.1). As a result, we obtain the following expression for the density of the energy emitted at angles φ and θ into a solid angle element do:

$$W = \frac{e^2 a_0^2 \omega_0^2}{8\pi^2 c^3} \left(\int n(\omega)\omega^2 \frac{1 - \cos(\omega - \omega\beta n \sin\theta\cos\varphi - \omega_0)t}{(\omega - \omega\beta n \sin\theta\cos\varphi - \omega_0)^2} F_\perp(\omega, \varphi, \theta) d\omega \right) do \,. \tag{43.3}$$

Taking the time derivative of expression (43.3), we find the radiation power:

$$\begin{aligned} P(\varphi, \theta) &\equiv \frac{dW}{dt} \\ &= \frac{e^2 a_0^2 \omega_0^2}{8\pi^2 c^3} \left(\int n(\omega)\omega^2 \frac{\sin(\omega - \omega\beta n \sin\theta\cos\varphi - \omega_0)t}{(\omega - \omega\beta n \sin\theta\cos\varphi - \omega_0)} F_\perp(\omega, \varphi, \theta) d\omega \right) do \,. \end{aligned} \tag{43.4}$$

The function that describes the frequency–angular radiation spectrum has the form

$$\begin{aligned} I(\omega, \varphi, \theta) &= n(\omega)\omega^2 \frac{\sin(\omega - \omega\beta n \sin\theta\cos\varphi - \omega_0)t}{(\omega - \omega\beta n \sin\theta\cos\varphi - \omega_0)} F_\perp(\omega, \varphi, \theta) \,, \\ I(\omega, \varphi, \theta) &\xrightarrow[t\to\infty]{} \pi n(\omega)\omega^2 \delta(\omega - \omega\beta n \sin\theta\cos\varphi - \omega_0) F_\perp(\omega, \varphi, \theta) \,. \end{aligned} \tag{43.5}$$

Calculating the integral in formula (43.4) at $n(\omega) = n = \text{const}$ for long time scales gives the power emitted by a moving dipole per unit time:

$$P(\varphi, \theta) \equiv \frac{dW}{dt} = \frac{W}{t} = \frac{e^2 a_0^2 \omega_0^4 n}{8\pi c^3 (1 - \beta n \sin\theta\cos\varphi)^3} F_\perp(\omega_0 (1 - \beta n \sin\theta\cos\varphi)^{-1}, \varphi, \theta) do \,. \tag{43.6}$$

The figures below present directional radiation patterns calculated from formulas (43.6). Figure 65 shows the directional pattern as a function of the angle φ at $\theta = \pi/2$. The dipole is at the origin of the coordinate axes x and y. The vector $\mathbf{a}_0$ is perpendicular to the plane of the figure. The velocity vector $\mathbf{u}$ points in the positive direction of the x axis. The directional patterns are plotted for different values of the parameter βn, which are indicated near the curves. Figure 66 shows the directional pattern as a function of the angle $\theta \in [0, \pi]$ at $\varphi = 0$ (the $x > 0$ half-plane) and $\varphi = \pi$ (the $x < 0$ half-plane). The vector $\mathbf{a}_0$ is parallel to the z axis. The dipole is at the origin of the coordinates. The velocity vector $\mathbf{u}$ again points in the positive direction of the x axis. The relevant values of βn are indicated near the curves. And finally, Fig. 67 shows the directional pattern as a function of θ for $\beta n = 0.2$ at different angles φ: curve 1 is for $\varphi = 0$ (forward radiation), curve 2 is for $\varphi = \pi/4$ (forward radiation at an angle of 45°), curve 3 is for $\varphi = \pi/2$ (radiation perpendicular to the velocity $\mathbf{u}$), curve 4 is for $\varphi = 3\pi/4$ (backward radiation at an angle of 45°), and curve 5 is for $\varphi = \pi$ (backward radiation).

From Figs. 65 and 66 we can see that the larger the parameter βn, the more forward the radiation and the smaller the solid angle around the direction ($\theta = \pi/2$, $\varphi = 0$) into which the radiation is emitted. Formulas (43.6) yield the following

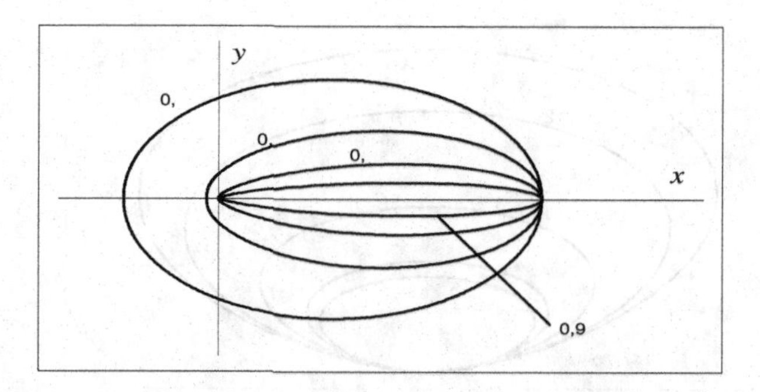

Fig. 65. Directional pattern of radiation from a transverse dipole as a function of the azimuthal angle φ at $\theta = \pi/2$.

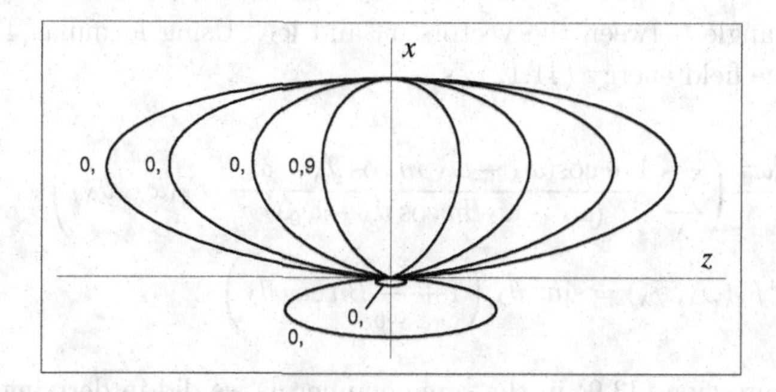

Fig. 66. Directional pattern of radiation from a transverse dipole as a function of the angle $\theta \in [0, \pi]$ at $\varphi = 0$ (the $x > 0$ half-plane) and $\varphi = \pi$ (the $x < 0$ half-plane).

familiar estimate for the range of angles into which radiation from a particle is emitted when $\beta n \sim 1$:

$$\sqrt{(\Delta\theta)^2 + (\Delta\varphi)^2} \sim \sqrt{1 - \beta n}. \tag{43.7}$$

Let us now consider radiation from a moving dipole under the assumption that an electron oscillates in the direction of motion, $\mathbf{a}_0 \| \mathbf{u}$. Such a dipole is called "longitudinal." We choose a spherical coordinate system (r, ϑ, ϕ) with the z axis directed along the velocity vector $\mathbf{u}$. Here, $\phi \in [0, 2\pi]$ is the azimuthal angle in a plane perpendicular to the vector $\mathbf{u}$ and $\vartheta \in [0, \pi]$ is the angle measured from the positive direction of the z axis. The vector $\mathbf{e}_{\lambda 1}$ lies in the azimuthal plane $\phi = \text{const}$, and the vector $\mathbf{e}_{\lambda 2}$ is perpendicular to this plane. Accordingly, for the right-hand sides of Eqs. (41.18b) (in which the first and third terms are again discarded for the moment), we have

$$\begin{aligned} &\mathbf{u} \neq 0\,, \quad \mathbf{k}_\lambda \mathbf{u} = \omega_\lambda \beta n \cos\vartheta_\lambda\,, \quad b_{\lambda 1} = -\sqrt{4\pi}(e/n) a_0 \omega_0 \sin\vartheta_\lambda\,, \quad b_{\lambda 2} = 0\,, \\ &\qquad c_{\lambda 1} = -\sqrt{4\pi}(e/n) a_0 \omega_\lambda \beta n \sin\vartheta_\lambda \cos\vartheta_\lambda\,, \quad c_{\lambda 2} = 0\,, \end{aligned} \tag{43.8}$$

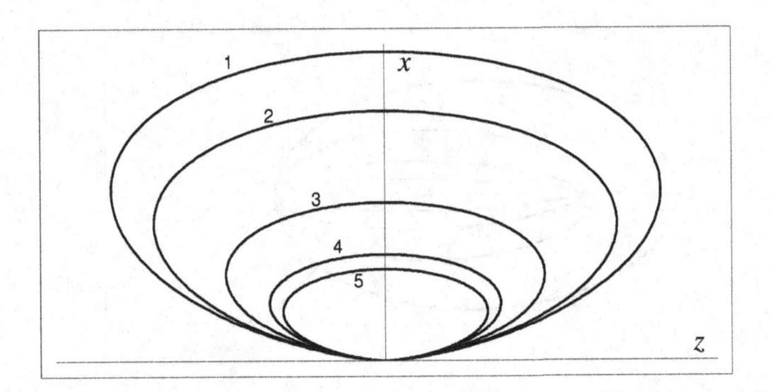

Fig. 67. Directional pattern of radiation from a transverse dipole as a function of θ for $\beta n = 0.2$ at different angles φ: 1 $\varphi = 0$, 2 $\varphi = \pi/4$, 3 $\varphi = \pi/2$, 4 $\varphi = 3\pi/4$, and 5 $\varphi = \pi$.

where ϑ_λ is the angle between the vectors $\mathbf{a}_0$ and $\mathbf{k}_\lambda$. Using formula (41.21), we can then write the field energy (41.12) as

$$W = \pi \frac{e^2 a_0^2 \omega_0^2}{n^2} \left(\sum_\lambda \frac{1 - \cos(\omega_\lambda - \omega_\lambda \beta n \cos\vartheta_\lambda - \omega_0)t}{(\omega_\lambda - \omega_\lambda \beta n \cos\vartheta_\lambda - \omega_0)^2} F_{\|}(\omega_\lambda, \vartheta_\lambda) \right),$$
$$F_{\|}(\omega_\lambda, \vartheta_\lambda) = \sin^2\vartheta_\lambda \left(1 + \frac{\omega_\lambda}{\omega_0} \beta n \cos\vartheta_\lambda \right)^2. \tag{43.9}$$

We transform expression (43.9) in the same manner as we did in deriving the field energy (42.6) from formula (42.1). As a result, we obtain the following expression for the energy emitted at angles ϕ and ϑ into a solid angle element $do = \sin\vartheta d\vartheta d\phi$:

$$W = \frac{e^2 a_0^2 \omega_0^2}{8\pi^2 c^3} \left(\int n(\omega)\omega^2 \frac{1 - \cos(\omega - \omega\beta n \cos\vartheta - \omega_0)t}{(\omega - \omega\beta n \cos\vartheta - \omega_0)^2} F_{\|}(\omega, \vartheta) d\omega \right) do. \tag{43.10}$$

Taking the time derivative of expression (43.10), we find the radiation power:

$$P(\vartheta) \equiv \frac{dW}{dt}$$
$$= \frac{e^2 a_0^2 \omega_0^2}{8\pi^2 c^3} \left(\int n(\omega)\omega^2 \frac{\sin(\omega - \omega\beta n \cos\vartheta - \omega_0)t}{(\omega - \omega\beta n \cos\vartheta - \omega_0)} F_{\|}(\omega, \vartheta) d\omega \right) do. \tag{43.11}$$

The function that describes the frequency–angular radiation spectrum has the form

$$I(\omega, \vartheta) = n(\omega)\omega^2 \frac{\sin(\omega - \omega\beta n \cos\vartheta - \omega_0)t}{(\omega - \omega\beta n \cos\vartheta - \omega_0)} F_{\|}(\omega, \vartheta),$$
$$I(\omega, \vartheta) \xrightarrow[t\to\infty]{} \pi n(\omega)\omega^2 \delta(\omega - \omega\beta n \cos\vartheta - \omega_0) F_{\|}(\omega, \vartheta). \tag{43.12}$$

Calculating the integral in formula (43.11) at $n(\omega) = n = \text{const}$ for long time scales gives the power emitted by a moving dipole per unit time:

$$P(\vartheta) \equiv \frac{dW}{dt} = \frac{W}{t} = \frac{e^2 a_0^2 \omega_0^4 n}{8\pi c^3 (1 - \beta n \cos\vartheta)^3} F_{\parallel}(\omega_0 (1 - \beta n \cos\vartheta)^{-1}, \vartheta) do$$
$$= \frac{e^2 a_0^2 \omega_0^4 n}{8\pi c^3} (1 - \beta n \cos\vartheta)^{-5} \sin^2\vartheta \, do \,. \tag{43.13}$$

By virtue of the axial symmetry of the system, formulas (43.10)–(43.13) do not contain the azimuthal angle ϕ.

Figures 68–70 display directional radiation patterns calculated from formula (43.13) as functions of the angle ϑ for different values of the parameter βn, which are indicated near the curves. The dipole moves in the positive direction of the z axis, with respect to which the vector $\mathbf{a}_0$ is parallel, too.

Integrating expression (43.13) over the angles, we arrive at the following formula for the total power radiated by a longitudinal dipole per unit time:

$$P_0 = \frac{dW_0}{dt} = \frac{W_0}{t} = \frac{e^2 a_0^2 \omega_0^4 n}{3c^3} S(\beta n) \,. \tag{43.14}$$

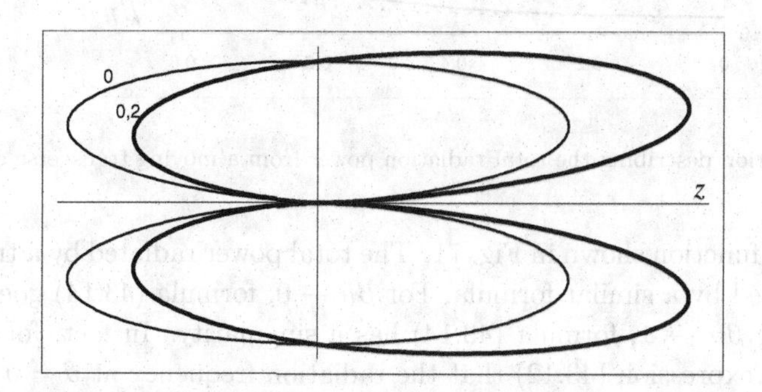

Fig. 68. Directional patterns of radiation from a longitudinal dipole for $\beta n = 0$ and $\beta n = 0.2$.

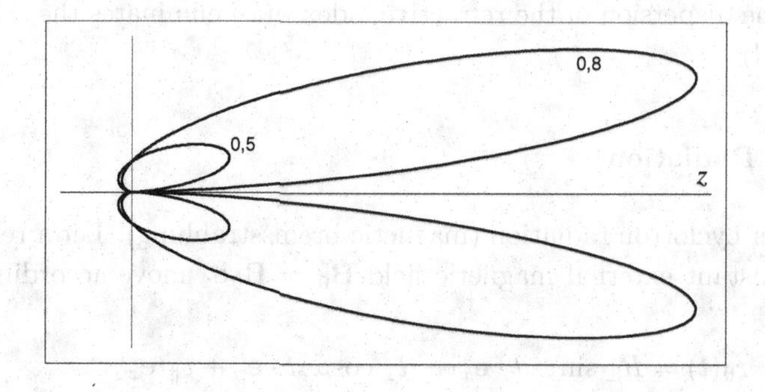

Fig. 69. Directional patterns of radiation from a longitudinal dipole for $\beta n = 0.5$ and $\beta n = 0.8$.

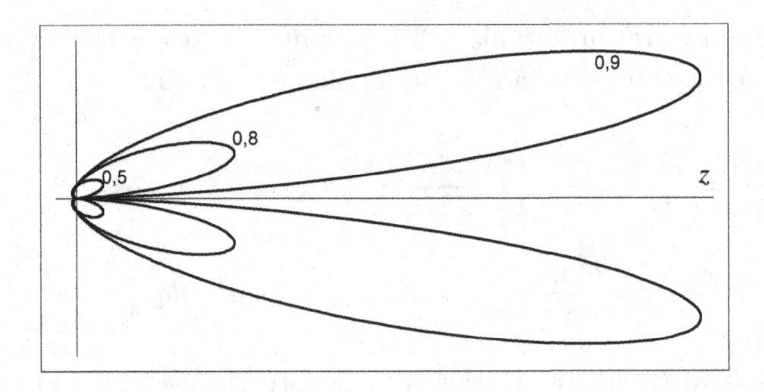

Fig. 70. Directional patterns of radiation from a longitudinal dipole for $\beta n = 0.5$, $\beta n = 0.8$, and $\beta n = 0.9$.

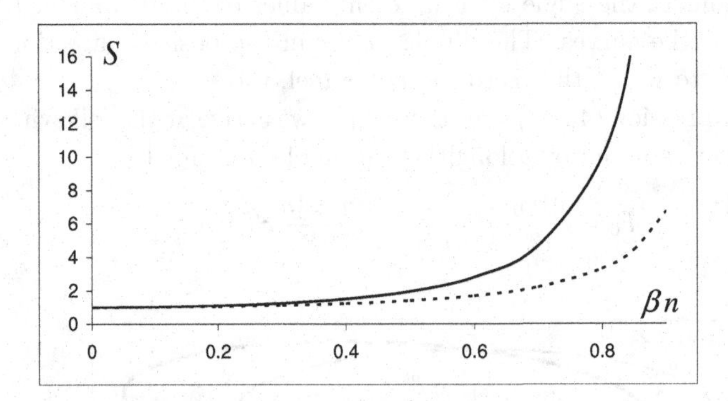

Fig. 71. Function describing the total radiation power from a moving transverse dipole.

Here, $S(\beta n)$ is a function shown in Fig. 71. The total power radiated by a transverse dipole is described by a similar formula. For $\beta n \to 0$, formula (43.14) goes over to (42.10), and, for $\beta n \to 1$, formula (43.14) has a singularity. In fact, for $\beta n = 1$, we can see from expression (43.12) that the radiation frequency at $\vartheta = 0$ becomes infinite. But as $\omega \to \infty$, the refractive index n should approach unity. That is why, at the radiation frequency, the case $\beta n = 1$ is impossible. In other words, accounting for the dispersion of the refractive index $n(\omega)$ eliminates the singularity in formula (43.14).

44. Cyclotron Radiation

Here, we consider cyclotron radiation (magnetic bremsstrahlung). Let a relativistic electron in a constant external magnetic field $\mathbf{B}_0 = \mathbf{B}_0\mathbf{e}_z$ move according to the law

$$\begin{aligned} \mathbf{r}_0(t) &= R_\perp \sin\omega_e t \cdot \mathbf{e}_x + R_\perp \cos\omega_e t \cdot \mathbf{e}_y + v_\parallel t\mathbf{e}_z\,, \\ \mathbf{v}_0(t) &= v_\perp \cos\omega_e t \cdot \mathbf{e}_x - v_\perp \sin\omega_e t \cdot \mathbf{e}_y + v_\parallel \mathbf{e}_z\,, \end{aligned} \tag{44.1}$$

where $v_\perp$ and $v_\parallel$ are the electron velocity components across and along the magnetic field, $R_\perp = v_\perp/\omega_e$ is the electron gyroradius, $\omega_e = eB_0/mc\gamma$ is the electron gyrofrequency, $\gamma = (1 - v_\perp^2/c^2 - v_\parallel^2/c^2)^{-1/2}$ is the electron relativistic factor, and $(\mathbf{e}_x,\ \mathbf{e}_y,\ \mathbf{e}_z)$ are unit vectors of a Cartesian coordinate system. The xy plane is perpendicular to the magnetic field vector $\mathbf{B}_0$.

Assume that the following inequality is satisfied:

$$\sqrt{k_{\lambda x}^2 + k_{\lambda y}^2} R_\perp \ll 1\,. \tag{44.2}$$

Substituting relationships (44.1) into Eqs. (41.15), we then obtain the equations

$$q_{\lambda 1,2}'' + \omega_\lambda^2 q_{\lambda 1,2} = \sqrt{4\pi}\frac{e}{n} v_\perp ((\mathbf{e}_{\lambda 1,2}\mathbf{e}_x)\cos\omega_e t - (\mathbf{e}_{\lambda 1,2}\mathbf{e}_y)\sin\omega_e t)\exp(-ik_{\lambda z}v_\parallel t)\,. \tag{44.3}$$

Let $\mathbf{k}_\lambda = \{k_{\lambda x}, 0, k_{\lambda z}\}$, which indicates that this vector lies in the xz plane. We introduce the unit polarization vectors of radiation: $\mathbf{e}_{\lambda 1} = \{-\cos\theta_\lambda, 0, \sin\theta_\lambda$ and $\mathbf{e}_{\lambda 2} = \{0, 1, 0\}$, where θ_λ is the angle between the vector $\mathbf{k}_\lambda = (\omega_\lambda/c)n\{\sin\theta_\lambda, 0, \cos\theta_\lambda\}$ and the positive direction of the z axis. As a result, we can write Eqs. (44.3) as

$$q_{\lambda 1}'' + \omega_\lambda^2 q_{\lambda 1} = -\sqrt{4\pi}\frac{e}{n} v_\perp \cos\omega_e t\cos\theta_\lambda \exp(-i\Omega_\lambda t)\,,$$

$$q_{\lambda 2}'' + \omega_\lambda^2 q_{\lambda 2} = \sqrt{4\pi}\frac{e}{n} v_\perp \sin\omega_e t\exp(-i\Omega_\lambda t)\,, \quad \Omega_\lambda = k_{\lambda z}v_\parallel = \omega_\lambda \beta_\parallel n\cos\theta_\lambda\,, \tag{44.4}$$

where $\beta_\parallel = v_\parallel/c$. Since both components of expansion (41.13) are nonzero, the electromagnetic field energy (41.12) reads

$$W = \sum_{\substack{\lambda \\ s=1,2}} (q_{\lambda s}' q_{\lambda s}'^* + \omega_\lambda^2 q_{\lambda s} q_{\lambda s}^*)\,. \tag{44.5}$$

In what follows, we are going to consider only resonances $\omega_\lambda \approx \Omega_\lambda + \omega_e$. Accordingly, we rewrite Eqs. (44.4) as

$$\begin{aligned} q_{\lambda 1}'' + \omega_\lambda^2 q_{\lambda 1} &= -\frac{1}{2}\sqrt{4\pi}\frac{e}{n} v_\perp \cos\theta_\lambda \exp(-i(\Omega_\lambda + \omega_e)t)\,, \\ q_{\lambda 2}'' + \omega_\lambda^2 q_{\lambda 2} &= \frac{1}{2} i\sqrt{4\pi}\frac{e}{n} v_\perp \exp(-i(\Omega_\lambda + \omega_e)t)\,. \end{aligned} \tag{44.6}$$

The initial-value problem for Eqs. (44.6) with zero initial conditions is formulated as problem (41.19). Therefore, using formula (41.21) with $\Omega = \Omega_\lambda + \omega_e$, we can rewrite the electromagnetic field energy (44.5) as

$$W = \pi\frac{e^2 v_\perp^2}{n^2}\left(\sum_\lambda \frac{1 - \cos(\omega_\lambda - \omega_\lambda\beta_\parallel n\cos\theta_\lambda - \omega_e)t}{(\omega_\lambda - \omega_\lambda\beta_\parallel n\cos\theta_\lambda - \omega_e)^2}(1 + \cos^2\theta_\lambda)\right). \tag{44.7}$$

Note that expression (44.7) has been averaged over the period of gyration in an external magnetic field. In expression (44.7), we switch from summation over λ to integration over frequencies (see (42.5)) to obtain the following expression for the electromagnetic energy of the cyclotron radiation:

$$W = \frac{e^2 v_\perp^2}{8\pi^2 c^3} \left(\int n(\omega)\omega^2 \frac{1 - \cos(\omega - \omega\beta_\| n \cos\theta - \omega_e)t}{(\omega - \omega\beta_\| n \cos\theta - \omega_e)} d\omega \right) (1 + \cos^2\theta) do\,. \quad (44.8)$$

Differentiating expression (44.8) with respect to time yields the radiation power:

$$P(\theta) \equiv \frac{dW}{dt}$$

$$= \frac{e^2 v_\perp^2}{8\pi^2 c^3} \left(\int n(\omega)\omega^2 \frac{\sin(\omega - \omega\beta_\| n \cos\theta - \omega_e)t}{(\omega - \omega\beta_\| n \cos\theta - \omega_e)} d\omega \right) (1 + \cos^2\theta) do\,. \quad (44.9)$$

The function that describes the frequency–angular radiation spectrum has the form

$$I(\omega, \varphi, \theta) = n(\omega)\omega^2 \frac{\sin(\omega - \omega\beta_\| n \cos\theta - \omega_e)t}{(\omega - \omega_\lambda\beta_\| n \cos\theta_\lambda - \omega_e)} (1 + \cos^2\theta)\,,$$
$$I(\omega, \theta) \xrightarrow[t\to\infty]{} \pi n(\omega)\omega^2 \delta(\omega - \omega\beta_\| n \cos\theta - \omega_e)(1 + \cos^2\theta)\,. \quad (44.10)$$

Calculating the integral in formula (44.9) at $n(\omega) = n = \text{const}$ for long time scales gives the power an electron gyrating in a magnetic field emits into a solid angle element do per unit time:

$$P(\theta) \equiv \frac{dW}{dt} = \frac{W}{t} = \frac{e^2 v_\perp^2 \omega_e^2 n}{8\pi c^3} (1 - \beta_\| n \cos\theta)^{-3} (1 + \cos^2\theta) do\,. \quad (44.11)$$

For $\beta_\| = 0$, expression (44.11) simplifies to

$$P(\theta) \equiv \frac{dW}{dt} = \frac{W}{t} = \frac{e^2 v_\perp^2 \omega_e^2 n}{8\pi c^3} (1 + \cos^2\theta) \sin\theta d\theta d\varphi\,. \quad (44.12)$$

Figures 72 and 73 show directional patterns of cyclotron radiation, calculated as functions of θ for different values of the parameter $\beta_\| n$, which are indicated near the curves.

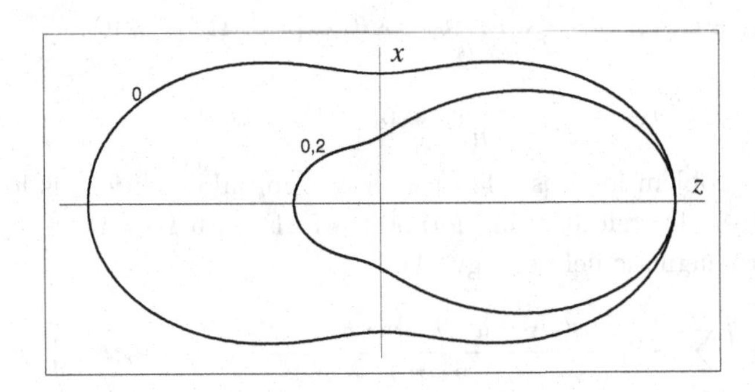

Fig. 72. Directional patterns of cyclotron radiation for $\beta_\| n = 0$ and $\beta_\| n = 0.2$.

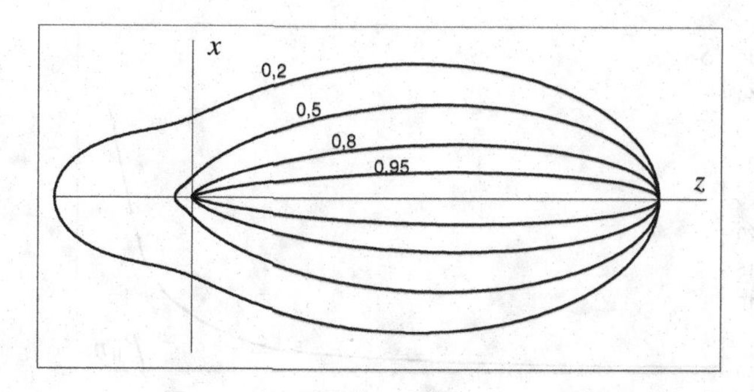

Fig. 73. Directional patterns of cyclotron radiation for $\beta_{\|}n = 0.2$, $\beta_{\|}n = 0.5$, $\beta_{\|}n = 0.8$, and $\beta_{\|}n = 0.95$.

Integrating expression (44.12) over θ from zero to π and over φ from zero to 2π, we arrive at the following familiar formula for the total radiation power emitted by an electron as it revolves along a Larmor circle in an external magnetic field:

$$P_0 = \frac{dW_0}{dt} = \frac{W_0}{t} = \frac{2e^2 v_\perp^2 \omega_e^2 n}{3c^3} = \frac{2e^4 v_\perp^2 B_0^2 n}{3m^2 c^5 (1 - v_\perp^2/c^2)} \,. \tag{44.13}$$

Let us clarify what other restrictions are imposed by inequality (44.2), which contains the component of the wave vector $\mathbf{k} = \{\mathbf{k}_\perp, k_z\}$ that is transverse to the magnetic field $\mathbf{B}_0$. Here, $k_\perp = \sqrt{k_x^2 + k_y^2}$. Radiation in the direction of the wave vector component $\mathbf{k}_\perp$ is that emitted at the angle $\theta = \pi/2$. Formula (44.10) implies that this radiation has the frequency ω_e. Since $\omega = kc/n$, we have $k_\perp = \omega_e n/c$ and inequality (44.2) becomes

$$n(v_\perp/c) \ll 1 \,. \tag{44.14}$$

Consequently, the transverse motion of an electron should be weakly relativistic. As for the longitudinal electron motion, no assumption is made on whether it is relativistic or not. The cyclotron radiation emitted under conditions in which inequality (44.14) fails to hold is called synchrotron radiation. An analysis of synchrotron radiation is rather involved and is not presented here.

Integrating expression (44.11) over the angles yields the following general formula for the total power of cyclotron radiation from an electron that moves with a relativistic longitudinal velocity and a weakly relativistic transverse velocity in an external magnetic field:

$$P_0 = \frac{dW_0}{dt} = \frac{W_0}{t} = \frac{2e^4 v_\perp^2 B_0^2 n}{3m^2 c^5 (1 - v_\perp^2/c^2)} S(\beta_{\|} n) \,. \tag{44.15}$$

Here, $S(\beta_{\|}n)$ is a function shown in Fig. 74. Since $S(0) = 1$, formula (44.15) contains as a particular case formula (44.13). For $\beta_{\|}n \to 1$, formula (44.15), like formula (43.14), has a singularity. Formulas (43.14) and (44.15) were obtained numerically.

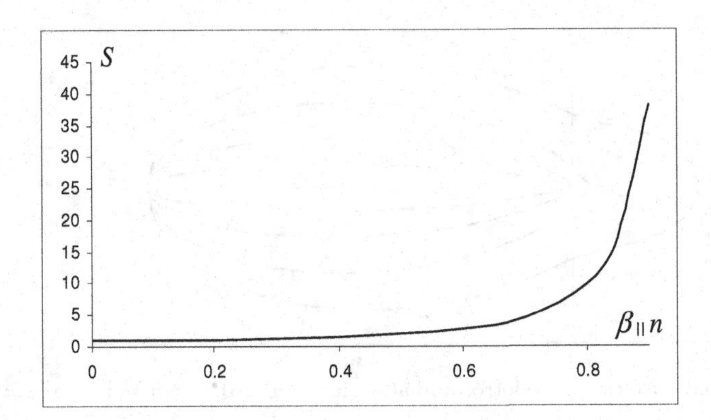

Fig. 74. Function describing the total cyclotron radiation power.

45. Cherenkov Effect. Anomalous and Normal Doppler Effects

Let us now consider radiation from an electron moving uniformly along a straight line. This radiation, called Cherenkov radiation, can occur only in a medium with a refractive index $n(\omega) > 1$. Setting $\mathbf{a}_0 = 0$ in formulas (41.16), we retain only the first term on the right-hand side of Eqs. (41.18b). We work in the same spherical coordinate system (ρ, ϑ, ϕ) as that used to consider radiation from a longitudinal dipole. Accordingly, we can write the following relationships:

$$\mathbf{u} \neq 0\,, \quad \Omega = \omega_\lambda \beta n \cos\vartheta_\lambda\,, \quad a_1 = -\sqrt{4\pi}\frac{e}{n}u\sin\vartheta_\lambda\,, \quad a_2 = 0\,, \tag{45.1}$$

where ϑ_λ is the angle between the velocity $\mathbf{u}$ and the vector $\mathbf{k}_\lambda$. With formula (41.21), the field energy (41.12) becomes

$$W = 4\pi\frac{e^2u^2}{n^2}\left(\sum_\lambda \frac{1-\cos\omega_\lambda(1-\beta n\cos\vartheta)t}{\omega_\lambda^2(1-\beta n\cos\vartheta)}\right)\sin^2\vartheta\,. \tag{45.2}$$

Switching to integration over frequency, we find

$$W = \frac{e^2u^2}{2\pi^2c^3}\int n(\omega)\frac{1-\cos[\omega(1-\beta n\cos\vartheta)t]}{(1-\beta n\cos\vartheta)^2}d\omega\sin^2\vartheta\,do\,. \tag{45.3}$$

Here, in the integrand, we must take into account dispersion, i.e., we must set $n = n(\omega)$. The cases considered above — those of dipole, undulator, and cyclotron radiation — do not require that dispersion be taken into account.

To find the radiation power, we differentiate expression (45.3) with respect to time:

$$P(\vartheta) \equiv \frac{dW}{dt} = \frac{e^2u^2}{2\pi^2c^3}\int n(\omega)\frac{\omega\sin[\omega(1-\beta n(\omega)\cos\vartheta)t]}{1-\beta n(\omega)\cos\vartheta)}d\omega\sin^2\vartheta\,do\,. \tag{45.4}$$

For $\omega t \gg 1$, expression (45.4) reduces to

$$\frac{dW}{dt} = \frac{W}{t} = \frac{e^2u^2}{2\pi c^3}\int n(\omega)\omega\delta(1-\beta n(\omega)\cos\vartheta)d\omega\sin^2\vartheta\,do\,. \tag{45.5}$$

In order to transform expression (45.5) further, we assume that the frequencies in the range $\omega \in (0, \omega_0)$, where ω_0 is a certain frequency (see below), satisfy the inequality $n(\omega) \geq 1$. The argument of the delta function in expression (45.5) varies between $1 - \beta n$ at $\vartheta = 0$ and $1 + \beta n$ at $\vartheta = \pi$. If, for all frequencies $\omega \in (0, \omega_0)$, the difference $1 - \beta n(\omega)$ is nonzero, then the integral over $\vartheta \in [0, \pi]$ in expression (45.5) equals zero for all $\vartheta \in [0, \pi]$. Consequently, if there is radiation, then it is emitted in the angular range

$$\frac{c}{un(\omega)_{\max}} < \cos\vartheta < \frac{c}{un(\omega)_{\min}},$$
$$0 < \arccos\left[\min\left(1, \frac{c}{un(\omega)_{\min}}\right)\right] < \vartheta < \arccos\frac{c}{un(\omega)_{\max}} < \frac{\pi}{2}. \tag{45.6}$$

Since $0 < \vartheta < \pi/2$, the radiation is emitted predominantly in the direction of electron motion. This radiation came to be called Cherenkov radiation. If the integral in expression (45.5) is first taken over solid angle, then subsequent integration over frequency should be restricted to the range of ω values that is determined by the inequality $c/n < u$.

In expression (45.5), we first integrate over $do = \sin\vartheta d\vartheta d\phi$. Let $\vartheta_0 = \vartheta_0(\omega)$ be a root of the equation $1 - \beta n(\omega)\cos\vartheta = 0$. We then have $\cos\vartheta_0 = (\beta n)^{-1}$, $\sin^2\vartheta_0 = 1 - (\beta n)^{-2}$, and

$$\delta(1 - \beta n(\omega)\cos\vartheta) = |\beta n \sin\vartheta_0|^{-1}\delta(\vartheta - \vartheta_0). \tag{45.7}$$

Taking into account the above relationships, we perform simple manipulations with the integral in expression (45.5) to obtain the following final result:

$$P_0 = \frac{dW_0}{dt} = \frac{W_0}{t} = \frac{e^2 u}{c^2}\int_{c/n<u} \omega\left(1 - \frac{c^2}{u^2 n^2(\omega)}\right) d\omega. \tag{45.8}$$

This is the familiar Tamm–Frank formula for the Cherenkov radiation power.

In addition to the Tamm–Frank formula (45.8), let us determine the directional pattern of the Cherenkov radiation. Denoting by ω^* the root of the equation

$$1 - \beta n(\omega)\cos\vartheta = 0, \tag{45.9}$$

we rewrite expression (45.5) as

$$\frac{dW}{dt} = \frac{W}{t} = \frac{e^2 u^2}{2\pi c^3}\int n(\omega)\omega\left|\beta\cos\vartheta\frac{dn}{d\omega}\right|^{-1}\delta(\omega - \omega^*)d\omega\sin^2\vartheta\, do. \tag{45.10}$$

Integration over frequency yields

$$\frac{dW}{dt} = \frac{W}{t} = \frac{e^2 u^2}{2\pi c^3}\int n(\omega^*)\omega^*\left|\beta\cos\vartheta\frac{dn(\omega^*)}{d\omega}\right|^{-1}\sin^2\vartheta\, do, \tag{45.11}$$

where the root ω^* is a function of the angle ϑ, as implied by Eq. (45.9). The directional radiation pattern is described by the integrand in expression (45.11), specifically,

$$D(\vartheta) = n(\omega^*)\omega^*\left|\beta\cos\vartheta\frac{dn(\omega^*)}{d\omega}\right|^{-1}\sin^2\vartheta. \tag{45.12}$$

For further analysis, we need to specify the dependence $n(\omega)$. We set

$$n^2(\omega) = 1 + \frac{\omega_p^2}{\omega_0^2 - \omega^2}, \tag{45.13}$$

where ω_p^2 and ω_0^2 are constants. Dependence (45.13) is characteristic of media with normal dispersion in the optical frequency range. Of course, one must keep in mind that, for $\omega \to \omega_0$, formula (45.13) fails to hold because it does not account for wave absorption in the medium. The solution to Eq. (45.9) has the form

$$\omega^*(\vartheta) = \omega_0 \sqrt{1 - \frac{\omega_p^2}{\omega_0^2} \frac{\beta^2 \cos^2 \vartheta}{1 - \beta^2 \cos^2 \vartheta}}. \tag{45.14}$$

Formula (45.14) implies that $\omega^* \leq \omega_0$. In this frequency range, we have $n(\omega) > 1$. In the frequency range $\omega > \omega_0$, the radiation under analysis is impossible because either $n^2 < 0$ or $n < 1$. Since, in the frequency range where radiation can occur, the refractive index is positive, we have $\cos \vartheta > 0$. Accordingly, from formula (45.14) we find the angular range in which Cherenkov radiation is emitted:

$$\arccos \left[\min \left(1, \frac{c}{u} \sqrt{\frac{\omega_0^2}{\omega_0^2 + \omega_p^2}} \right) \right] \leq \vartheta \leq \frac{\pi}{2}. \tag{45.15}$$

Condition (45.15) is a particular case of general condition (45.6). For $\omega_0 = 0$ — the case of a cold isotropic plasma — the angular range (45.15) becomes zero, indicating that Cherenkov emission of transverse waves in an isotropic plasma is impossible.

Substituting formulas (45.13) and (45.14) into expression (45.12) and performing simple manipulations, we arrive at the following directional pattern:

$$D(\vartheta) = \omega_p^2 \frac{\beta \cos \vartheta \sin^2 \vartheta}{(\beta^2 \cos^2 \vartheta - 1)^2}. \tag{45.16}$$

Formula (45.16) is only valid in the angular range (45.15) (as well as in an angular range that is symmetric with respect to zero); outside this range, we must set $D(\vartheta) \equiv 0$. The directional patterns (45.16) calculated for two β values are depicted in Fig. 75.

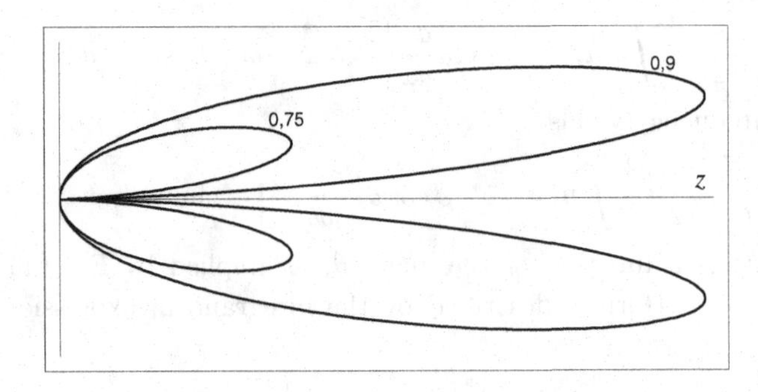

Fig. 75. Directional patterns of Cherenkov radiation for $\beta = 0.75$ and $\beta = 0.9$.

In the nonrelativistic case (such that $\beta^2 \ll 1$), the maximum of the directional pattern is at $\vartheta = \arctan(\sqrt{2}) \approx 0.96$ and, in the ultrarelativistic limit, the maximum is at an angle of $\vartheta \approx \sqrt{1-\beta^2} \ll 1$.

Let us also apply the Tamm–Frank formula to calculate the Cherenkov radiation power in a medium with a refractive index given by formula (45.13). For simplicity, we restrict ourselves to the particular case

$$\frac{c}{u}\sqrt{\frac{\omega_0^2}{\omega_)^2 + \omega_p^2}} = 1\,, \tag{45.17}$$

in which, according to formula (45.14) and condition (45.15), radiation is emitted at all frequencies from zero to ω_0 in the entire angular range from zero to $\pi/2$. It is a simple matter to take the integral in formula (45.8), with the result that

$$P_0 = \frac{dW_0}{dt} = \frac{W_0}{t} = \frac{e^2 u}{c^2}\omega_0^2 F(\omega_p^2/\omega_0^2)\,. \tag{45.18}$$

Here, $F(x) = (x/2)((1+x)\ln(1+x^{-1}) - 1)$. The plot of this function is presented in Fig. 76. For $x \to \infty$, we have $F(x) \to 1/4$.

The relationships between the frequencies of the emitted waves and the angles at which they are emitted in the case of Cherenkov radiation from an electron in a medium with the refractive index (45.13) is explained in Fig. 77, which shows the dispersion curves of waves in the medium under consideration — the solution $\omega(k)$ to the dispersion relation

$$\omega^2 = k^2c^2/n^2(\omega)\,. \tag{45.19}$$

There are two types of waves: optical (curve b) and acoustic (curve a). Cherenkov radiation is possible only at acoustic frequencies. We can easily see that the left-hand side of relationship (45.17) is the ratio $V_{ph}(0)/u$, where $V_{ph}(0) = c/n(0)$ is the phase velocity of an acoustic wave at $\omega \to 0$.

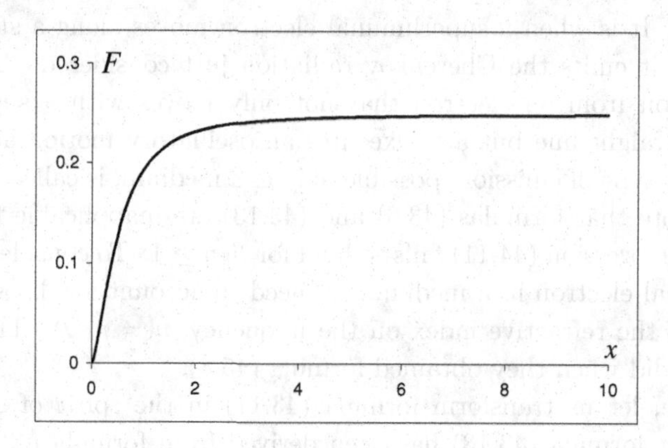

Fig. 76. Function $F(x)$ (45.18), describing the Cherenkov radiation power.

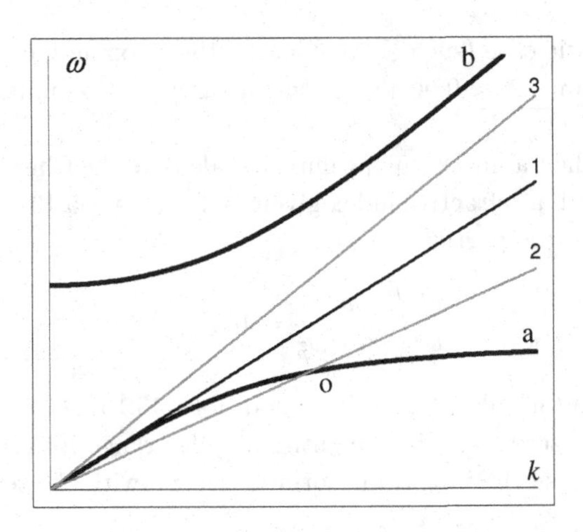

Fig. 77. Determination of the frequencies and angles at which Cherenkov radiation is emitted in a medium with the refractive index (45.13).

In Fig. 77, we also draw the straight lines $\omega = ku = \beta kc$. Equality (45.17) refers to line 1. Straight lines $\omega = ku\cos\vartheta$ lie below (see, e.g., line 2). Consequently, under conditions corresponding to equality (45.17), radiation is emitted at all frequencies from zero to ω_0 and at all angles from zero to $\pi/2$ (of course, in accordance with the directional pattern). If the left-hand side of equality (45.17) is greater than unity, then we have $V_{ph}(0) > u$ and the line $\omega = ku$ is line 2. That is why radiation does not occur at frequencies below point o in the figure, and radiation at the frequency corresponding to this point is emitted at all angles from zero to $\pi/2$. And finally, if the left-hand side of equality (45.17) is less than unity, then we have $V_{ph}(0) < u$ and the line $\omega = ku$ is line 3. Accordingly, radiation is emitted at all frequencies from zero to ω_0 and there is no emission at small angles.

An electron moving with a superluminal velocity in a medium can also emit another type of radiation. By superluminal motion is meant a motion with a velocity $u > c/n(\omega)$. It is when a superluminal electron moves along a straight line in a medium that it emits the Cherenkov radiation just considered. Let us now investigate radiation from an electron that not only moves with a superluminal velocity along a straight line but also executes an oscillatory motion at a certain frequency ω_0. This type of emission, possible only in a medium, is called anomalous Doppler effect. Note that formulas (43.6) and (43.13) are inapplicable for $\beta n \geq 1$ or $u \geq c/n$. Also, expression (44.11) fails to hold for $\beta_{\|} n \geq 1$. To consider radiation from a superluminal electron in a medium, we need to account for dispersion, i.e., the dependence of the refractive index on the frequency, $n = n(\omega)$. This is what Tamm and Frank did when they obtained formula (45.8).

As an example, let us transform formula (43.11) in the spirit of Tamm and Frank. Recall that formula (43.13) has been derived from formula (43.11) in the case $n(\omega) = n = \text{const}$. For $\beta n \geq 1$, formula (43.11) is inapplicable. Let us write formula (43.11) on long time scales t as

$$P(\vartheta) \equiv \frac{dW}{dt} = \frac{e^2 a_0^2 \omega_0^2}{4c^3} \int n(\omega)\omega^2 \delta(\omega - \omega\beta n(\omega)\cos\vartheta - \omega_0) F_{\|}(\omega, \vartheta) \sin\vartheta d\omega\, d\vartheta\,. \tag{45.20}$$

Here, integration over ω is carried out from zero to ∞ and integration over ϑ, from zero to π. As for integration over the azimuthal angle, it has already been performed.

From formula (45.20) we can see that the frequency of a wave emitted at an angle ϑ is given by the expression

$$\omega = \frac{\omega_0}{1 - \beta n \cos\vartheta}\,. \tag{45.21}$$

When $\beta n > 1$, in the cone of angles such that $\cos\vartheta > (\beta n)^{-1}$, we have $\omega < 0$. It might seem that negative frequencies do not contribute to the integral in formula (45.20) because they are beyond the interval of integration. This is not the case, however: in transforming Eqs. (41.18b), we have retained the second term on their right-hand side and have discarded the third term. In other words, for the resonant interaction between an electron and the emitted waves, we have used the condition $\omega_\lambda = \mathbf{k}_\lambda \mathbf{u} + \omega_0$. But we can also use another condition, namely, $\omega_\lambda = \mathbf{k}_\lambda \mathbf{u} - \omega_0$. Consequently, we must put the $\pm$ sign in front of ω_0 in formula (45.20) or must put absolute value bars around the denominator in formula (45.21). Another apparent drawback of formula (45.21) is that, for $\beta n \cos\vartheta = 1$, the frequency is infinite. This is also not the case, however: as $\omega \to \infty$, the refractive index approaches unity, $n(\infty) \to 1$. We will consider anomalous Doppler effect by using formula (45.20) with the plus sign in front of ω_0.

In formula (45.20), we first integrate over ϑ. Let $\vartheta_0 = \vartheta_0(\omega)$ be a root of the equation $\omega - \omega\beta n(\omega)\cos\vartheta + \omega_0 = 0$. Representing the delta function in the integrand as

$$\delta(\omega - \omega\beta n(\omega)\cos\vartheta + \omega_0) = |\omega\beta n(\omega)\sin\vartheta_0|^{-1}\delta(\vartheta - \vartheta_0)\,, \tag{45.22}$$

we integrate over the angle to obtain

$$P_0 \equiv \frac{dW_0}{dt} = \frac{W_0}{t} = \frac{e^2 a_0^2}{4uc^2} \int_{c/n<u} \omega^3 \left[1 - \left(\frac{\omega + \omega_0}{\omega\beta n(\omega)}\right)^2\right] d\omega\,. \tag{45.23}$$

This formula describes the total radiation power from an oscillating electron under anomalous Doppler effect conditions.

For $\beta n < 1$, the anomalous Doppler effect is impossible. The emission that then occurs is called normal Doppler effect. But the normal Doppler effect can also occur for $\beta n > 1$, accompanying the anomalous Doppler effect. From the condition that frequency (45.21) is positive we can see that the normal effect occurs within a cone of angles such that $\cos\vartheta < (\beta n)^{-1}$. Consequently, for $\beta n > 1$, the anomalous effect takes place inside the Cherenkov cone, while the normal effect, outside the cone. For $\beta n < 1$, only the normal Doppler effect is possible.

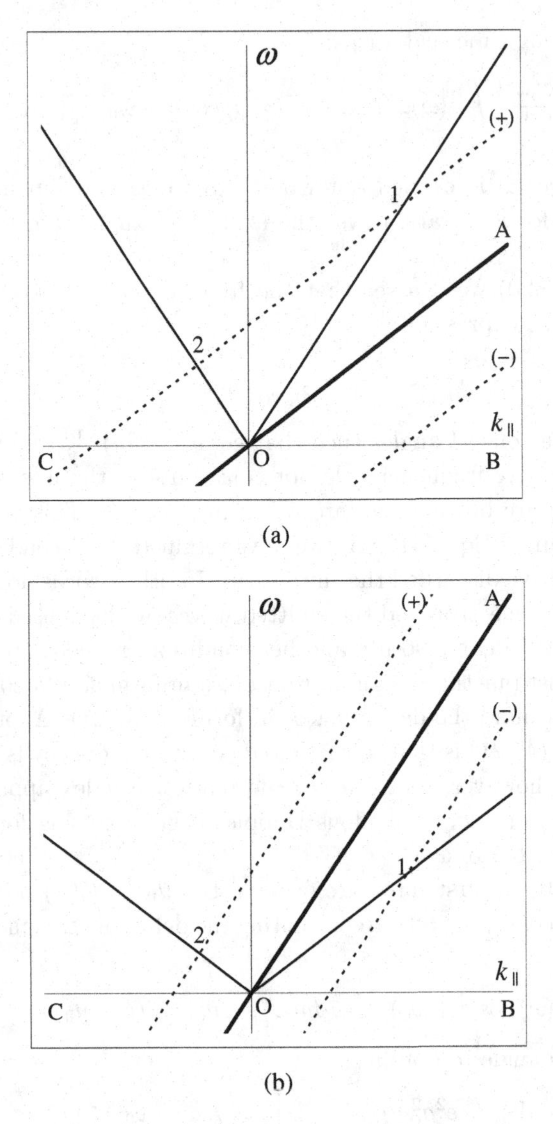

Fig. 78. (a) Normal Doppler effect for $\beta n < 1$ (the quantities ω and $k_{\parallel}$ are determined from points *1* and *2*). (b) Anomalous (point *1*) and normal (point *2*) Doppler effects for $\beta n > 1$.

Figure 78 illustrates the resonance conditions and the relationship between the frequencies of the waves emitted under normal and anomalous Doppler effect conditions. The illustration refers only to forward ($k_{\parallel} > 0$ or $\cos\vartheta = 1$) and backward ($k_{\parallel} < 0$ or $\cos\vartheta = -1$) radiation with $\mathbf{k} = \{0, 0, k_{\parallel}\}$, emitted from an electron moving with the velocity $\mathbf{u}$ in the positive direction of the z axis. The radiation spectra are taken to have the simplest form

$$\omega = \mathrm{sign}(k_{\parallel}) \cdot k_{\parallel} \frac{c}{n}, \tag{45.24}$$

where $n = \text{const.}$[1] The resonance conditions are written as

$$\omega = k_{\parallel} u \pm \omega_0 \,, \tag{45.25}$$

where the plus sign is for the normal Doppler effect and the minus sign is for the anomalous effect.

Figure 78(a) illustrates the case $\beta n < 1$. The lighter straight lines show the radiation spectra (45.24), the heavier line is the Cherenkov resonance line $\omega = k_{\parallel} u$, the dotted line (+) is the resonance line $\omega = k_{\parallel} u + \omega_0$ of the normal Doppler effect, and the dotted line (−) is the resonance line $\omega = k_{\parallel} u - \omega_0$ of the anomalous Doppler effect. The radiation frequencies follow from spectra (45.24) and conditions (45.25); they are given by the formulas

$$\omega_{1,2} = \frac{\omega_0}{1 \mp \beta n} \,. \tag{45.26}$$

Formulas (45.26) have been obtained using condition (45.25) with the upper sign. Under condition (45.25) with the lower sign, radiation at frequencies $\omega > 0$ is not emitted. In Fig. 78(a), the resonant frequencies (45.26) are designated by points 1 and 2. At the frequency ω_1, waves are emitted in the forward direction, and, at the frequency ω_2, radiation from an electron is emitted backward. The region of the anomalous Doppler effect is a region within the acute angle AOB. Since, in this region, there are no electromagnetic eigenmodes (i.e., no dispersion curves), anomalous Doppler radiation is not emitted there. The region of the normal Doppler effect is a region within the obtuse angle AOC. In this region, there are electromagnetic eigenmodes and normal Doppler radiation is emitted at the resonant frequencies (45.26).

Figure 78(b) illustrates the case $\beta n > 1$, showing the same lines in the same notation as in Fig. 78(a). The radiation frequencies follow from spectra (45.24) and conditions (45.25); they are given by the formulas

$$\omega_{1,2} = \frac{\omega_0}{\beta n \mp 1} \,. \tag{45.27}$$

Formulas (45.27) have been obtained using both conditions (45.25). The condition with the lower sign leads to the frequency ω_1, and the condition with the upper sign, to the frequency ω_2. Point 1 lies within the acute angle AOB, i.e., within the region of the anomalous Doppler effect. Consequently, forward radiation at the frequency ω_1 is an anomalous Doppler effect. Point 2 lies within the obtuse angle AOC, i.e., within the region of the normal Doppler effect, thereby indicating that backward radiation at the frequency ω_2 is a normal Doppler effect.

Let us also obtain an analogue of formula (45.23) for cyclotron emission from an electron under anomalous Doppler effect conditions. To do this, we rewrite formula (44.9) on long time scales as

$$P(\theta) \equiv \frac{dW}{dt} = \frac{e^2 v_\perp^2}{4c^3} \int n(\omega)\omega^2 \delta(\omega - \omega\beta_{\parallel} n(\omega)\cos\theta \pm \omega_e) d\omega (1 + \cos^2\theta) \sin\theta d\theta \,. \tag{45.28}$$

[1]Dispersion should be taken into account in calculating the integrals in formulas (45.8) and (45.23). In considering radiation emitted at a certain frequency, dispersion can be ignored.

Here, the plus sign is to be chosen for the anomalous Doppler effect and the minus sign corresponds to the normal Doppler effect. Going through the same manipulations as in deriving (45.23), we arrive at the formula

$$P_0 \equiv \frac{dW_0}{dt} = \frac{W_0}{t} = \frac{e^2 v_\perp^2}{4uc^2} \int \omega \left[1 + \left(\frac{\omega + \omega_e}{\omega \beta n(\omega)} \right)^2 \right] d\omega \,. \tag{45.29}$$

46. Application of the Hamiltonian Method to the Problem of the Excitation of Longitudinal Waves

And finally, we consider how the Hamiltonian method can be applied to the problem of the excitation of longitudinal waves by an electron, in particular, longitudinal waves in an isotropic plasma. In the previous section, we have shown that the Cherenkov emission of transverse waves in such a plasma is impossible. That this conclusion is not valid for longitudinal waves in an isotropic plasma has already been shown in Sec. 32. Now, we demonstrate this once again, but in another way. We begin with the following differential equations (see Eqs. (6.2)):

$$\begin{aligned} &\frac{\partial \mathbf{E}}{\partial t} + 4\pi \mathbf{j}_p = -4\pi e \mathbf{v}_0(t)\delta(\mathbf{r} - \mathbf{r}_0(t)) \,, \\ &\frac{\partial \mathbf{j}_p}{\partial t} - \frac{\omega_p^2}{4\pi}\mathbf{E} = 0 \,. \end{aligned} \tag{46.1}$$

The energy of longitudinal perturbations in a plasma is described by the formula

$$W = \int \left(\frac{E^2}{8\pi} + \frac{2\pi}{\omega_p^2} j_p^2 \right) dV \,. \tag{46.2}$$

In accordance with the Hamiltonian method, we use the following expansion in plane waves:

$$\mathbf{j}_p = \frac{\omega_p^2}{\sqrt{4\pi}} \sum_\lambda (q_\lambda \mathbf{e}_\lambda \exp(i\mathbf{k}_\lambda \mathbf{r}) + q_\lambda^* \mathbf{e}_\lambda \exp(-i\mathbf{k}_\lambda \mathbf{r})) \,. \tag{46.3}$$

Here, $\mathbf{e}_\lambda = \mathbf{k}_\lambda / k_\lambda$ is the unit polarization vector of a longitudinal wave. From the second of Eqs. (46.1) we then find

$$\mathbf{E} = \sqrt{4\pi} \sum_\lambda (q'_\lambda \mathbf{e}_\lambda \exp(i\mathbf{k}_\lambda \mathbf{r}) + q'^*_\lambda \exp(-i\mathbf{k}_\lambda \mathbf{r})) \,. \tag{46.4}$$

With expansion (46.3) and expression (46.4), the energy of plasma waves (46.2) becomes

$$W = \sum_\lambda (\mathbf{q}'_\lambda \mathbf{q}'^*_\lambda + \omega_p^2 \mathbf{q}_\lambda \mathbf{q}^*_\lambda) \,. \tag{46.5}$$

Eliminating the field **E** in the first of Eqs. (46.1) and substituting expansion (46.3) yields the following equations for the excitation of plasma oscillators — plasmons:

$$q''_\lambda + \omega_p^2 q_\lambda = \sqrt{4\pi} e (\mathbf{e}_\lambda \mathbf{v}_0) \exp(-i\mathbf{k}_\lambda \mathbf{r}_0) \,. \tag{46.6}$$

In formula (46.5) and Eqs. (46.6), the volume of the region occupied by the plasma has been set equal to unity, $L^3 = 1$. Formula (46.5) and Eqs. (46.6) differ from formula (41.12) and Eqs. (41.15) in that they contain only one frequency, ω_p. But when the dispersion of plasma waves is taken into account, we must replace ω_p^2 in (46.5) and (46.6) with $\omega_{p\lambda}^2$. Note that the quantities $\omega_{p\lambda}^2$ for plasmons and ω_λ^2 for photons depend on $\mathbf{k}_\lambda$ in a radically different manner.

Assume that an electron in a plasma moves uniformly along a straight line: $\mathbf{v}_0 = \mathbf{u}$ and $\mathbf{r}_0 = \mathbf{u}t$. Comparing Eqs. (46.6) and (41.19) and using formula (41.21), we can write the energy of plasma waves as

$$W = 4\pi \sum_\lambda \left(\frac{\mathbf{k}_\lambda \mathbf{u}}{k_\lambda}\right)^2 \frac{1 - \cos(\omega_p - \mathbf{k}_\lambda \mathbf{u})t}{(\omega_p - \mathbf{k}_\lambda \mathbf{u})^2}. \tag{46.7}$$

Differentiation of (46.7) with respect to t leads to the following formula for the power of the emitted plasmons:

$$\frac{dW}{dt} = 4\pi e^2 \sum_\lambda \left(\frac{\mathbf{k}_\lambda \mathbf{u}}{k_\lambda}\right)^2 \frac{\sin(\omega_p - \mathbf{k}_\lambda \mathbf{u})t}{(\omega_p - \mathbf{k}_\lambda \mathbf{u})}. \tag{46.8}$$

We switch from summation to integration according to the rule (see (42.4), (42.5))

$$\sum_\lambda \cdots \to \frac{1}{(2\pi)^3}\int \cdots k^2 dk\, do \tag{46.9}$$

to convert (46.8) into the form

$$\frac{dW}{dt} = \frac{e^2}{2\pi^2}\int (\mathbf{ku})^2 \frac{\sin(\omega_p - \mathbf{ku})t}{(\omega_p - \mathbf{ku})} dk do. \tag{46.10}$$

For long time scales, formula (46.10) reduces to

$$\frac{dW}{dt} = \frac{W}{t} = \frac{e^2}{2\pi}\int (\mathbf{ku})^2 \delta(\omega_p - \mathbf{ku}) dk do. \tag{46.11}$$

We introduce a spherical coordinate system in $\mathbf{k}$ space, with the z axis pointing along the velocity $\mathbf{u}$ and with the angle θ being the angle between the vectors $\mathbf{k}$ and $\mathbf{u}$. Integration of formula (46.11) over the azimuthal angle φ yields

$$\frac{dW}{dt} = \frac{W}{t} = e^2 u^2 \int k^2 \cos^2\theta \sin\theta\, \delta(\omega_p - ku\cos\theta) dk d\theta. \tag{46.12}$$

From formula (46.12) it follows that the minimum wavenumber of the emitted plasmons is $k_{\min} = \omega_p/u$. A plasmon with this wavenumber is emitted exactly in the forward direction.

Integrating (46.12) over the angle, we obtain the following formula for the total energy an electron loses per unit time as it excites longitudinal waves in an isotropic plasma:

$$\frac{dW_0}{dt} = \frac{W_0}{t} = \frac{e^2\omega_p^2}{2u}\int_{k_{\min}}^{k_{\max}} (1/k^2) dk^2. \tag{46.13}$$

In this formula, the lower limit of integration is nonzero; it has been determined earlier. As for the upper limit of integration, $k_{\max}$, we have no grounds for choosing its value. With $k_{\max} = \infty$, the integral in formula (46.13) diverges logarithmically. We can use the familiar result of the plasma theory: plasma waves with wavenumbers $k > \omega_p/V_{Te}$ are strongly damped (here, V_{Te} is the thermal velocity of the plasma electrons). This effect is known as Landau damping (for more detail on the properties of longitudinal waves in a plasma, see Secs. 6, 10). Accordingly, we can set $k_{\max} = \omega_p/V_{Te}$ to simplify formula (46.13) to

$$\frac{dW_0}{dt} = \frac{W_0}{t} = \frac{e^2\omega_p^2}{u} \ln \frac{u}{V_{Te}} . \tag{46.14}$$

This formula coincides with formula (32.11) and describes the power of longitudinal waves emitted via the Cherenkov mechanism by an electron in its uniform motion along a straight line in an isotropic plasma. It might seem strange that we apply the term Cherenkov emission to the process of the excitation of longitudinal waves. In electrodynamics, by emission is commonly meant the excitation of transverse waves. But the terminology we are using here is quite justified. The reason is twofold. First, the theories of the excitation of transverse and longitudinal waves are mathematically identical. And second, in anisotropic media, there are no purely transverse and purely longitudinal waves, because they are coupled to each other (see, e.g., Sec. 13). Under conditions such that

$$\omega(\mathbf{k}) = \mathbf{ku} , \tag{46.15}$$

the emission process is the Cherenkov emission of transverse–longitudinal electromagnetic waves in a highly anisotropic medium. Analogously, under conditions such that

$$\omega(\mathbf{k}) = \mathbf{ku} \pm \omega_0 , \quad \omega > 0 , \tag{46.16}$$

the emission process is either the normal or the anomalous Doppler effect, regardless of the polarization of the emitted waves and of their nature.

APPENDICES

Appendix 1

Absorption of the Energy of a Localized Source

In Sec. 25, we have analyzed the problem of how the energy of an external source is absorbed in a dissipative medium. We have assumed that perturbations produced by the source in the medium are spatially harmonic, i.e., are proportional to $\sim \exp(ikz)$, and as such are not localized in space. In this Appendix, we generalize the analysis to the case of a nonharmonic source.

Let the right-hand side of Eq. (3.12) have the form

$$F(t,z) = \left[\sum_k f_0(k)\exp(ikz)\right]\exp(-i\omega_0 t). \tag{A1.1}$$

Formula (A1.1) describes a typical experimental situation in which an external source is harmonic in time but nonharmonic in the spatial coordinate. In a medium, such a source produces a spatially nonuniform, localized perturbation at a prescribed frequency (such as perturbations from an RF discharge, laser radiation, etc.). If the spatial modes in formula (A1.1) are independent, then, according to formula (25.11), the total energy loss of the source is described by the expression

$$Q = \sum_k \left(\frac{\partial D_0'(k)}{\partial\omega}\right)^{-1} \frac{D_0''(k)}{[D_0'(k)]^2 + [D_0''(k)]^2}|f_0(k)|^2, \tag{A1.2}$$

where $D_0(k) = D(\omega_0, k)$ is the complex dispersion function. If the characteristic length of a medium (or a perturbation) in the Z direction is L, then, in the wavenumber range from zero to k, there are $N = (2\pi)^{-1}kL$ oscillations, i.e., we have $dN = (2\pi)^{-1}Ldk$. With this circumstance in mind, we go over from summation to integration in expression (A1.2) to obtain the following formula for the energy loss density:

$$\frac{Q}{L} = \frac{1}{2\pi}\int \left(\frac{\partial D_0'(k)}{\partial\omega}\right)^{-1} \frac{D_0''(k)}{[D_0'(k)]^2 + [D_0''(k)]^2}|f_0(k)|^2 dk. \tag{A1.3}$$

A generalization of formula (A1.3) to the three-dimensional case is obvious: the scalar k is replaced with the wave vector $\mathbf{k} = \{k_x, k_y, k_z\}$ and the integration is over $dk = dk_x dk_y dk_z$.

In calculating the integral in formula (A1.3), one must distinguish between the cases of a nonresonant and a resonant source. In the nonresonant case, we have $\omega_0 \neq \omega_m(k)$ for all eigenmodes of the medium over the entire integration region. Here, $\omega_m(k)$ are the eigenfrequencies determined from the equation $D'(\omega, k) = 0$ and the subscript runs through the values $m = 1, 2, \ldots, N$, where N is the number of eigenmodes. Since the dissipation is assumed to be weak, the integral in formula (A1.3) in the nonresonant case reduces to

$$\frac{Q}{L} = \frac{1}{2\pi} \int \left(\frac{\partial D_0'(k)}{\partial \omega} \right)^{-1} \frac{D_0''(k)}{[D_0'(k)]^2} |f_0(k)|^2 dk \,. \tag{A1.4}$$

Formula (A1.4) describes the energy loss due to conventional Joule heating of the medium.

When the source is resonant, for some of the eigenmode branches, the frequency ω_0 of the external source coincides with the eigenfrequency. We denote by k_{0mj} the real roots of the equation

$$\omega_0 = \omega_m(k) \,. \tag{A1.5}$$

Here, m is the number of the branch and j is the number of the root (for each fixed m, Eq. (A1.5) can have more than one real root k). It is obvious that, near the root, the derivatives of the dispersion function can be represented as

$$D_0'(k) = (k - k_{0mj}) \frac{\partial D_0'}{\partial k}(k_{0mj}) \,, \quad D_0''(k) = D_0''(k_{0mj}) \,. \tag{A1.6}$$

With these representations, formula (A1.3) can be cast into the form

$$\frac{Q}{L} = \frac{1}{2\pi} \sum_{m,j} \left\{ \int \left(\frac{\partial D_{0mj}'}{\partial \omega} \right)^{-1} \frac{D_{0mj}''}{(\partial D_{0mj}'/\partial k)^2 (k - k_{0mj})^2 + (D_{0mj}'')^2} |f_0(k)|^2 dk \right\} , \tag{A1.7}$$

where the derivatives are taken at the point ($\omega = \omega_0$, $k - k_{0mj}$).

When the dissipation is weak, the integrand in formula (A1.7) has a pronounced resonant structure. In fact, using the limiting relationship

$$\delta(x) = \lim_{\alpha \to 0} \frac{1}{\pi} \frac{\alpha}{x^2 + \alpha^2} \,, \tag{A1.8}$$

we take the integral in (A1.7) to obtain the following formula for the resonant energy loss of an external source in a weakly dissipative medium:

$$\frac{Q}{L} = \frac{1}{2} \sum_{m,j} \left[\left(\frac{\partial D_{0mj}'}{\partial \omega} \right)^{-2} \frac{|f_0(k_{0mj})|^2}{|V_{g0mj}|} \right] , \tag{A1.9}$$

where

$$V_{g0mj} = \left. \frac{d\omega_m(k)}{dk} \right|_{k = k_{0mj}} \tag{A1.10}$$

are the group velocities of the waves excited by the source in the medium. Note that formula (A1.9) has been derived by using relationship (29.32). It is the energy losses due to the resonant excitation of the eigenmodes of the medium that are described by formula (A1.9). That the denominators in this formula contain the group velocities stems from the spatial nonuniformity of an external source. Indeed, from the regions where it is deposited, the energy is transported away with the group velocity. Consequently, the lower the group velocity, the greater the amount of energy stored in the region where the external action is localized.

Appendix 2

On the Theory of Electromagnetic Wave Scattering by a Free Electron

1. In order to apply the results obtained in Secs. 42 and 43 to the theory of electromagnetic wave scattering by a freely moving electron, it is necessary to relate the amplitude $\mathbf{a}_0$ and frequency ω_0 of electron oscillations in formulas (41.16) to the corresponding parameters of the scattered electromagnetic wave. In this way, the ratios of the radiation power given by formulas (42.9), (42.10), (43.6), (43.13), and (43.14) to the absolute value of the Poynting vector can be treated as the cross sections for scattering into a solid angle do or as the total scattering cross sections.

Let a linearly polarized plane electromagnetic wave be described by the formulas

$$\mathbf{E} = \mathbf{E}_0 \sin(\Omega t - \boldsymbol{\kappa}\mathbf{r}) ,$$
$$\mathbf{B} = \frac{c}{\Omega}[\boldsymbol{\kappa}\mathbf{E}_0] \sin(\Omega t - \boldsymbol{\kappa}\mathbf{r}) , \tag{A2.1}$$

where $\Omega > 0$ is a frequency and $\mathbf{E}_0$ is a constant vector orthogonal to the wave vector $\boldsymbol{\kappa}$. We can also write the relationship

$$\kappa = \pm\frac{\Omega}{c}n(\Omega) , \tag{A2.2}$$

where $n(\Omega)$ is the refractive index. The Poynting vector of the wave (A2.1), averaged over the period $2\pi/\Omega$, has the form

$$\mathbf{P} = \frac{1}{8\pi}cn(\Omega)E_0^2\frac{\boldsymbol{\kappa}}{|\kappa|} . \tag{A2.3}$$

The motion of an electron with the velocity $\mathbf{v}$ is described by the relativistic equation

$$\frac{d\mathbf{v}}{dt} = \frac{e}{m}\sqrt{1-\frac{v^2}{c^2}}\left\{\mathbf{E} + \frac{1}{c^2}[\mathbf{v}\mathbf{B}] - \frac{1}{c^2}\mathbf{v}(\mathbf{v}\mathbf{E})\right\} . \tag{A2.4}$$

Let an electron have an unperturbed velocity $\mathbf{u}$, which can be zero or nonzero. Assuming that the amplitude $\mathbf{E}_0$ of an electromagnetic wave (A2.1) is small, we insert the relationships $\mathbf{v} = \mathbf{u}$ and $\mathbf{r} = \mathbf{u}t$ into the right-hand side of Eqs. (A2.4). As a result, performing simple integration, we arrive at formulas (41.16) in which

$$\mathbf{a}_0 = \frac{e}{m\gamma\omega_0^2}\left(\mathbf{E}_0 + \frac{1}{\Omega}[\mathbf{u}[\boldsymbol{\kappa}\mathbf{E}_0]] - \frac{1}{c^2}\mathbf{u}(\mathbf{u}\mathbf{E}_0)\right), \qquad \text{(A2.5)}$$
$$\omega_0 = \Omega - \boldsymbol{\kappa}\mathbf{u},$$

and $\gamma = (1 - u^2/c^2)^{-1/2}$ is the relativistic factor of the electron. A general analysis of the problem requires lengthy calculations. That is why we restrict ourselves here to two particular examples.

2. The first example is scattering by an immobile electron. For $\mathbf{u} = 0$, we have $a_0 = eE_0/m\Omega^2$ and $\omega_0 = \Omega$. We substitute these quantities into formula (42.9) and divide the result by the absolute value of the Poynting vector (A2.3) to obtain the following expression for the cross section for scattering into a solid angle do:

$$d\sigma = \left(\frac{e^2}{mc^2}\right)^2 \sin^2\theta \, do\,. \qquad \text{(A2.6)}$$

In this case, the frequency of the scattered waves is independent of the scattering angle and coincides with the frequency Ω of the incident wave.

Integrating expression (A2.6) over the angles leads to the familiar Thomson formula for the total cross section for scattering by an immobile electron:

$$\sigma = \frac{8\pi}{3}\left(\frac{e^2}{mc^2}\right)^2. \qquad \text{(A2.7)}$$

This same formula can also be derived by dividing expression (42.10) by the absolute value of the Poynting vector (A2.3).

3. The second example is scattering by a relativistic electron. We direct the coordinate axis X along the electron velocity, such that $\mathbf{u} = \{u, 0, 0\}$, and consider only waves propagating exactly along the X axis. We set

$$\boldsymbol{\kappa} = \{\kappa, 0, 0\}, \quad \mathbf{E}_0 = \{0, 0, E_0\}, \qquad \text{(A2.8)}$$

where κ is one of the quantities (A2.2). For $u > 0$, the case $\kappa > 0$ corresponds to the scattering of the forward wave (i.e., a wave propagating in the direction of electron motion) and the case $\kappa < 0$, to the scattering of the backward wave.

Taking into account (A2.8), from relationships (A2.5) we find

$$\mathbf{a}_0 = \{0, 0, a_0\}, \quad a_0 = \frac{eE_0}{m\gamma\Omega^2}[1 \mp \beta n(\Omega)]^{-1}, \qquad \text{(A2.9)}$$
$$\omega_0 = \Omega|1 \mp \beta n(\Omega)|\,.$$

Here, the lower sign is for the scattering of the backward wave and the upper sign refers to the scattering of the forward wave. Relationships (A2.9) imply that $\mathbf{a}_0 \perp \mathbf{u}$, i.e., that the scattering electron is a transverse dipole. Consequently, using formulas (43.6) and (A2.3), we obtain the following expression for the cross section for scattering into a solid angle do:

$$d\sigma = \left(\frac{e^2}{mc^2\gamma}\right)^2 \frac{n(\omega)}{n(\Omega)} \frac{[1 \mp \beta n(\Omega)]^2}{[1 - \beta n(\omega)\sin\theta\cos\varphi]^3} \sin^2\theta \, do\,. \qquad \text{(A2.10)}$$

In contrast to (A2.6), the scattering cross section (A2.10) is generally frequency-dependent, because, for a dispersive medium, the refractive index depends on frequency, $n = n(\Omega)$. Moreover, in accordance with formula (43.5), the frequency of the scattered wave depends on the scattering angle,

$$\omega = \Omega \frac{|1 \mp \beta n(\Omega)|}{1 - \beta n(\omega) \sin\theta \cos\varphi} . \tag{A2.11}$$

It should be stressed that formulas (A2.10) and (A2.11) contains the refractive indices at two frequencies: the frequency of the incident wave, $n(\Omega)$, and the frequency of the scattered wave, $n(\omega)$. The frequency Ω is fixed. And the frequency ω is a complicated function of the scattering angles, as can be seen from formula (A2.11). The refractive index $n(\Omega)$ arises from relationship (A2.2), formula (A2.3), and relationships (A2.5). As for the refractive index $n(\omega)$, it comes from the fact that Eqs. (41.11) contain the frequencies of the field oscillators, ω_λ.

For a nondispersive medium, the total cross section can be found by integrating formula (A2.10) over the angles:

$$\sigma = \frac{8\pi}{3} \left(\frac{e^2}{mc^2\gamma} \right)^2 (1 \mp \beta n)^2 S(\beta n) . \tag{A2.12}$$

The function $S(\beta n)$ is shown in Fig. 71.

Appendix 3

Problem with a Source for the Wave Equation in Spontaneous Emission Theory

1. In the Hamiltonian approach, spontaneous emission from an electron has been described as resonant excitation of the field oscillators. Another approach is a straightforward solution of the wave equation for the electromagnetic field potentials, with the right-hand side accounting for the charge and current densities of an electron executing a given motion. In this case, the emitted waves are identified by the behavior of their amplitudes at large distances from the source. The solution to the wave equation with a source has been considered in Sec. 32. But in that section, the radiation field has been calculated at the point charge. Let us now consider the fields at large distances from the source.

We begin by solving a more general problem of determining the field created by an arbitrary source with the charge and current densities $\rho_0(t, \mathbf{r})$ and $\mathbf{j}_0(t, \mathbf{r})$, related to one another by the continuity equation

$$\frac{\partial \rho_0}{\partial t} + \operatorname{div} \mathbf{j}_0 = 0\,. \tag{A3.1}$$

The electromagnetic field equations (41.1) with prescribed sources $\rho_0(t, \mathbf{r})$ and $\mathbf{j}_0(t, \mathbf{r})$ can be transformed by using the potentials $\Phi(t, \mathbf{r})$ and $\mathbf{A}(t, \mathbf{r})$ defined by expressions (41.3). In the Lorentz gauge

$$\nabla \cdot \mathbf{A} + \frac{\varepsilon}{c}\frac{\partial \Phi}{\partial t} = 0\,, \tag{A3.2}$$

from Eqs. (41.1) we obtain the following equations for the potentials:

$$\begin{aligned} \Delta \mathbf{A} - \frac{\varepsilon}{c^2}\frac{\partial^2 \mathbf{A}}{\partial t^2} &= -\frac{4\pi}{c}\mathbf{j}_0\,, \\ \varepsilon\left(\Delta \Phi - \frac{\varepsilon}{c^2}\frac{\partial^2 \Phi}{\partial t^2}\right) &= -4\pi\rho_0\,. \end{aligned} \tag{A3.3}$$

Here, ε is the permittivity of the medium in material equation (41.2). For a medium with frequency dispersion, the permittivity ε is an operator, $\varepsilon(\hat{\omega})$. Note also that, in the Coulomb gauge $\nabla \cdot \mathbf{A} = 0$, the equations for the potentials are Eqs. (41.4).

Applying the Fourier transformation in time and in spatial coordinates to the vector potential,

$$\mathbf{A}(t,\mathbf{r}) = \int d\omega \int d\mathbf{k}\,\mathbf{A}(\omega,\mathbf{k})\exp(-i\omega t + i\mathbf{kr})\,,$$
$$\mathbf{A}(\omega,\mathbf{k}) = \frac{1}{(2\pi)^4}\int dt \int d\mathbf{r}\,\mathbf{A}(t,\mathbf{r})\exp(i\omega t - i\mathbf{kr})\,, \tag{A3.4}$$

and, in the same manner, to the scalar potential $\Phi(t,\mathbf{r})$, we represent the solutions to Eqs. (A3.3) as

$$\mathbf{A}(t,\mathbf{r}) = \frac{4\pi}{c}\int d\omega \int d\mathbf{k}\,\mathbf{j}_0(\omega,\mathbf{k})\frac{\exp(-i\omega t + i\mathbf{kr})}{k^2 - \varepsilon(\omega)\omega^2/c^2}\,,$$
$$\Phi(t,\mathbf{r}) = 4\pi\int d\omega \int d\mathbf{k}\rho_0(\omega,\mathbf{k})\frac{\exp(-i\omega t + i\mathbf{kr})}{\varepsilon(\omega)(k^2 - \varepsilon(\omega)\omega^2/c^2)}\,. \tag{A3.5}$$

where, in accordance with continuity equation (A3.1), the Fourier transforms $\rho_0(\omega,\mathbf{k})$ and $\mathbf{j}_0(\omega,\mathbf{k})$ of the charge and current densities $\rho_0(t,\mathbf{r})$ and $\mathbf{j}_0(t,\mathbf{r})$ are related by the relationship

$$\omega\rho_0(\omega,\mathbf{k}) = \mathbf{kj}_0(\omega,\mathbf{k})\,. \tag{A3.6}$$

2. Let us consider several problems that are not only very illustrative but also are of great interest in themselves. We begin by calculating the field of a point charge e moving uniformly along a straight line with the velocity $\mathbf{u} = \{0,0,u\}$. We have

$$\rho_0(t,\mathbf{r}) = e\delta(\mathbf{r} - \mathbf{u}t)\,,$$
$$\mathbf{j}_0(t,\mathbf{r}) = e\mathbf{u}\delta(\mathbf{r} - \mathbf{u}t) = \mathbf{u}\rho_0(t,\mathbf{r})\,. \tag{A3.7}$$

Applying the Fourier transformation (A3.4) and using the integral representation (28.6) for the delta function, we find

$$\rho_0(\omega,\mathbf{k}) = \frac{e}{(2\pi)^3}\delta(\omega - \mathbf{ku})\,, \quad \mathbf{j}_0(\omega,\mathbf{k}) = \frac{e\mathbf{u}}{(2\pi)^3}\delta(\omega - \mathbf{ku})\,, \tag{A3.8}$$

in which case relationship (A3.6) is obviously satisfied identically.

We substitute expressions (A3.8) into formulas (A3.5) and, using the properties of the delta function, integrate the result over ω to obtain

$$\mathbf{A}(t,\mathbf{r}) = \frac{e}{2\pi^2}\frac{\mathbf{u}}{c}\int d\mathbf{k}\frac{\exp[i\mathbf{k}(\mathbf{r} - \mathbf{u}t)]}{k^2 - \varepsilon(\mathbf{ku})(\mathbf{ku})^2/c^2}\,,$$
$$\Phi(t,\mathbf{r}) = \frac{e}{2\pi^2}\int d\mathbf{k}\frac{\exp[i\mathbf{k}(\mathbf{r} - \mathbf{u}t)]}{\varepsilon(\mathbf{ku})(k^2 - \varepsilon(\mathbf{ku})(\mathbf{ku})^2/c^2)}\,. \tag{A3.9}$$

The integrals in formulas (A3.9) are identical. In order to calculate them, we switch to cylindrical coordinates $\mathbf{r} = \{\mathbf{r}_\perp, z\}$ and $\mathbf{k} = \{\mathbf{k}_\perp, k - z\}$ such that $d\mathbf{k} = k_\perp dk_\perp d\varphi dk_z$, $\mathbf{ku} = k_z u$, $\mathbf{kr} = \mathbf{k}_\perp\mathbf{r}_\perp + k_z z$, and $\mathbf{k}_\perp\mathbf{r}_\perp = k_\perp r_\perp\cos\varphi$, where φ is the angle between the vectors $\mathbf{r}_\perp$ (which determines the observation point) and $\mathbf{k}_\perp$

(over which the integration is carried out). In these coordinates, formulas (A3.9) — say, the second of them — can be converted into the form

$$\Phi(t, \mathbf{r}) = \frac{e}{2\pi^2} \int k_\perp dk_\perp dk_z d\varphi \frac{\exp(ik_\perp r_\perp \cos\varphi) \exp[ik_z(z - ut)]}{\varepsilon(k_z u)[k_\perp^2 + k_z^2(1 - \varepsilon(k_z u)u^2/c^2)]} . \quad \text{(A3.10)}$$

Using the familiar integral representations

$$\int_0^{2\pi} \exp(ia\sin\varphi) d\varphi = \int_0^{2\pi} \exp(ia\cos\varphi) d\varphi = 2\pi J_0(a) , \quad \text{(A3.11)}$$

where $J_0(a)$ is the Bessel function, we integrate in formula (A3.10) over φ. The result is

$$\Phi(t, \mathbf{r}) = \frac{e}{\pi} \int k_\perp dk_\perp dk_z \frac{J_0(k_\perp r_\perp) \exp[ik_z(z - ut)]}{\varepsilon(k_z u)[k_\perp^2 + k_z^2(1 - \varepsilon(k_z u)u^2/c^2)]} . \quad \text{(A3.12)}$$

It is then expedient, first, to integrate over k_Z from minus to plus infinity (the integral over $k_\perp$ is taken from 0 to ∞). The relevant integration contour — the real axis of the complex plane k_z — is closed by a semicircle of infinite radius in the upper (lower) half of the plane k_z for $z - ut > 0$ ($z - ut < 0$). The result of the integration is determined by the poles of the integrand in formula (A3.12). In turn, the poles in question are the roots of the equations

$$\begin{aligned} &\varepsilon(k_z u) = 0 , \\ &k_\perp^2 + k_z^2(1 - \varepsilon(k_z u)u^2/c^2) = 0 . \end{aligned} \quad \text{(A3.13)}$$

The left-hand sides of Eqs. (A3.13) are, respectively, the dispersion functions of longitudinal and transverse waves in an isotropic medium at the Cherenkov resonance frequency $\omega = \mathbf{ku} = k_z u$ (see Secs. 45, 46). The asymptotic behavior of the potentials $\Phi(t, \mathbf{r})$ and $\mathbf{A}(t, \mathbf{r})$ at large distances from the charge is largely governed by the positions of the poles on the complex plane k_z, i.e., by whether or not they are on the real axis.

In order to complete the calculation of the integral in formula (A3.12), it is necessary to specify the function $\varepsilon(\omega)$. This issue, however, requires serious analysis and is not worth treating in the appendix, the more so because Cherenkov emission of transverse and longitudinal waves has been considered in detail in Secs. 45 and 46. So, we restrict ourselves to the simplest case of a medium without frequency dispersion. That is, we set

$$\varepsilon(\omega) = \varepsilon_0 = \text{const.} \quad \text{(A3.14)}$$

We first assume that the velocity of the charge is less than the speed of light in the medium:

$$u^2 < c^2/\varepsilon_0 . \quad \text{(A3.15)}$$

Under inequality (A3.15), the integrand has two poles

$$k_{z1,2} = \pm i \frac{k_\perp}{\sqrt{1 - \varepsilon_0 u^2/c^2}} , \quad \text{(A3.16)}$$

which do not lie on the real axis. In this case, formula (A3.12) becomes

$$\Phi(t, \mathbf{r}) = \frac{e}{\pi\varepsilon_0}\int \frac{J_0(k_\perp r_\perp)\exp[ik_z(z-ut)]k_\perp dk_\perp dk_z}{(1-\varepsilon_0 u^2/c^2)(k_z-k_{z1})(k_z-k_{z2})}$$

$$= \frac{e}{\varepsilon_0\sqrt{r_\perp^2(1-\varepsilon_0 u^2/c^2)+(z-ut)^2}}\,. \tag{A3.17}$$

Note that integration over k_z in formula (A3.17) has been performed by applying the methods of residue theory (since the poles (A3.16) are symmetric about the real axis, it does not matter in which of the half-planes — upper or lower — the integration contour is closed) and then by using the formula

$$\int_0^\infty J_0(ax)\exp(-bx)dx = (a^2+b^2)^{-1/2}. \tag{A3.18}$$

In the case at hand, i.e., under the assumption (A3.14), we can also use the expression

$$\mathbf{A}(t, \mathbf{r}) = \frac{\mathbf{u}}{c}\varepsilon_0\Phi(t, \mathbf{r})\,. \tag{A3.19}$$

Knowing the potentials $\Phi(t, \mathbf{r})$ and $\mathbf{A}(t, \mathbf{r})$, from formulas (41.3) we can find the electric and magnetic fields, $\mathbf{E}(t, \mathbf{r})$ and $\mathbf{B}(t, \mathbf{r})$, of a moving charge. At large distances from the source, potentials (A3.17) and (A3.19) behave as $1/r$ and the fields, as $1/r^2$. Consequently, the electromagnetic energy flux vector — the Poynting vector (A2.3) — behaves as $\mathbf{P} \sim 1/r^4$. The total energy flux through a sphere of radius r around the charge is proportional to $|P|\pi r^2 \sim 1/r^2$ and vanishes at $r \to \infty$. This means that the fields with potentials (A3.17) and (A3.19) are confined to the charge: they do not propagate as radiation away from the charge and, consequently, do not form a free electromagnetic wave. As expected, for $\mathbf{u} \to 0$, the vector potential (A3.19) vanishes and the scalar potential becomes a Coulomb potential, $\Phi(t, \mathbf{r}) = e/\varepsilon_0 r$. Hence, under inequality (A3.15), a charge moving uniformly along a straight line does not emit radiation.

Let now the inequality opposite to inequality (A3.15) hold. In this case, the poles (A3.16) are on the real axis of the complex plane k_z:

$$k_{z1,2} = \pm\frac{k_\perp}{\sqrt{\varepsilon_0 u^2/c^2 - 1}}\,. \tag{A3.20}$$

Nevertheless, it is again legitimate to use the residue theorem to calculate the integral over k_z in formula (A3.17). Moreover, in the spirit of the Laplace transformation, the integral over ω in formula (A3.5) is to be taken along a straight line lying in the upper half of the complex plane ω (see Fig. 1). Consequently, from expressions (A3.8) we can see that, for $u > 0$, the wave vector component k_z also has a positive imaginary part. That is, in formula (A3.17), integration over k_z is carried out along the straight path $L = (-\infty < k_z' < \infty,\ k_z'' = i\alpha,\ \alpha \to +0)$ which passes above the real axis.

For $z - ut > 0$, the integration path L is closed by a semicircle $R^{(+)}$ in the upper half of the complex plane k_z. Since the poles (A3.20) lie outside the contour $L + R^{(+)}$, the integration yields zero. Hence, we have

$$\Phi(t,\mathbf{r})|_{z>ut} = \mathbf{A}(t,\mathbf{r})|_{z>ut} \equiv 0\,. \tag{A3.21}$$

Relationships (A3.21) imply that, under the inequality opposite to (A3.15), the electromagnetic field does not manage to reach the region $z > ut$.

For $z - ut < 0$, the integration path L is closed by a semicircle $R^{(-)}$ lying in the lower half of the complex plane k_z. Since the poles lie inside the contour $L + R^{(-)}$, the result of integration in formula (A3.17) is nonzero. Calculating the residues at the poles (A3.20) and taking into account the direction of integration along the contour, we reduce formula (A3.17) to

$$\Phi(t,\mathbf{r})|_{z<ut} = \frac{2e}{\varepsilon_0\sqrt{\varepsilon_0 u^2/c^2 - 1}} \int_0^\infty dk_\perp J_0(k_\perp r_\perp) \sin\frac{k_\perp(ut - z)}{\sqrt{\varepsilon_0 u^2/c^2 - 1}}\,. \tag{A3.22}$$

Using the familiar relationships

$$\int_0^\infty J_0(ax)\sin(bx)dx = \begin{cases} 0, & b < a \\ (b^2 - a^2)^{-1/2}, & b > a \end{cases}, \tag{A3.23}$$

we then convert the scalar potential (A3.22) into the form

$$\Phi(t,\mathbf{r})|_{z<ut} = \begin{cases} \dfrac{2e}{\varepsilon_0}\dfrac{1}{\sqrt{(z-ut)^2 - r_\perp^2(\varepsilon_0 u^2/c^2 - 1)}}, & ut - z > r_\perp\sqrt{\varepsilon_0 u^2/c^2 - 1} \\ 0, & ut - z < r_\perp\sqrt{\varepsilon_0 u^2/c^2 - 1} \end{cases}. \tag{A3.24}$$

The vector potential $\mathbf{A}(t,\mathbf{r})$ is described by formula (A3.19).

Hence, under the inequality opposite to (A3.15), the field of a point charge moving uniformly along a straight line is nonzero not over the entire region $z < ut$ but only within the region

$$z < ut - r_\perp\sqrt{\varepsilon_0\frac{u^2}{c^2} - 1} \equiv Z(r_\perp, t)\,. \tag{A3.25}$$

The surface $z = Z(r_\perp, t)$ is a cone, because $r_\perp$ is a cylindrical coordinate and z is the symmetry axis. Moreover, we are dealing with a single napped cone (see Fig. 79), because from relationships (A3.21) we can see that, for $z > ut$, the field potentials are identically zero. The vertex angle of the cone is determined by the expression

$$\mathrm{tg}^2\theta' = \frac{r_\perp^2}{(ut - Z)^2} = \frac{1}{\varepsilon_0 u^2/c^2 - 1}\,. \tag{A3.26}$$

The angle θ' varies from $\pi/2$ for $\varepsilon_0 u^2/c^2 = 1$ to zero for $\varepsilon_0 u^2/c^2 \to \infty$.

Using formulas (41.3), (A3.24), and (A3.19), we arrive at the following expressions for the nonzero components of the electromagnetic field:

$$E_z = \left(\varepsilon_0\frac{u^2}{c^2} - 1\right)\frac{\partial\Phi}{\partial z}\,, \quad E_r = -\frac{\partial\Phi}{\partial r_\perp}\,, \quad B_\varphi = \varepsilon_0\frac{u}{c}E_r\,. \tag{A3.27}$$

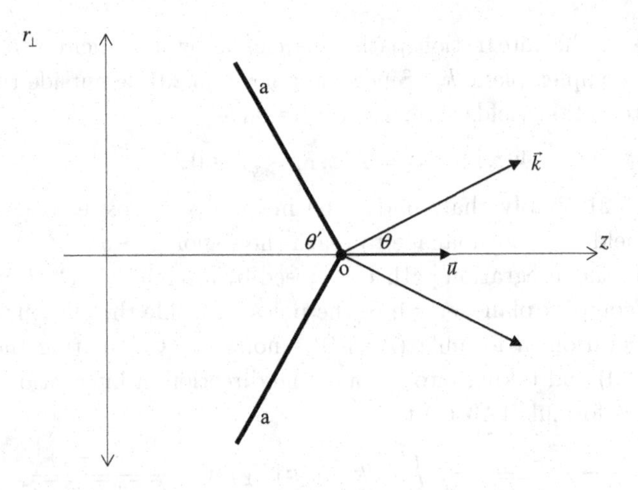

Fig. 79. Cherenkov radiation cone in a nondispersive medium: point o is the position of the charge ($z = ut,\ r_\perp = 0$) and heavy lines oa show the wave front.

With formulas (A3.27) and (A3.24), the Poynting vector can be cast into the form

$$\mathbf{P} = \frac{u}{c}\varepsilon_0 \frac{c}{4\pi r_\perp}\left(\frac{\partial\Phi}{\partial r_\perp}\right)^2 \begin{pmatrix} (ut - z)\mathbf{n}_r \\ 0 \cdot \mathbf{n}_\varphi \\ r_\perp \mathbf{n}_z \end{pmatrix}, \tag{A3.28}$$

where $(\mathbf{n}_r, \mathbf{n}_\varphi, \mathbf{n}_z)$ are the orthogonal unit vectors of the cylindrical coordinate system. The fields (A3.27) include the components confined to the charge and also free radiation. For $z = Z(r_\perp, t) - 0$, potential (A3.24) becomes infinite and, for $z = Z(r_\perp, t) + 0$, it vanishes, so there is an infinite jump in the potential. That is why, on the cone $z = Z(r_\perp, t)$, the fields are infinite, as well as the components of the Poynting vector (A3.28). It is the infinity of the fields on the one-sided infinite cone (not at a point!) that describes free radiation. For $\varepsilon_0 u^2/c^2 > 1$, the charge moves with a velocity higher than the speed of light in the medium. Consequently, the radiation just considered is Cherenkov emission. Integration of the Poynting vector (A3.28) over a surface around the charge obviously leads to a singularity. However, for $n(\omega) = n_0 = \text{const}$, the Tamm–Frank formula (45.8) leads to a singularity as well.

The direction in which radiation is emitted is determined from the relationship (see Fig. 79):

$$\mathrm{tg}^2\theta = \frac{k_\perp^2}{k_z^2} - \varepsilon_0 u^2/c^2 - 1 = \mathrm{ctg}^2\theta'\,. \tag{A3.29}$$

The angle θ is independent of the frequency ω. That is, in a nondispersive medium, waves at all frequencies are emitted with the same velocity in the same direction. Consequently, on the conical emission front — the Cherenkov cone — the electromagnetic energy density is infinite. It can be said that, during Cherenkov radiation in a nondispersive medium, a shock electromagnetic wave forms. Of course, a correct description of Cherenkov radiation requires that the frequency dispersion of the medium, $\varepsilon(\omega)$, be taken into account.

3. Let us consider one more problem that can be solved based on formulas (A3.5) without allowance for dispersion. We mean the problem of the scattering of an electromagnetic wave by an electron. The electron motion in the field of an electromagnetic wave (A2.1) is described by formulas (41.16). We restrict ourselves to the case $\mathbf{u} = 0$, i.e., to the scattering by an immobile electron, or, more precisely, by an electron oscillating about a fixed position. In this case, the charge and current densities of the electron are given by the expressions

$$\begin{aligned} \rho_0(t, \mathbf{r}) &= e\delta(\mathbf{r} - \mathbf{a}_0 \sin\omega_0 t)\,, \\ \mathbf{j}_0(t, \mathbf{r}) &= e\omega_0 \mathbf{a}_0 \cos\omega_0 t\delta(\mathbf{r} - \mathbf{a}_0 \sin\omega_0 t)\,. \end{aligned} \tag{A3.30}$$

The Fourier transformed charge and current densities have the form

$$\begin{aligned} \rho_0(\omega, \mathbf{k}) &= \frac{e}{(2\pi)^3} \sum_n J_n(\mathbf{k}\mathbf{a}_0)\delta(\omega - n\omega_0)\,, \\ \mathbf{j}_0(\omega, \mathbf{k}) &= \frac{e}{(2\pi)^3} \frac{\omega_0}{\mathbf{k}\mathbf{a}_0} \mathbf{a}_0 \sum_n nJ_n(\mathbf{k}\mathbf{a}_0)\delta(\omega - n\omega_0)\,. \end{aligned} \tag{A3.31}$$

It is obvious that relationship (A3.6) is satisfied identically. In deriving expressions (A3.21), we have used the formula

$$\exp(\pm ia\sin\varphi) = \sum_n J_n(a)\exp(\pm in\varphi)\,. \tag{A3.32}$$

Note that the frequency ω has a positive imaginary part. Consequently, in accordance with expressions (A3.21), the frequency ω_0, too, has a nonzero imaginary part, which is positive for $n > 0$ and is negative for $n < 0$. Since the correction to the frequency should be of a certain sign, we assume that $\mathrm{Im}\omega_0 = \sigma > 0$. Recall that introducing the imaginary part of the frequency makes it possible to account for the causality principle — the one that is automatically taken into account in the Laplace transform method. In the final result, we will take the limit $\sigma \to +0$.

We substitute expressions (A3.31) into formulas (A3.5), ignore frequency dispersion, and integrate over ω by using the properties of the δ-function. As a result, we obtain the formulas

$$\begin{aligned} \mathbf{A}(t, \mathbf{r}) &= \frac{e}{2\pi^2} \frac{\omega_0 \mathbf{a}_0}{c} \int d\mathbf{k} \frac{1}{\mathbf{k}\mathbf{a}_0} \sum_n nJ_n(\mathbf{k}\mathbf{a}_0) \frac{\exp(i\mathbf{k}\mathbf{r} - in\omega_0 t)}{k^2 - \varepsilon_0 n^2\omega_0^2/c^2}\,, \\ \Phi(t, \mathbf{r}) &= \frac{e}{2\pi^2\varepsilon_0} \int d\mathbf{k} \sum_n J_n(\mathbf{k}\mathbf{a}_0) \frac{\exp(i\mathbf{k}\mathbf{r} - in\omega_0 t)}{k^2 - \varepsilon_0 n^2\omega_0^2/c^2}\,. \end{aligned} \tag{A3.33}$$

Further analysis will be carried out in the dipole approximation,

$$|\mathbf{k}\mathbf{a}_0| \ll 1\,. \tag{A3.34}$$

Inequality (A3.34), which means that the amplitude of electron oscillations is small in comparison with the radiation wavelength, allows us to calculate the potentials by accounting for only the terms with $n = \pm 1$. The term with $n = 0$ describes the Coulomb component of the scalar potential, i.e., the component that is unrelated to the problem under consideration here. The terms with $|n| \geq 2$ make a contribution of higher order in the small parameter (A3.34). Since $\mathrm{Im}\omega_0 > 0$, the main contribution to potentials (A3.33) on long time scales comes from the term with $n = 1$. Accounting for this term alone, we obtain

$$\begin{aligned} \mathbf{A}(t, \mathbf{r}) &= \frac{e}{(2\pi)^2} \frac{\omega_0}{c} \mathbf{a}_0 \exp(-i\omega_0 t) F(\mathbf{r}) , \\ \Phi(t, \mathbf{r}) &= -i \frac{e}{(2\pi)^2 \varepsilon_0} \exp(-i\omega_0 t) \left(\mathbf{a}_0 \frac{\partial}{\partial \mathbf{r}} \right) F(\mathbf{r}) . \end{aligned} \tag{A3.35}$$

Here,

$$F(\mathbf{r}) = \int d\mathbf{k} \frac{\exp(i\mathbf{kr})}{k^2 - \varepsilon_0 \omega_0^2/c^2} = \frac{4\pi}{r} \int_{-\infty}^{\infty} dk \frac{k \sin kr}{k^2 - k_0^2} , \tag{A3.36}$$

with $k_0 = \sqrt{\varepsilon_0}\omega_0/c$ and $r = |\mathbf{r}|$. In transforming the integral in formula (A3.36), we have used spherical coordinates in $\mathbf{k}$ space such that $d\mathbf{k} = k^2 \sin\theta d\theta d\varphi dk$ and $\mathbf{kr} = kr\cos\theta$ and then have taken into account the evenness of the integrand of the integral over dk. Taking the integral with infinite limits yields

$$\int_{-\infty}^{\infty} dk \frac{k \sin kr}{k^2 - k_0^2} = \frac{1}{2i} \left[\int_{-\infty}^{\infty} dk \frac{k \exp(ikr)}{k^2 - k_0^2} - \int_{-\infty}^{\infty} dk \frac{k \exp(-ikr)}{k^2 - k_0^2} \right] = \pi \exp(ik_0 r) . \tag{A3.37}$$

Here, each of the integrals in the square brackets has been calculated in accordance with Jordan's lemma by closing the integration path by a semicircle of infinite radius in the corresponding half of the complex plane k $(r > 0)$. The positive imaginary part of the frequency ω_0 has been taken into account.

Substituting (A3.37) and (A3.36) into (A3.35) yields the following formulas for the potentials:

$$\begin{aligned} \mathbf{A}(t, \mathbf{r}) &= e \frac{\omega_0}{c} \mathbf{a}_0 \frac{1}{r} \exp(-i\omega_0 t + ik_0 r) , \\ \Phi(t, \mathbf{r}) &= e k_0 \frac{\mathbf{a}_0 \mathbf{r}}{\varepsilon_0 r} \frac{1}{r} \exp(-i\omega_0 t + ik_0 r) . \end{aligned} \tag{A3.38}$$

Here, in the formula for the scalar potential, we have retained only the term of the leading order in $1/r$ — the one that determines the potential Φ at large distances from the source and describes free radiation. The term proportional to $\sim 1/r^2$ has been discarded. Solutions (A3.38) correspond to divergent spherical waves of the form

$$\frac{1}{r} \exp[ik_0(r - c_0 t)] , \tag{A3.39}$$

where $c_0 = c/\sqrt{\varepsilon_0}$ is the speed of light in the medium.

From formulas (41.3) we then calculate the electromagnetic field vectors $\mathbf{E}(t,\mathbf{r})$ and $\mathbf{B}(t,\mathbf{r})$, again retaining only the terms of the leading order in $1/r$:

$$\begin{aligned} \mathbf{E}(t,\mathbf{r}) &= ie\frac{\omega_0^2}{c^2}\left(\mathbf{a}_0 - \frac{(\mathbf{a}_0\mathbf{r})\mathbf{r}}{r^2}\right)\frac{1}{r}\exp(-i\omega_0 t + ik_0 r)\,, \\ \mathbf{B}(t,\mathbf{r}) &= ie\frac{\omega_0^2}{c^2}\sqrt{\varepsilon_0}\frac{[\mathbf{a}_0\mathbf{r}]}{r}\frac{1}{r}\exp(-i\omega_0 t + ik_0 r)\,. \end{aligned} \tag{A3.40}$$

And finally, calculating the Poynting vector

$$\mathbf{P} = \frac{c}{4\pi}\frac{1}{4}\langle[(\mathbf{E}+\mathbf{E}^*)(\mathbf{B}+\mathbf{B}^*)]\rangle\,, \tag{A3.41}$$

where the angle brackets stand for averaging over the period $2\pi/\omega_0$, yields the formula

$$\mathbf{P} = \frac{1}{r^2}\frac{e^2 a_0^2 \omega_0^4 \sqrt{\varepsilon_0}}{8\pi c^3}\sin^2\theta\frac{\mathbf{r}}{r}\,, \tag{A3.42}$$

with θ being the angle between the vectors $\mathbf{r}$ and $\mathbf{a}_0$. The flux of the vector field $\mathbf{P}$ through a surface element $r^2 do$, where do is a solid angle element, coincides as expected with the flux (42.9), obtained by the Hamiltonian method.

In using the Hamiltonian method, we did not appeal to the causality principle because we began with the solution to an oscillator equation with zero initial conditions, i.e., to problem (41.19). In the method described in this appendix, the causality principle is accounted for by introducing the imaginary part of the frequency. If the Laplace transformation were used instead of the Fourier transformation in time in formulas (A3.4), then the imaginary part of the frequency would appear automatically from the requirement for the Laplace transforms to be analytic functions in the upper half of the complex plane ω.

References

1. A.N. Tikhonov and A.A. Samarskii. Equations of Mathematical Physics. Translated in English, Courier Dover Publications, 1990. Russian version, Moscow, Nauka, 1972.
2. A.G. Tikhonov, A.N. Sveshnikov. The Theory of Functions of a Complex Variable. Moscow, MIR, 1973. Russian version, Moscow, Nauka, 1970.
3. V.S. Vladimirov. Generalized functions in mathematical physics, Moscow, MIR, 1979. Russian version, Moscow, Nauka, 1976.
4. V.A. Ilyin and E.G. Poznyak. Linear algebra. Moscow, MIR, 1986. Russian version, Moscow, Nauka, 1970.
5. F.W. Olver. Asymptotics and Special Functions. Publisher: Academic Pr, Publication, 1974. F. W. Olver. Asymptotics and Special Functions. A K Peters Ltd. Publication, 1997. Translated in Russian, Moscow, Nauka, 1978.
6. L.S. Pontryagin. Ordinary Differential Equations. Addison-Wesley, 1962. Russian version, Moscou, Nauka, 1974.
7. A.A. Sokolov, I.M. Ternov, and V.Ch. Zhukovskii. Quantum Mechanics. Imported Pubn. 1986. Russian version, Moscow, Nauka, 1979.
8. S. Flugge. Practical Quantum Mechanics. Springer Study Edition, Springer Verlag, 1994; S. Flugge. Practical Quantum Mechanics (Classics in Mathematics). Springer Verlag, 1998. Translated in Russian, v. 1, Moscow, MIR Publishers, 1974.
9. A.S. Davydov. Quantum Mechanics, Pergamon Pr. 2nd edn., 1976. Russian version, Moscow, Nauka, 1973.
10. D.I. Blokhintsev. Quantum Mechanics. Springer Verlag, 1st edition, 1964. Russian version, Moscow, Gostechizdat, 1963.
11. F. Aleksandrov, L.S. Bogdankevich, and A.A. Rukhadze. Principles of Plasma Electrodynamics. Springer Verlag, 1984. Russian version, Moscow, Vysshaya Shkola, 1988.
12. V.L. Ginzburg. Propagation of Electromagnetic Waves in Plasma. Gordon and Breach, 1962. Russian version, Moscow, Nauka, 1967.
13. V.L. Ginzburg. Applications of Electrodynamics in Theoretical Physics and Astrophysics. 2nd edn., Gordon and Breach Sci. Publ., 1989. Russian version, Moscow, Nauka, 1981.
14. A.B. Mikhailovskii. Theory of Plasma Instabilities. Consultants Bureau, 1974. Russian version, Moscow, Atomizdat, v/2, 1972.
15. E.M. Lifshitz and L.P. Pitaevskii. Physical Kinetics. Oxford: Pergamon Press, 1981. Russian version, Moscow, Nauka, 1979.
16. L.M. Brekhovskikh and V.V. Goncharov. Introduction to Mechanics of Continuous Media. Moscow: Nauka, 1982. Translated in English 1st edn., Springer Verlag, 1985, 2nd edn., Springer Verlag, 1994.

17. L.M. Brekhovskikh. Waves in Layered Media. New York: Academic Press, 1980. Russian version, Moscow, Nauka, 1973.
18. A.I. Akhiezer and I.A. Akhiezer. Electromagnetism and Electromagnetic Waves. Moscow: Vyshshaya Shkola, 1985 (in Russian).
19. M.V. Kuzelev and A.A. Rukhadze. Electrodynamics of Dense Electron Beams in Plasma. Moscow: Nauka, 1990. Translated in English, Plasms Free Electron Lasers, Eddition Frontier, 1995.
20. A.S. Il'inskii, V.V. Kravtsov, and A.G. Sveshnikov. Mathematical Models ofElectrodynamics. Moscow: Vysshaya Shkola, 1991 (in Russian).
21. B.Z. Katsenelenbaum. High-Frequency Electrodynamics. Wiley InterScience, 2006. Russian version, Moscow, Nauka, 1966.
22. V.V. Nikol'skii and T.I. Nikol'skaya. Electrodynamics and Radio Wave Propagation. Moscow: Nauka, 1989 (in Russian).
23. Max Born and Emil Wolf. Principles of Optics: Electromagnetic Theory of Propagation, Interference and Diffraction of Light. Cambridge University Press; 6th edn., 1997. Tranlated in Russian, Moscow, Nauka, 1970.
24. Plasma Electrodynamics, Vol. 1: Nonlinear theory. A.I. Akhiezer, I.A. Akhiezer, R.V. Polovin, A.G. Sitenko, and K.N. Stepanov. Pergamon Press, 1975. Plasma Electrodynamics, Volume 2: Nonlinear Theory and Fluctuations. A.I. Akhiezer, I.A. Akhiezer, R.V. Polovin, A.G. Sitenko, and K.N. Stepanov. Pergamon Press, 1975. Russian version, Moscow, Nauka, 1974.
25. N.A. Krall and A.W. Trivelpiece. Principles of Plasma Physics. McGraw-Hill, 1973. Translated in Russian, Moscow, MIR, 1975.
26. R.J. Briggs. Electron-Stream Interaction with Plasmas (Research Monograph). The MIT Press, 1964. Translated in Russian, Moscowm MIR, 1971.
27. N.S. Erokhin, M.V. Kuzelev, S.S. Moiseev, and A.A. Rukhadze. Nonequilibrium and Resonant Processes in Plasma Radiophysics. Moscow: Nauka, 1982 (in Russian).
28. L.A. Vainshtein and V.A. Solntsev. Lectures on Microwave Electronics. Moscow: Sovetskoe Radio, 1973 (in Russian).
29. W.H. Louisell. Coupled Mode and Parametric Electronics. Wiley & Sons, 1960. Translated in Russian, Moscow, MIR, 1963.
30. F.S. Crawford, Jr. Waves. Berkeley Physics Course. Vol. 3. McGraw-Hill College, 1968. Tramslated in Russian, Moscow, Nauka, 1984.
31. A.M. Fedorchenko and N.YR. Kotsarenko. Absolute and Convective Instability in Plasma and Solids. Moscow: Nauka, 1981 (in Russian).
32. P.S. Landa, Self-Oscillations in Distributed Systems. Moscow: Nauka, 1983 (in Russian).
33. M.B. Vinogradova, O.V. Rudenko, and A.P. Sukhorukov. Theory of Waves. Moscow, Nauka, 2nd edn., 1990 (in Russian).
34. Nonlinear Waves. Ed. by A.V. Gaponov and L. A. Ostrovskii. Moscow: Nauka, 1977 (Translated from English).
35. G. Dech. Handbook on the Practical Use of Laplace Transforms and Z-Transforms. Moscow: Nauka, 1971 (in Russian).
36. E. Titchmarsh. Introduction to the Theory of Fourier Integrals, Chelsea Publishing Company, 3rd edn. 1986. Translated in Russian, Moscow, Gostechizdat, 1948.